WATER CONFLICTS IN INDIA

WATER CONFLICTS IN INDIA
A MILLION REVOLTS IN THE MAKING

Editors

K.J. Joy
Biksham Gujja
Suhas Paranjape
Vinod Goud
Shruti Vispute

Routledge
Taylor & Francis Group

LONDON AND NEW YORK

First published 2008
by Routledge

2 Park Square, Milton Park, Abingdon, Oxfordshire OX14 4RN
711 Third Avenue, New York, NY 10017

Routledge is an imprint of the Taylor & Francis Group, an informa business

First issued in paperback 2018

Transferred to Digital Printing 2008

This volume has been supported by the WWF Project, 'Dialogue on Water, Food and Environment'. The views expressed in the individual articles are those of the respective authors, and the Editors or the institutions they represent do not necessarily endorse them.

Typeset in 10/12 Calisto MT by
Nikita Overseas Pvt. Ltd.,
1410 Chiranjiv Tower,
43 Nehru Place,
New Delhi 110 019.

British Library Cataloguing-in-Publication Data
A catalogue record for this book is available from the British Library.

ISBN 978-0-415-42411-0 (hbk)
ISBN 978-1-138-37675-5 (pbk)

Forum for Policy Dialogue on Water Conflict in India

PARTNER ORGANISATIONS

Centre for World Solidarity
(CWS),
Hyderabad

Chalakudi River
Samrakshan Samithi,
Trichur

IWMI Tata
Water Policy Program,
Anand

Pragathi,
Bangalore

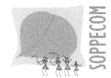

Society for Promoting
Participative Ecosystem
Management
(SOPPECOM), Pune

Vikram Sarabhai Centre for
Development Interaction
(VIKSAT),
Ahmedabad

World Water Institute
(WWI),
Pune

WWF Dialogue on Water,
Food and Environment

WWF International

Steering Committee Members

Contents

Acknowledgements

Working with over ninety authors to bring out this compendium on behalf of the Forum for Policy Dialogue on Water Conflicts in India was an interesting experience, though not without its headaches. It was possible to see it through to its completion because it was truly a collaborative effort of the institutions and individuals who comprise the Forum: the Centre for World Solidarity (CWS), Hyderabad; the Chalakudi River Samrakshana Samithi, Trichur; the IWMI–Tata Water Policy Program, Anand; Pragathi, Bangalore; the Society for Promoting Participative Ecosystem Management (SOPPECOM), Pune; the Vikram Sarabhai Centre for Development Interaction (VIKSAT), Ahmedabad; the World Wide Fund for Nature (WWF)–ICRISAT Dialogue Project, Patancheru; the World Water Institute (WWI), Pune; and some independent researchers. The WWF funded the preparation of this compendium as part of its project 'Dialogue on Water, Food and Environment'. We are thankful to WWF and all other collaborating institutions and individuals comprising the Forum for their support.

The preparation of this volume was overseen by a Steering Committee and facilitated by some members and others who had volunteered to function as State Coordinators. They include: Bhaskar Rao, a progressive farmer from Karnataka; Biksham Gujja from WWF International; Ganesh Pangare from WWI; K. J. Joy of SOPPECOM; K. V. Rao, a progressive farmer from Andhra Pradesh; Latha A. of the Chalakudi River Samrakshana Samithi; Malavika Chauhan, an independent researcher in Delhi; N. J. Rao from the Indian Institute of Science, Bangalore; R. Doraiswamy of Pragathi; Rakesh Tiwary from the IWMI–Tata Water Policy Program; S. K. Anwar from CWS; Srinivas Mudrakartha from VIKSAT; Uma Maheshwari R., a researcher and freelance journalist from Hyderabad; Vinod Goud from WWF–ICRISAT Dialogue Project; and R. S. Bhalla from FERAL, Pondicherry. We gratefully acknowledge the efforts of the Steering Committee members and the State Coordinators; without their interface we would never have been able to manage the complex interactions needed with the ninety-plus authors from all corners of India.

The case studies included in the compendium have been reviewed by well-known experts: Ajit Patnaik, Amita Shah, Indira Hirway, K. V. Raju, Kanchan Chopra, N. J. Rao, Paul Appasamy, R. K. Patil, R. Parthasarathy, Ramaswamy Iyer, Ratna Reddy, S. Janakarajan, S. N. Lele, Sara Ahmed, V. B. Eswaran and Vishwa Ballabh. In spite of their extremely busy schedules, they found the time to review the case studies within the time given to them. The comments and suggestions received from them have greatly improved the quality of many of the case studies and we gratefully acknowledge their efforts. Similarly, the feedback we received from the participants of the two-day conference on 'Policy Dialogue on Water Conflicts in India', organised by the Forum in March 2005, have helped immensely in the planning of this volume, and the final shape it has taken. We are thankful to the participants of the conference – activists, judges, lawyers, farmers, academics, media persons and, not to forget, politicians – for their valuable inputs and, as importantly, their encouragement.

Approached with the idea of publishing this volume as a book, Routledge India readily agreed to do it. Since then Omita Goyal and her team at Routledge, especially Sayantani and Nilanjan, have shown great patience and persistence and it is thanks to them that the book has come out as such an excellent production. Revathi Suresh did the first round of copyediting, and Marion Jhunja did the design and layout of the first draft. Raja Mohanty has done the cover and Sraban Dalai of WWF–ICRISAT Dialogue Project has done the map that shows the case-study sites. Our sincere thanks to all of them.

The *Economic and Political Weekly* (*EPW*) brought out a Special Issue on water conflicts in India on February 18, 2006, which contained 18 of the compendium case-studies along with an overview article. The special issue helped give visibility to the issue of water conflicts and also created widespread interest in the book. We are thankful to *EPW* and its Editor, C. Ramamanohar Reddy, for bringing out the special issue on water conflicts.

Most of all, our sincere thanks go to all our case study and thematic review writers. Without their cooperation, especially in meeting deadlines and patiently responding to our repeated queries and demands, this book would never have been possible. We could not have wished for a better person than Professor A. Vaidyanathan to write the Foreword to the volume, and we sincerely thank him, not only for having found the time to do so, but also for having put it in a larger context.

Last, but not the least, we are grateful to all our colleagues in our respective institutions—SOPPECOM and the WWF–ICRISAT Dialogue Project—for all the hard work that they have put in at all stages of the production of this volume. The usual disclaimers apply; if any mistakes and errors remain they are solely our responsibility.

Considering the resource base of the various contributors, some visuals reproduced in this volume may not be very sharp. The Editors nonetheless decided to use these visuals in the hope that they would give the readers a general idea of the location of the cases and the impacts discussed.

We sincerely hope that the book will contribute to an informed public debate on water conflicts in India, and possibly even facilitate their resolution.

K. J. Joy, Biksham Gujja, Suhas Paranjape,
Vinod Goud and Shruti Vispute

Foreword

Conflicts have always been, and remain, an endemic feature of the water sector. It arises whenever and wherever the available supplies from a given source (aquifer, diversion works, surface storage) fall short of demand. Conflicts between head-reach and tail-end, and over sharing of water in times of shortage, were quite widespread and common in traditional water supply systems. These have become more and more widespread and intense over time.

The demand for water for domestic use, irrigation and non-agricultural uses has been growing rapidly. Growth of population, production and incomes has increased the requirements for domestic, agricultural and non-agricultural uses. The relatively high private returns on irrigated (compared to un-irrigated land), due partly to higher productivity of irrigated land and partly government policy of heavy subsidisation of water and electricity, have added to the demand pressures.

Advances in civil engineering and pumping technology have made it possible to increase supplies vastly, and make it accessible to wider areas and larger number of users. Since Independence, these technologies have been used to increase the utilisation of surface and groundwater on an unprecedented scale. With utilisable water potential being limited and the actual demand and utilisation going up, on the one hand the scope for further expansion progressively diminishes, and on the other adverse environmental impacts increase.

In point of fact, in several parts of the country, utilisation of both surface and groundwater has been reached or even exceeded the limit of sustainable use. Wasteful and inefficient use of harnessed water increases the gap between effective supply and demand at the users' end. As a result, supply has not kept pace with demand and the gap has grown. This has led to intensified competition and conflicts over access to and allocation of available supplies between uses and users.

The essays in this collection give numerous specific examples of conflicts between uses and users at different levels, ranging from individual systems to sharing of waters among riparians in inter-state and international river basins. They highlight the varied nature of the conflicts between uses and users and between beneficiaries of projects and those adversely affected. The reasons for scarcity and conflicts are numerous, including growing demand and its changing composition, overestimation of water availability, defective project design compounded by poor water management, deterioration in water quality, and mining of sand from river beds.

The papers also give an idea of the ways in which the conflicts at different levels are attempted to be resolved, and the fact that these are far from being effective. Several reasons are identified including lack of well-defined principles for determining allocations between different claimants, non-involvement of users in making or changing rules, absence of transparency and clarity regarding their rights and entitlements and for changing them in light of changing demand-supply conditions, the ineffectiveness of judicial processes, and laxity in their enforcement due to weak or deteriorating institutional mechanisms.

The problem of scarcity, conflicts and their resolution is complex. Achieving an acceptable balancing of the claims of different uses and users in different regions to a limited supply, the claims of immediate demands and ensuring long-term sustainability is a very difficult task both technically and politically. It is more difficult when both supply and demand conditions are changing over time. The technical complexity arises from the fact that one has to handle a complex and multi-layered intermeshing of engineering, hydrology, legal and institutional aspects. Economic policy instruments can be used to influence demand for and returns to water for different uses in such a way that it conforms to the socially preferred state.

These decisions are now made entirely by the State and its agencies. There is ample evidence to suggest that these decisions are not based on a proper study and analysis of facts. Such information and analyses are seldom made public. This precludes informed public discussion of issues involved or the rationale for the decisions. Partly because of this, critics of government policy tend to underestimate the complexity of the problem and its solutions. It is wrong to pose the issue in terms of dichotomous alternatives relating to specific aspects — such as large versus small projects; centralised versus decentralised planning and management by local communities; state management versus management by user representatives and the like. Public discourse also tends to sidestep or make light of the reforms needed to correct distortions in the allocation and use of water arising from the failure of governance, perverse government policies which encourage its inefficient and wasteful use, and the unavoidable conflicts arising from the fact that water is scarce in relation to demand.

It is very important to promote more informed public discussion with due appreciation of the complexity of the water problem. An essential condition for this is to force the government to place all relevant information relating to water resources, their utilisation and allocation, and the rationale of decisions freely accessible in the public domain. Civil society has to be proactive in pointing out gaps and inconsistencies in the data, and contribute actively to improving the database. But more and better information is not sufficient. It is also to develop the capacity to analyse the information and come up with technically sound alternative solutions (rather along the lines attempted in the case of Sardar Sarovar Project); institutional mechanisms for democratic management by representatives of all stakeholders and for settlement of disputes based on clearly defined principles; and economic incentives combined with transparent but strict governance to induce efficient, equitable and sustainable use of available water. The present collection is a modest but important step in this direction.

September 2007   A. Vaidyanathan
Chennai

Introduction
A Million Revolts in the Making:
Understanding Water Conflicts in India[*]

K. J. Joy, Biksham Gujja, Suhas Paranjape,
Vinod Goud and Shruti Vispute

The fingers of thirst wither away while probing for edible beans,
And man seeks compassion in a devastated body,
The earth's mouth opens wide as a begging bowl,
What a marvellous sight it is to watch your secular regimes wagging their tail!
You will draw water upstream,
And we downstream,
Bravo! Bravo! How you teach *chaturvarnya*[1] even to the water in your sanctified style!

Namdeo Dhasal, *Golpitha*, 1972
Translated from Marathi by Dilip Chitre[2]

Water Conflicts: The Context

'Rivers should link, not divide us' said Dr. Manmohan Singh, Prime Minister of India, while inaugurating the conference of the state Irrigation Ministers on December 1, 2005.[3] He expressed concern over inter-state disputes and urged the state governments to show 'understanding and consideration, statesmanship and an appreciation of the other point of view'. Ponnala Laxmaiah, Irrigation Minister of Andhra Pradesh, returned from the meeting to be hauled over the coals the next day by Janardhan Reddy, his own party senior, over the so-called Pothireddy Padu diversion planned to divert water to the Chief Minister Y. S. Rajashekar Reddy's native district.[4] MLAs from his own party in the Telengana region declared that they would oppose this water diversion to the end. On the issue of the controversial Polavaram dam on the Godavari river—an inter-state river—in Andhra Pradesh, the two Chief Ministers of the neighbouring states of Orissa and Chhattisgarh decided to jointly oppose the state of Andhra Pradesh. They did not even participate in the meeting of the three Chief Ministers called by the Union Minister for Water Resources.[5] The same Polavaram dam is also being opposed by political parties within Andhra Pradesh, particularly from the Telengana region. Water conflicts, not water, seem to be percolating faster to the grassroots! So no wonder that the previous Union Minister for Water Resources described himself more as a 'Minister of Water Conflicts' than a 'Minister of Water Resources'![6]

Water conflicts in India now reach every level and divide every segment of our society—political parties, states, regions and sub-regions within states, districts, castes and individual farmers. Water conflicts within and between many developing countries are taking a serious turn. Fortunately, the 'water wars', a chance remark by the UN Secretary General that later became a media phrase, forecast by so many, have not materialised. War did take place, but over oil, not water, though it is true that some of the literature available globally, especially on the Middle East, sees water conflicts more in terms of wars, peace and survival.[7] Though water wars may not have taken place, water is radically altering and affecting political boundaries all over the world, between as well as within countries. In India, water conflicts are likely to get worse before they begin to be resolved. Till then they pose a significant threat to economic growth, social stability, security and ecosystem health. And under threat are the poorest of the poor as well as the very sources of our water—rivers, wetlands and aquifers.

Conflicts may be taken to be bad or negative, but they are logical developments in the absence of proper democratic, legal and administrative mechanisms to handle issues that are at the root of water conflicts. Part of the problem stems from the specific nature of water as a resource: for example: (a) water is divisible and amenable to sharing; (b) contrarily, it is a common pool resource so that a unit of water used by someone is a unit denied to others; (c) it has multiple uses and users and involves resultant trade-offs; (d) excludability is an inherent problem and exclusion costs involved are often very high; (e) it requires a consideration and understanding of nested expanding scales and boundaries from the local watershed to inter-basin transfers; and (f) the way water is planned, used and managed causes externalities—both positive and negative, and many of them are unidirectional and asymmetric.

These characteristics have a bearing on water-related institutions[8] and have the potential not only to trigger contention and conflict and become instruments of polarisation and exclusion, but also to become instruments of equitable and sustainable prosperity for all those who directly or indirectly depend on them for their livelihood.

There is also a relative paucity of frameworks, policies and mechanisms that deal with water resources. In contrast, policies, frameworks, legal set-ups and administrative mechanisms to deal with immobile natural resources have greater visibility and have a greater body of experience backing them, however contested they may be. Take land, for example. Many reformist as well as revolutionary movements are rooted in issues related to land. Several political and legal interventions have addressed the issue of equity and social justice in respect of land. Most countries have gone through land reforms of one type or the other. Similarly, there is comprehensive literature on forest resources and rights. This is not to say that conflicts over them have necessarily been effectively or adequately resolved, but they have received much more serious attention, have been studied in their own right, and practical as well as theoretical means of dealing with them have been sought. In contrast, water conflicts have not received the same kind of attention.

This is not to say that there has not been any attempt to academically engage with water conflicts in India. Vandana Shiva has published two books on water conflicts. The first one (1991) deals with sub-themes like conflicts over river waters, large dams and conflicts in the Krishna basin, mining and water conflicts, and fisheries and conflicts at sea. It also has a small section on people's alternatives in the context of water scarcity. The second one (2002) takes more of a historical, global perspective and analyses issues like international water trade, damming, mining, aquafarming, and the historical erosion of communal water rights, showing how 'water privatisation is threatening cultures and livelihoods worldwide'. She also calls for a movement to preserve water access for all, and offers a blueprint for global resistance based on examples of successful campaigns like the one in Cochabamba, Bolivia, where citizens fought for and retained their water rights.

Another important contribution to the academic discourse on water conflicts in India is from A. Vaidyanathan and H. M. Oudshoom.[9] Their edited volume (2004) discusses the sources and nature of water scarcity and conflicts and also the mechanisms used to address them. The book strongly argues for a 'transdisciplinary' dialogue as water and water conflicts cannot be fully understood and mechanisms to resolve them cannot be designed within disciplinary boundaries and a coherent, multi-pronged action on several inter-related fronts like technological, legal, institutional and economic.

The Book: Background and Process

As may be seen, the earlier work on water conflicts is sporadic and narrowly focused in terms of the nature of conflict. It was felt that there is a need to look at the whole range of water conflicts and start a process of dialogue on such conflicts. This book is an attempt in that direction; a modest but first attempt to capture in one compendium a large number of cases representing a wide variety of water conflicts across India. It is premised on the belief that understanding and documenting the variety of conflicts in terms of scale and nature with the help of illustrative cases in all their complexity would contribute to initiating informed public debate and facilitating their resolution.

This introductory chapter, the eight thematic reviews that initiate each theme, and the sixty-three case studies (see Map 1 for the list of cases and their locations) that comprise the book have their roots in a process initiated as part of the World Wide Fund (WWF) project, 'Dialogue on Water, Food and Environment'.[10] Discussions in civil society forums led to an awareness of the need to look at water conflicts, and some information on a small number of relatively better known water conflicts in south India was collected and a summary of the cases published as a small booklet (Doraiswamy and Gujja 2004). During a meeting in Bangalore towards the end of 2004, organised to discuss this booklet, participants described many more varied conflicts and it was felt that there is a need to pay more attention to water conflicts in India. It soon became clear that information on water conflicts was scattered, unorganised, documented inadequately or not at all. The initial set of institutions and organisations that participated in the discussions also decided to form the Forum for Policy Dialogue on Water Conflicts in India (hereafter Forum).[11]

It was decided that one of the first steps should be to bring out a volume on water conflicts in India. A small group was formed to discuss the action plan, and a core group and a steering committee were set up to carry out and guide the activities. Since the Forum had strengths more in peninsular India, it was decided to concentrate mainly on that region, but also include a few representative cases from the rest of India. The case studies are not full-fledged research papers but a summarised account of the conflict, the issues it involves and its current status. In most cases, the authors have taken care to capture the differing perceptions of the conflicting parties. Each case study was sent to an expert for review and the reviewers' comments were treated as issues to be addressed, and so long as they were adequately addressed, there was no attempt to modify the case study to bring it in line with reviewers' opinions. On March 21–22, 2005, the Forum organised the Policy Dialogue on Water Conflicts in India, a two-day meeting in which nearly 120 people drawn from various walks of life—politics, judiciary, activism, farming community, academia, and media—participated. The inputs from the Policy Dialogue meeting have helped improve the quality of this volume. We believe that the process of preparing the volume has been as important as the product itself.

The cases presented here are not, and cannot be, a comprehensive account of water conflicts in India. They are more an attempt to illustrate the wide diversity of water conflicts in India. Some cases like the more than 30-year-old Cauvery dispute[12] have not been included; first, because they are very well known, and second, they would probably require separate volumes to do them justice. In spite of many limitations, it was felt that bringing out this volume would be an important first step, mainly because it gives us a glimpse into 'the million revolts' that are brewing around water. We hope that it begins a process of serious reflection on water conflicts within an evolving comprehensive framework.

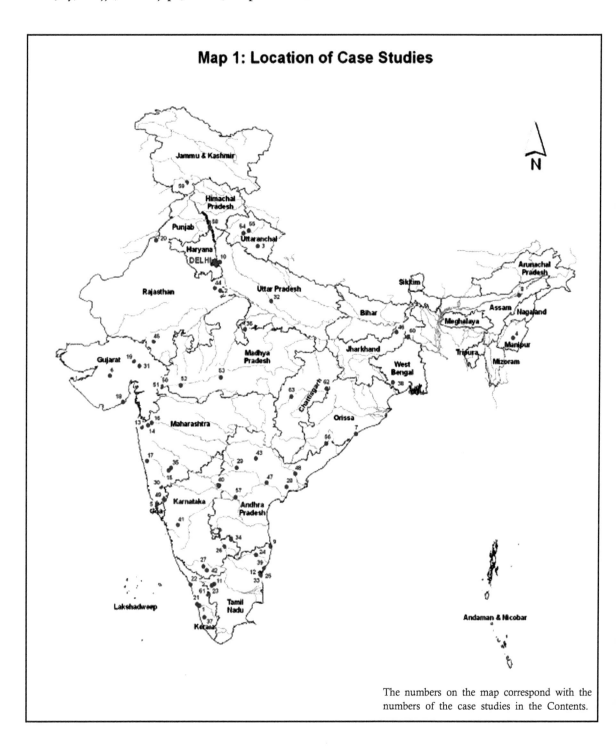

Map 1: Location of Case Studies

The numbers on the map correspond with the numbers of the case studies in the Contents.

Map 1: Location of Case Studies

Case Studies and Themes

After being reviewed by the experts, a total of sixty-three case studies were selected for the volume. Many of these cases have been, or are being, fought in courts. Many others involve agitations and grassroots action. Organising these studies for publication involved adopting a thematic principle for grouping and presenting them, even though each fits into more than one theme. Since water conflicts are often a multi-faceted microcosm of wider conflicts which would make it difficult to identify any one aspect as the dominant one, it was not possible to make the themes mutually exclusive. After much discussion, it was decided to organise the cases into eight broad themes described briefly in the following sections.[13] In the book, invited thematic review pieces introduce the theme and discuss some of the case studies covered under each.[14]

Contending Water Uses

Water, as noted earlier, is a common pool resource and hence when the same unit is demanded for different kinds of uses we have a contestation and a potential conflict. The ten case studies included under this broad theme deal with conflicts that are related to contention between *different* kinds of uses and range from a contention of water for wetlands versus agriculture to contention over building a bridge and its impacts on an island ecosystem.

The case of the Keoladeo National Park (Chauhan, in this volume) is a classic one of contested water use. The contestation is between the needs of the Bharatpur wetlands, a spectacular bird sanctuary which is a World Heritage Ramsar site, and the irrigation needs of the local farmers. The conflict between the needs of the wetlands and irrigation continues to be resolved temporarily every year, in an *ad hoc* manner. Two other Ramsar sites—Kuttanadu and Loktak—illustrate how structures may benefit one group of users while adversely impacting others and the ecosystems. Rice farmers and fishermen are in conflict in Kuttanadu (Thampuran), whereas in Loktak (Maitra) unintended effects are more important. The Bogibeel Bridge case deals with the unintended and unanticipated effects of a bridge across the mighty Brahmaputra (Mahanta). Lack of an integrated approach to ecosystem interventions conditioned by unifocal thinking is at the root of this lack of anticipation.

In the well-known Chilika case (Deshpande & Bedamatta), another Ramsar site, the conflict is many-sided and involves fishermen and settlers, while the Gagas basin in the Himalyas is a microcosm in which urban demands, state policy and increasing pressure endanger the source itself (Paul & Paul).

The case of Vadali (Prakash and Sama) in Chotila taluk, Gujarat, shows that in the absence of grassroots regulation, watershed development may even lead to accentuation of drinking water problems since more water may be diverted to irrigation by the relatively rich farmers. Another case is that of Chennai (Janakarajan) where there is a growing conflict in peri-urban areas between those who want to mine the groundwater to supply it to the cities and those who want to use it for irrigation. A similar case is Sangolda village (Dongre & Poteker) in Goa, and although not on a scale, on a large it illustrates how deep it has reached. The Ganga canal (Das) for Delhi is another case of conflict between large urban needs and rural livelihoods.

Three salient points emerge from these case studies: (*a*) structures built to improve the ecosystems may have unintended effects that actually harm people and ecosystems; (*b*) improving water resources through rainwater harvesting at the micro level might improve water availability but sharpen conflicts if equity is not addressed; and (*c*) in the conflict between rural and urban uses, rural needs are steadily losing out.

Equity, Access and Allocations

This broad theme focuses mainly on equity issues between different users but *within the same kind of use*, unlike the first theme which deals with *different* contending uses. The ten cases under this theme cover a wide variety of equity and access issues including contestation over and between old and new water rights, old and new projects, tail-enders and head-reachers, inter-basin transfers, Dalits' and upper castes, and so on. The case of Bhavani river (Rajagopal & Jayakumar), tributary to the Cauvery in Tamil Nadu, shows how competing demands between old and new settlers were further aggravated by growing demands of industry and drought, and how a multi-stakeholder dialogue attempted to resolve the issue. Not all old rights holders have been fortunate in having prior rights over water. The 'Phad' systems (Sane & Joglekar) in the Tapi Basin in Maharashtra, which had inbuilt equitable access and cropping system management, have all but collapsed as they were absorbed into modern irrigation structures. The Palkhed project (Lele & Patil) in Maharashtra is a classic case of tail-ender double discrimination, both in terms of norm of allocation as well as the actual allocation they finally receive. Almost eighty years ago, Dr. B. R. Ambedkar launched a water satyagraha in Mahad, Maharashtra, by marching to the Chavadar tank to open up all public watering places to the Dalits.[15] The Mahad to Mangaon (Paranjape *et al.*) case shows how, in a drought year, centuries of caste-based oppression and prejudice, deep-rooted cultures and traditions, rear their head once again to deny water to the Dalits. In contrast, the Jholapuri river (Ahmed and UTTHAN) case in coastal Gujarat shows how a gender-based dialogue that addressed water management went some way towards resolving the conflict.

The Tembu Lift Irrigation Scheme (TLIS) (Kavde-Datye) lifts water 300 m from the Krishna and makes it available to a severely drought-prone rain-shadow region in Maharashtra. The TLIS case discusses issues related to equitable access to water as well as energy that are involved in the conflict around the scheme. The proposed Nar-Par diversion (Desai) diverts the water of west-flowing rivers into other basins in Maharashtra. That, along with the famous Indira Gandhi canal case (Das), illustrate how so-called interlinking and diversions may widen and sharpen conflict rather than resolve it. Two other cases, the Pondicherry case of Keezhaparikkalpet (Raghunath and Vasanthan) and the north Gujarat case of Mehsana district (Mc Kay and Diwakara), relate to the conflict between those who own wells and those who do not, but from totally different perspectives and contexts.

As Paranjape and Joy point out in their thematic review on equity-related conflicts, the core issue is the absence of clear-cut norms of equitable water allocation and distribution. Allocation norms have evolved according to local situations, size and nature of project, and historical socio-political relations. To tackle the conflicts over allocations and access we need a better concept of a right or an entitlement to water. How much water should a person or a household be entitled to as a right? Here we need a livelihood needs framework that sees assurance of minimum livelihood needs and the corresponding water requirement as an associated right. Attached to this is the need to share shortages and surpluses in a principled manner. It also entails doing away with the obstacles that deny the disadvantaged sections of our society their rights.

Conflicts around Water Quality

Issues related to water quality or pollution are fast emerging in various parts of India. Earlier these issues were treated as inevitable consequences of growth and industrial development, and therefore largely ignored. However, growing scale, increased awareness and active civil society engagement has brought water quality conflicts more and more to the forefront. The main issue here is how and in what form do users return water to the ecosystem. Polluted water returned by users causes problems to 'downstream users', and decreased freshwater availability causes economic loss, social distress and ill health. Sadly, deterioration in quality becomes apparent only after adverse impact is strongly felt, and in the last instance ecosystems are the major losers.

A dozen cases, drawn from different parts of the country, are included in the book under this theme. The Ganga Action Plan (GAP) case (Singh) in Kanpur traces the complex interaction between the leather industry, the Jajmau villages and government agencies like the Jal Nigam and the Pollution Control Board around the issue of the pollution of the Ganga waters. The Kolleru Sanctuary case (Rao *et al.*) shows the complex interaction between livelihood needs and water quality. However, the Khari river case (Mudrakartha *et al.*) in Gujarat describes a relatively rare example where stakeholder dialogue has brought about a significant improvement in pollution control. In the Noyyal (Jayakumar and Rajagopal) and Palar (Janakarajan) basin cases in Tamil Nadu, multi-stakeholder processes seem to have led to some concrete measures to check pollution caused mainly by the booming textile industry. The other cases include: the pollution of the Periyar (Suchitra) and Chaliyar (George and Krishnan), both in Kerala; Pandasozhanallur (Larroquette and Appavou) in Pondicherry; Arkavati sub-basin (Dominic), Hootgalli (Manasi and Deepa) near Mysore, both in Karnataka; Musi (Jairath *et al.*) in Andhra Pradesh; Ghodawat Industries (Das and Panagare) in Kolhapur, Maharashtra—all these bring out how water pollution impacts both the ecosystems and the peoples' lives and livelihoods. The corollary is that any possible solution to conflicts related to water quality needs to address both the ecosystem needs as well as people's livelihoods.

Most water quality conflicts have been continuing for a decade or more[16] and involve loss of health and livelihoods, reduction in fish catch, loss of species, loss of crops, and even loss of lives.[17] However, the conflicts are complex because there is 'only a thin line separating the victims and the beneficiaries (George and Krishnan). In almost all these cases, except Kolleru and Musi, point source pollution has spread to nearby rivers and groundwater, and the industry is clearly the culprit.[18]

In his thematic review on water quality conflicts, Paul Appasamy raises three questions: (*a*) whether closure of the factories is the solution; (*b*) whether industries can coexist with agriculture and other water users; and (*c*) what is the long-term solution? In spite of considerable civil society initiative, several legislations and pollution control boards at the state and central levels, on the whole we have failed in evolving a long-term answer that can protect our rivers and aquifers from contamination.

Perhaps we need a three-pronged approach. First, a legal framework based on rapidly enforced criminal and civil penalties. Strict but non-implementable legal frameworks appear good only on paper. Second, environmental mediation, a pragmatic direction to settle issues quickly and amicably. Third, encouraging voluntary compliance. This last is a long way from becoming effective in India, since consumers/users in particular are still focused mainly on price rather than on quality and safety.

To these we should add another concern, the ecosystem. Ecosystems have no voice, no votes, and some important ecosystem issues have never entered the agenda of water conflicts. For example, concepts of ecological flows, minimum ecosystem requirements and preservation of ecosystem services are not even being explored. Yet, our long-term future will finally be decided by whether we tackle these issues before we poison the wellspring of life on this planet.

Sand Mining

The book has four cases of conflicts over sand mining. The Vade Agraram (Larroquette and Appavou) case, where sand is being mined from the village stream because of high silica content, shows how industries siphon vital village resources. The Papagani catchment (Rao) case shows how inter-state streams are affected. Karnataka issues sand-mining permits, while the downstream farmers in Andhra Pradesh protest the consequent reduction in groundwater. In the Baliraja dam (Vispute) case on the Yerala river in Maharashtra, villagers took control of sand mining and used the proceeds to help build the dam after a protracted struggle. The Nandanvara dam (Chowla) case in Madhya Pradesh involves the general problem of mining and its impact on water bodies.

Uncontrolled sand mining has a deleterious effect on stream flows. Apart from its ecological impacts— to name a few, impact on stream flows and sandy acquifers, deepening of riverbeds, sub-surface intrusion of saline seawater in coastal areas, erosion of the banks—it also impacts on the livelihoods of the local people, for example, through decreased availability of water for both domestic and irrigation purposes as the wells near the banks go dry. Sand is also a building material, and local people depend on it for the construction of their homes and other local uses. It provides seasonal employment to the local labourers. In many states it is one of the major sources of revenue for the gram *panchayats*. The contractor-bureaucrat-politician nexus further complicates the situation and the conflicts very often take the form of conflict between this nexus and the local people.

Sand is primarily a local resource and it is time we looked at sand mining in the context of rights over local resources and their management and the loss to local communities. P. B. Sahasranaman, in his thematic review suggests, 'it is important to determine the quantity of sand that can be safely mined by taking into account annual rate of accrual or replenishment keeping a long time span of about 25 years or so.' He also suggests, among other things, that 'the value of sand must incorporate in it the cost to riparian health' and argues for an increase in the price of sand so that there is an incentive to look for alternative building materials, given the fact that the issue of sand mining is likely to get worse in future due to escalating urban growth and building activities.

Micro-Level Conflicts

Ten case studies have been included under this theme that comprises conflicts on a truly micro scale—within a village, a community or around a small tank. The thousands of such micro-level conflicts that exist in India are more varied and, contrary to expectation, often much more complex to understand, and involve a very wide range of issues. No volume can ever aspire to do justice to them. This sample is only illustrative of such cases.

The case of the gravity dam in Paschim Midnapur district (Singh and Sinha) in West Bengal brings out the need for coordination between the different government agencies involved at the village level. While a project like Polavaram, which submerges thousands of hectares of forest, gets clearance within a fortnight, this small gravity dam, submerging at best 12 ha, failed to get clearance for years! The Lava Ka Baas (Kashwan) case investigates the complex issue of inclusions and exclusions in the notion of community around this structure built by Tarun Bharat Sangh under the leadership of Magsaysay award winner Rajendra Singh. It also brings out the complexity of water conflicts and the need to critically evaluate even the best of efforts. Two other case studies involve conflicts between the Dalits and the rich upper caste farmers in the village. In the case of Seliamadu village (Raghunath and Vasanthan) in Pondicherry the conflict was over tank usufruct rights. In the Somadevi Kere (Bhende) case in Yadgir taluka of Gulbarga district, the Dalit Madars had customary right to cultivate the tank bed and this right became a bone of contention when the issue of de-silting and renovation of the tank came up. It benefited the Lingayats, a politically powerful caste in Karnataka who owned most of the land in the command area.

In another interesting case of a tank in Karnataka, villagers resolved a two-decade-old conflict by introducing 'tail-end first' principle in water distribution (Shah). But the similar 'head-reachers and tail-enders' conflict in Shapin basin (Lal *et al.*), Jharkhand still continues except when there is plenty of water during a good monsoon year. In the Kudregundi Halla tank (Menon and Patil) case in Mysore district in Karnataka, the conflict over a most unlikely resource, seepage water from the tank, merged with other caste and class issues in the village.

The Dharmasagar tank (Rao and Murali) case near Warangal in Andhra Pradesh has a complex history going back seventy years, illustrating a classic case of urban centres gradually taking over rural water without any compensation. The Ubeshwarji Ka Nala[19] (Prasad) case study involves a complex 300-year-old history of conflict between two adjacent villages over a 'water source that is considered as a blessing from Lord Shiva'. This case helps lay to rest any romantic notion of community and shows that left to themselves, 'communities' may sustain rather than resolve their fights over very long periods. The Morathodu Irrigation Scheme (Shahjan *et al.*) describes how the scheme in the Chalakudy river basin in Kerala, planned and implemented under the now famous 'People's Planning Campaign', has actually become a nightmare for the local people because of a number of oversights.

These cases show that local-level water conflicts are increasing and spilling into many other issues, and although there are instances of successful resolution of conflicts, what stands out is the lack of mechanisms to mediate, to provide platforms for dialogue and contestation between rights and stakeholders.

Dams and Displacement

Conflicts over dams and displacement are relatively well publicised and better documented. There is a lot of material already available on many of the cases. In the thematic review, Bharat Patankar and Anant Phadke raise several important issues, challenging the major technical flaws, lack of transparency in sharing the data, the reluctance to look at serious alternatives including alternative designs, and, more importantly, they argue for rehabilitation 'that preserves self-respect'. They also argue that in many drought-prone regions in India, exogenous water from large and medium dams may be needed to supplement and strengthen local water harvesting, and that their integration is the way to avoid dividing the poor and pitting them against each other as drought-affected beneficiaries versus displaced victims.[20]

On the Sardar Sarovar project, the book includes two of the three intended articles: one describing the Narmada Bachao Andolan (Sangvai) and its stand, and another that suggests an alternative restructuring of the project (Paranjape and Joy) that could maintain or increase the benefit while greatly reducing the costs of human displacement, thus having the potential to bring the contending parties around a negotiated settlement. The third intended piece presenting a critical viewpoint from the proponents of the project unfortunately did not materialise. Two cases in Madhya Pradesh are of interest here from the point of view of an alternative approach to rehabilitation and livelihoods of the project-affected persons. The Haribad case shows how even in a small dam context there may be an overlap of conflicts similar to the overlap in big dams (Mansuri and Dharmadhikary). However, it also shows a possible way out as 'many beneficiary families pledged portions of their land in the project command to create an 80 ha land pool for the rehabilitation of the project-affected persons. The case of the fishing cooperative in Tawa (Vikash Singh) has the potential to become an important component of a policy of rehabilitation with self-respect.

The theme includes some of the other important cases of dams and displacement: the Tehri dam (Vimal Bhai) in Uttaranchal, the Pulichintala (Rama Mohan) and Polavaram (Rama Mohan) dams in Andhra Pradesh,[21]. It also covers some of the local cases, like the Uchangi dam (Adagale & Pomane) in Maharashtra where the government agreed partially to an alternative that reduced the dam height, or the Bhilangana small hydro-power project (Rawat & Kaintura) that seems to be more tractable. However, here again, the issue of bio-diversity and loss of forest lands and change in river ecology are often ignored or simply submerged. As said earlier, there are no guidelines on minimum ecological flows and in their absence it is difficult to arrive at consensus on such issues.

Transboundary Water Conflicts

Conflicts between countries are generally classed as transboundary conflicts. However, in India, constituent states themselves are often very large and since water is a state subject, enjoy considerable autonomy in this respect. For this reason, both inter-state and inter-country disputes have been included in this theme. The thematic article by Ramaswamy Iyer provides an excellent overview of the subject. Of the five cases included in this volume, two involve conflicts between countries and three between states in India.

The first inter-country case sparked by the Baghlihar dam (Sinha) extends to the Indus Treaty between India and Pakistan. The other case, sharing the Ganga (Sen), is that of inter-basin diversions of the Himalayan rivers and their implications for Indo–Bangladesh relations. In both cases, there is a need to look beyond political expediency to long-term durable understanding on the issues involved.

In India, practically all major rivers cross several state boundaries rendering many inter-state disputes as complex as inter-country disputes! The Cauvery and the Krishna water disputes are too complex, too large and quite well documented and have not thus been included in the book.[22] However, the case of the Telugu Ganga Project (Ramadevi and Nikku) which has been included shows how a project that exemplified 'major inter-state cooperation' has turned into a major inter-state conflict over two decades. The Sutlej Yamuna Link Canal (Khurana) is another example of politics, litigations and hysterical propaganda. While an Indo–Pak agreement over sharing waters has withstood hostile political relations and wars, similar agreements have led to bitter conflicts between Indian states.[23] Even small projects are not immune, as shown by the example of the three-decades-old dispute between Andhra Pradesh and Orissa over a small dam on the Jhanjhavati river (Rama Mohan).

Privatisation

Privatisation of water is an important upcoming arena of conflict not only in India but in many other countries in Asia, Latin America and Africa as well. Three important cases of water conflicts due to water privatisation are included in the book under this theme. The famous Plachimada case[24] involving Coca-Cola versus the local communities and the *panchayat* is an important example of a protracted struggle waged by local farmers. The Sheonath river (Das and Pangare) and the Kelo river (Kashwan and Sharma) cases, both from Chhattisgarh, bring out clearly what is in store if there is no vigilance exercised in respect of the kind and extent of privatisation, or in respect of whether or not privatisation of rights and entitlements takes place under the garb of privatising services.

Sunita Narain, in her thematic review of water privatisation and private–public partnerships, the latest buzzword in water management, calls for a nuanced understanding of the issue of water privatisation (source privatisation and privatisation of service delivery cannot be equated) and also brings to the forefront the underlying issues of equity and assurance that are most threatened by privatisation, or what goes on under the rubric of public–private partnerships. The current debate about water privatisation is highly polarised between two well-entrenched positions of 'for' and 'against', and there seems to be very little attempt to explore the middle ground of seeing water as both a social and economic good.[25] This has implications for issues like ownership, rights and allocations, pricing and cost recovery, and the regulatory framework. According to Sunita Narain, 'the water issue is not about privatisation. It is about the governance and regulatory framework to secure the rights and access of all to clean water. It is about the right to life. It is also about the rights to water for all.'

The Possible Way Ahead

Water conflicts are symptoms of larger issues in water resources governance. The book, mainly a pre-analytical effort, does not aim at a detailed analysis of water conflicts, their root causes, and the ways ahead. However, implicit in these 'million revolts' is a demand for change, first in the many ways we think about water, and second, in the many ways we manage it. And many isolated insights can already be gleaned from the different case studies and thematic reviews in the book. In this concluding section we briefly enumerate some of these insights.

First, we need to get out of the thinking that sees water flowing out to the sea as water going waste. This thinking, still prevalent in the country, led to a water management strategy centred on dams. It is also important to have a historical perspective and not demonise dams and earlier dam builders from today's vantage point. While questioning the wisdom of selling the same technology approach that was valued in that era, we need to look ahead.

The lesson is that water is a resource embedded within ecosystems; we cannot treat it as a freely manipulatable resource. For example, too many of our mega projects, whether big dams, or diversions or interlinking schemes treat it as freely manipulatable and do harm to the long-term viability and sustainability of the resource itself. Our wetlands and rivers are already in bad shape. It is time we took them into account on their own, and not simply as a resource to be mined. Otherwise we will end up spending more in managing conflicts than what we get from our projects!

We need to change our thinking in respect of the role of large systems and dams. We need to see local water resources as the mainstay of our water system and need to see large-scale irrigation as a stabilising and productivity-enhancing *supplement* feeding into it. For this we must deliver water in a dispersed manner to local systems, rather than in concentrated pockets, creating ecosystem islands dependent fully on exogenous water that can only be maintained at great economic and social cost.

Then there is the vexed question of who pays how much for water. We need to realise, first and foremost, that so far it is the urban poor, the rural areas and the ecosystems who have paid a much higher cost, directly as well as indirectly, for the water that the rich and the middle classes in the country enjoy, especially from public sources. More than anything, we have here a case of reverse subsidy. We need to ensure that full costs are recovered from the rich and the middle classes. They have the capacity to pay, as the profits to bottled water manufacturers show. Without this it will not be possible for cities to maintain adequate quality for the water they return to downstream ecosystems and communities.

Two of the most important issues that have emerged are those of rehabilitation and pollution. Though some progress has been made in states like Maharashtra in respect of 'rehabilitation with self-respect', there is an urgent need for a policy and enactment at the national level for the rehabilitation of all project-affected people. In respect of pollution, as already discussed, we need to move to a mix of civil and criminal penalties and introduce environmental mediation as an active method of addressing pollution issues.

What is also evident is the total lack of effectiveness of the so-called river basin organisations in tackling water conflicts. What is sorely needed is a system of nested institutions that start from the micro level, may be a village, and proceed upwards to a basin-level board or authority. Water is a highly dispersed and local resource even while it is an interconnected resource. Centralised basin-level authorities alone will never be able to take care of the complex problems that arise at all levels.[26] It is also important that these micro-level institutions do not automatically follow the boundaries of a presumed community, since it is clear from many cases that intra-community divisions enter decisively into water conflicts.

The case studies clearly bring out that struggles and viewpoints around water issues in India are highly polarised. The richness and diversity of bio-physical, social, economic and political contexts in India itself creates a tendency of fragmentation and polarisation rather than synthesis, leading to long-drawn-out wars of attrition in which the losers are invariably the vulnerable and weaker sections. It is important in this respect to look at multi-stakeholder platforms (MSPs) or similar processes that bring stakeholders together. The case studies also show that MSPs have resulted in better outcomes than polarised wars of attrition.[27]

However, there are a few aspects that need urgent attention if MSPs are to become meaningful and stable instruments of water governance. They will first need to take into account and give proper attention to the heterogeneity of stakeholders, prior rights and the context of MSP formation. But more importantly, they will also have to be informed by an innovative approach to water-sector reform that will allow accommodation of different stakeholder interests, will need to be supported by access to reliable data, information and decision support systems, and be based on an acceptable normative framework.[28]

Such a framework, Rogers and Hall (2002) point out, needs to be an inclusive framework (institutional and administrative) within which strangers or people with different interests can practically discuss and agree to cooperate and coordinate their actions.[29] This is all the more important in the water sector where opinions are sharply divided on crucial issues: for example, is water a social good and part of the human rights framework or an economic good like any other. There is a similarly sharp difference of opinion about source creation, about large vs. small systems, equitable access and entitlements. The framework adopted will therefore be of critical importance.

The framework needs to be capable of creating space for a dialogue if an MSP is to be initiated. For example, a framework that inherently sees large and small as mutually exclusive and opposed alternatives leaves little scope for dialogue between the dam-affected and the drought-affected: large dam votaries would tend to either invoke the 'greater common good' to ignore the suffering and displacement of already marginalised communities like the *adivasis* (tribals), and opponents would invoke that very suffering to deny the possibility of reliable water supply to severely drought-affected areas. However, if the framework is based on the need to *integrate* the small and the large, several possibilities emerge— destructive centralised submergence behind the dam could be reduced by diverting and storing as much as possible of the water flows in the small systems within the command/service area instead of storing them behind the dam and open up space for a joint exploration by the two important stakeholders, the would-be project-affected and the beneficiaries. The conventional framework governing water resource planning, source development, norms of access and service delivery is also responsible for many types of conflicts amongst the direct stakeholders and a highly polarised discourse on water. The challenge is to evolve a consensual framework that will be inclusive enough even as it takes into account crucial concerns like equity and sustainability.

The beginning is likely to be modest. Recently Wang Shucheng, China's Minister of Water Resources, said in his keynote speech[30] to the International Commission on Irrigation and Drainage (ICID) Congress that 'by constructing water saving society, China will upgrade its resources, use efficiency, improve its eco-environment, enhance its capability for sustainable development, and push the entire society towards the civil development path that features better production development, affluent life for the people, and a sound ecology. Our objective is to prevent aggregate agricultural water consumption from further increasing and ensure that water for grain security will be satisfied through agricultural water saving and enhancement of water use efficiency'.[31] We may not necessarily adopt this Chinese formulation, but it is an example of the kind of focus and precision we need.

In conclusion, we refer to Dr. Manmohan Singh's advice to the state governments of 'understanding and consideration, statesmanship, and appreciation for other points of view'. These are applicable to all the actors in the water sector—the central government, state governments, courts, media, civil society, industry, and farmers. Unless we come together and evolve a consensual framework in India and go beyond the polarised discourses, rivers will continue to divide us, emotionally and politically, leading to a million revolts, the efforts at physical interlinking notwithstanding.

Notes

*   An earlier version of this article along with eighteen case studies was published in the *Economic and Political Weekly*, vol. XVI (7), February 18–24, 2006. Map 1 prepared by Sraban Kumar Dalal (s.dalal@cgiar.org), WWF Int'l–ICRISAT Project.

1.   *Chaturvarnya* refers to the overarching division of castes into a hierarchy of four varnas: the brahmin on the top, followed by kshatriya, vaishya and shudra varnas, and also functions as an ideological justification of the caste system and its hierarchy.

2.   *Infochange Agenda*, 'The Politics of Water', Issue 3, October 2005.

3.   *The Hindu*, December 2, 2005.

4.   The Andhra Pradesh state assembly was stalled for several days on this issue. Opposition and the electoral partner TRS (Telangana Rashtriya Samiti) joined the protest demanding that the Government Order diverting water from the Krishna be withdrawn.

5.   'In a strategic move, Chhattisgarh decided to join hands with Orissa to oppose the Polavaram project coming up in Andhra Pradesh when the Chief Ministers of the three states meet in New Delhi on Wednesday', *Deccan Chronicle*, October 4, 2006.

6.   *The Hindu*, January 13, 2005.

7.   See, for example Allan (1996); Bulloch and Adel (1993); Murakami (1995); Myles (1996); Starr (1995). Periodicals like *Studies in Conflict and Terrorism* have published articles on water conflicts more or less from the same standpoint. Gleick (1993) discusses the history of water-related disputes and water resources systems as offensive and defensive weapons.

8. There is a considerable amount of literature available on some of these, especially about common pool resources, their defining characteristics and the 'fit' between these characteristics and the institutions to manage them. Lele (2004) summarises some of these discussions and debates.

9. This book emerged from a seminar on 'Managing Water Scarcity: Experiences and Prospects' organised at Amersfoort (The Netherlands) in October 1997 by the Indo-Dutch Programme on Alternatives in Development (IDPAD). The book also has a few chapters which deal with experiences from river basins outside India.

10. 'Dialogue on Water, Food and Environment' was set up by ten international organisations. More information on the project is available at *www.iwmi.org/dialogue*

11. The meeting in Bangalore and the subsequent interactions led to the formation of the 'Forum for Policy Dialogue on Water Conflicts in India'. The Forum presently consists of the Centre for World Solidarity (CWS), Hyderabad; Chalakudi River Samrakshana Samithi, Trichur; IWMI–Tata Water Policy Program, Anand; Pragathi, Bangalore; Society for Promoting Participative Ecosystem Management (SOPPECOM), Pune; Vikram Sarabhai Centre for Development Interaction, Ahmedabad; WWF International, Hyderabad; and World Water Institute (WWI), Pune, as also a few independent researchers. Apart from preparing this volume, the Forum also organised media campaigns in five states and a two-day conference on water conflicts on March 21–22, 2005. K. J. Joy of SOPPECOM has been the nodal person for the Forum. For more information about the Forum write to Joy at *joykjjoy@gmail.com* or *soppecom@vsnl.com*

12. An inter-state dispute primarily between Karnataka and Tamil Nadu.

13. John Briscoe and R. P. S. Malik have organised the water conflicts under the following themes: conflicts at the international level, conflicts at the inter-state level, conflicts between upstream and downstream riparians in intra-state rivers, conflicts between communities and the state, conflicts between farmers and the environment, and conflicts within irrigation projects. For details see Briscoe and Malik (2006: 22–26).

14. Thematic review authors include Biksham Gujja, Suhas Paranjape, K. J. Joy, Paul Appasamy, P. B. Sahasranaman, K. V. Raju, Bharat Patankar, Anant Phadke, Ramaswamy Iyer and Sunita Narain.

15. Dalits are also called 'Scheduled Castes' and were 'Untouchables' under the caste system.

16. Grasim's (Chaliyar) is perhaps one of the longest and most bitter stories of water quality conflicts in the country. After nearly thirty-two years, the Grasim factory in Mavoor near Kozhikkode was closed in 2001, leaving the Chaliyar river polluted and more than 200 dead due to cancer and thousands losing much else.

17. Paul Appasamy, in his thematic review on water quality-related conflicts, discusses in detail the various sectors that have been affected because of water pollution.

18. Grasim Industries in Kerala was closed in 2001 and Ghodawat Industries in Kolhapur, Maharashtra was closed after a major riot on New Year's day, 2005.

19. 'Ubeshwarji Ka Nala Divides Two Villages: Inter-village Water Conflict in Rural Rajasthan' by Eklavya Prasad in this volume traces the 300-year-old history and fifty-year-old bitter conflict over sharing water between Dar and Ghelotaon Ka Waas, and suggests that the 'localised intervention in the form of direct communication between stakeholders might be a most effective and sustainable approach.'

20. This is important in the context of the polarised debate on 'large vs. small' or the 'no dam' stand of some environmental movements. 'In some drought-prone areas, for example in many areas in Maharashtra, millions of people are also uprooted due to drought and the lack of supplementary irrigation from dams. They migrate to large cities like Mumbai and live in slums under the most degraded conditions. Second, an extreme assumption that people would never want to move and should always stay where they have their roots also contradicts historical experience' (quoted from Bharat Patankar and Anant Phadke's thematic review on dams and displacement).

21. The *Deccan Chronicle*, October 12, 2006 reported, 'The Central Empowered Committee (CEC) constituted by the Supreme Court has stayed the construction of seven irrigation projects in the State on the ground that they do not have required forest clearances .... In the case of the Indirasagar (Polavaram) major irrigation project, too, the state government had stopped works on April 30, following a CEC direction, which pointed out that the project lacked required clearances.'

22. The Cauvery and Krishna disputes are also covered in an earlier effort by R. Doraiswamy and Biksham Gujja (2004).

23. Ramasway Iyer, in his review article on Transboundary Conflicts, has explained this apparent paradox of India entering agreements with Pakistan, while not being able to resolve the dispute between Punjab, Haryana and Rajasthan.

24. See C. Surendranath in this volume. Following the recent disclosures by the Centre for Science and Environment (CSE), New Delhi about the high levels of pesticide contents in soft drinks in India, the Government of Kerala has banned production and sale of soft drinks in the state, including Coca-Cola at Plachimada.

25. For a detailed discussion on this see Gleick (2002).

26. 'Governments have to follow letter and sprit of their own guidelines ... [they] should not be just rituals'; C.H. Hanumantha Rao in his Foreword in Gujja et al. (2006).

27. In this context it is important to mention that shifting water from the State List to the Union List, a solution that is being suggested by some experts as a way out, especially for inter-state water conflicts, is also not going to work given the complexities involved.

28. The cases of Palar and Noyal basins in Tamil Nadu, the case of Khari river in Gujarat, and cases like the Uchangi dam and Tembu Lift Irrigation Scheme in Maharashtra, all point to this.

29. The details of this normative framework are discussed in Joy, Paranjape and Kulkarni (2004).

30. Speech of H.E. Mr. Wang Shucheng at the opening ceremony of the 19th International Commission on Irrigation and Drainage Congress (ICID) on September 15, 2005.

31. For details see 'Alternative Restructuring of the Sardar Sarovar Project: Breaking the Deadlock' by Suhas Paranjape and K. J. Joy in this volume, and Paranjape and Joy (1995).

References

Allan J. A. (ed.). 1996. *Water, Peace and the Middle East: Negotiating Resources in the Jordan Basin*. New York: St. Martin's Press.

Briscoe John and R.P.S. Malik. 2006. *India's Water Economy: Bracing For a Tubulent Future*. Delhi: Oxford University Press. pp. 22-26.

Bulloch, John and Adel Darwish. 1993. *Water Wars: Coming Conflicts in the Middle East*. London: Gollancz.

Doraiswamy R. and Biksham Gujja. 2004. *Understanding Water Conflicts: Case Studies from South India*. WWF-ICRISAT Project, Patancheru (Andhra Pradesh), and *Pragathi*, Bangalore.

Gleick, Peter. 1993. 'Water and Conflict: Fresh Water Resources and International Security', *International Security*, vol. 18: 79-112.

———. 2002. *The New Economy of Water: The Risks and Benefits of Globalisation and Privatisation of Fresh Water*. Oakland, CA: Pacific Institute for Studies in Development, Environment and Security.

Gujja, Biksham, S. Ramakrishna, Vinod Goud and Sivaramakrishna. 2006. *Perspectives on Polavaram: A Major Irrigation Project on Godavari*. New Delhi: Academic Foundation.

Myles, James R. 1996. *U. S. Global Leadership: The U.S. Role in Resolving Middle East Water Issues*. Carlisle Barracks, PA: US Army War College.

Murakami, Masahiro. 1995. *Managing Water for Peace in the Middle East: Alternative Strategies*. New York: United Nations University Press.

Paranjape, Suhas and K.J. Joy. 1995. *Sustainable Technology: Making the Sardar Sarovar Project Viable*. Ahmedabad: Centre for Environment Education.

Paranjape, Suhas, K.J. Joy and Seema Kulkarni. 2004. 'Multi-stakeholder Participation and Water Governance: A Suggested Normative Framework', Paper presented at the IRMA Silver Jubilee Symposium on 'Governance in Development: Issues, Challenges and Strategies', December 14-19, 2004, IRMA, Anand.

Rogers, Peter and Alan W. Hall. 2002. *Effective Water Governance*. Stockholm: Global Water Partnership.

Sharachchandra, Lele. 2004. 'Beyond State-Community and Bogus "Joint"ness: Crafting Institutional Solutions for Resource Management' in Max Spoor (ed.), *Globalisation, Poverty and Conflict: A Critical "Development" Reader*, Dordrecht: Kluwer Academic Publishers: pp 283-303.

Shiva, Vandana. 1991. *Ecology and the Politics of Survival: Conflicts over Natural Resources in India*. New Delhi: Sage Publications India Pvt Ltd.

———. 2002. *Water Wars: Privatizatiion, Pollution, and Profit*. Cambridge: South End Press.

Starr, Joyce Shira. 1995. *Covenant Over Middle Eastern Waters: Key to World Survival*. New York: H. Holt.

Vaidyanathan, A. and H. M. Oudshoom. 2004. *Managing Water Scarcity: Experiences and Prospects*. New Delhi: Manohar Publishers and Distributors.

PART 1

Contending Water Uses: A Review

Biksham Gujja

It is important to emphasise that the case studies included in this volume elude simple categorisation, and do not easily fit the labels on the boxes we put them in. Each of the themes exhibits several dimensions of the same conflict. It is with this caveat in mind that one has to approach the ten papers dealing with contending water uses. They provide a distinct flavour of the type of conflicts that prevail in the country. Each paper tells a story—a story that might relate not only to contending uses, but many other themes as well; they are not mutually exclusive. Some stem from new claims on old resources, some from the unplanned consequences of structural interventions, some from claims over improved and amplified resources or services, some from the rural–urban divide, and some from plain mistrust.

Old Resources, New Claims

Water allocation for different kinds of uses in the pre-Independence era evolved in the context of the then prevailing social and political environment. Over the years, the context has changed, the balance of power has shifted, the relationship between different kinds of uses has altered, and the groups which perceived that their rights were being neglected are now asserting their right to a 'fair' share of water.

One of the case studies centres around the fight over water allocated to an ecosystem of global significance—the famous Keoladeo National Park (KNP). Also known as the Bharatpur wetlands, KNP is well known as a Ramsar and World Heritage site. A unique sanctuary for birds, the wetlands are entirely man-made and have attained global significance for their biodiversity and bird populations. Close to 380 species of birds are found in this 29 km² stretch, approximately 10 of which comprise marshes and bogs. It is a classic example of a society which has stood by its responsibility to protect unique ecosystems by allocating adequate water for them. However, the question now being raised is: at whose cost?

The park requires about 10 Mm³, and given its importance, this quantity is almost inconsequential in the context of allocations at the national level. It is, however, not insignificant for the twenty-one villages in the area that have lost several customary rights since the park was established. The present conflict is the culmination of a long history of mistrust between the park and the people. The proposed solution to the long-drawn-out and bitter conflict over Keoladeo is a structural one—i.e. to bring water from the river Chambal exclusively for the park.

The conflict over Keoladeo's water is not just about competing demands of ecosystems and irrigation; it reflects different perceptions within our society. While urban opinion concentrates on trying to protect the unique ecosystem, rural perspective sees it as people's water being allocated to the park. The disconnect between the two is the larger issue and can be seen to some extent in every national park. The solution involves not just water, but also how the park management system establishes a dialogue based on mutual respect with the people of the area.

Keoladeo also presents another interesting and associated issue—sharing the benefits of tourism. Unique natural sites such as KNP offer opportunities for tourism, but the local communities who have to make sacrifices and adjustments for this often receive no benefits or compensation.

This well-documented case presents various dimensions of this important conflict, provides all relevant facts and also suggests a way out. The suggested long-term solution to the dispute involves water diversion from a different river basin. While such diversions might help, it is likely to raise another debate related to the dilemmas of long-distance and inter-basin water transfers.

Structural Interventions and Unplanned Consequences

Two cases—Kuttanadu and Loktak—illustrate how structures can benefit one group of users while adversely impacting others. These cases are not so much about changes in quantity or quality of water, but about change in the natural flow of water. Interestingly, both cases are linked to Ramsar sites.

The Kuttanadu case demonstrates how investments meant to result in benefits for local people can go wrong, benefiting one section over another. The Vembanadu–Kole wetland is one of the largest, brackish ecosystems in the country. It is fed by ten rivers and is also home to a large waterfowl population during winters. Over ninety species of resident birds and fifty species of migratory birds are found in the area. In Kuttanadu, the government invested in structures, intending to divert and/or alter the flow of rivers with the stated objective of aiding paddy farmers. There were two major structural interventions and these have resulted in unforeseen consequences. First, the Thottappally spillway was completed in 1955 to 'flush' floodwater from three rivers—Pampa, Minimal and Chenoil. Second, the Kuttanadu–Thannirmukkam bund was constructed to regulate saline water intrusion from Vembanadu lake. The bund was planned at the point where the lake had minimum width (1,250 m). The system was supposed to keep areas upstream of Vembanadu fit for cultivation by regulating saline water through a bund and shutters. However, the shutters were not properly designed and soon suffered severe damage. As a result, the salt water content of the lake increased which in turn reduced the productivity of the waters. The movement of organisms upstream from the sea was blocked and this drastically brought down the fish population. The social balance between fishermen and paddy farmers was also disturbed.

The poor design and construction of the structures totally ignored natural processes as well as users other than the primary beneficiaries. The result has been loss of income and livelihoods without, as it became clear later, benefiting the farmers either. In the final analysis these structures have damaged the social fabric as well as the ecosystem. The country, which has signed an international treaty to protect wetlands, has failed to keep its obligation. Kuttanadu could well be turned into a win-win project if a dialogue is established between the stakeholders and additional investments based on a thorough understanding of the ecosystem are made.

The structural interventions at Loktak, Manipur—another Ramsar site—offer other insights. The size and depth of the lake was shrinking; the solution offered was a barrage downstream to maintain the water level in Loktak, a structural intervention intended to help the lake as well as produce electricity. But the construction of the barrage has only created more problems including the flooding of productive areas. Land in Manipur, particularly land along riverbanks, is precious. Nearly 20,000 ha, 20 per cent of the total cultivable land along the Manipur river, was permanently flooded. Increase in lake levels caused yet another 6,000 ha to be submerged.

The interventions did not serve their primary purpose either. The lake ecosystem did not improve; on the contrary, natural flow was altered and disturbances in the fine balance between nutrients, silt and floating islands or *phumdis* resulted in the deterioration of the ecosystem. Fish biodiversity was affected and four species disappeared. The thinning of the *phumdis* is beginning to threaten the Sanghi deer in the Keibul Lamjao National Park.

Loktak exemplifies situations where structural interventions, though well-intended, need to be evaluated and implemented properly. They need to be made on a scientific basis, after analysing long-term data and monitoring effects, after establishing proper water management systems and coordination between various authorities.

Both these Ramsar sites are unique ecosystems. India, as a signatory to the Ramsar convention, has an obligation to protect these wetlands. These cases warn us of the dangers of undertaking structural interventions without a proper understanding of the consequences.

Claims over Improved Water Resources

The case of Vadali in Chotila *taluk*, Gujarat, makes clear that improvement of water resources could lead to conflicts in the absence of clear guidelines and the implementation of those guidelines. Normally check dams and percolation structures are built with state investment. In this case, the farmers who wanted the benefits of de-silting a check dam did not want other communities to share the water by building wells close to the source. A community-based organisation, WASMO, aided the poorer sections; they were able to construct a well and benefit from de-silting the dam.

Several NGOs have been working hard at improving water resources at the local level by introducing various methods such as rainwater harvesting and tank restoration. However, even after resources have been improved, the benefits might not reach the poorest of the poor. In fact the drinking water crisis in rural areas might actually provide an opportunity for the better-off to exploit the situation by landing government contracts to supply drinking water, which they do by transporting water from farm wells and supplying it to the villages.

It is important that community-based programmes to improve water resources also come with a policy framework which ensures that water does reach the poor. Lack of such a policy and clarity on the subject could lead to new conflicts over sharing improved resources.

Rural–Urban Conflicts

Increasingly, cities are driving the water agenda. Urban centres demand huge quantities of water and large-scale, long-distance diversions are being driven by this demand. This obviously has implications for the rural economy and livelihoods. For instance, Chennai faces acute water shortage. To meet this demand, water markets have developed around the city. The water is mostly brought from peri-urban areas in tankers and sold for domestic use. While some in peri-urban areas milk the situation, others lose out.

In the absence of guidelines on the maximum depth of wells and at what width they can be drilled, there is a great deal of confusion and consequently, rising tensions. Those who make a living from supplying water to the city are rapidly depleting groundwater, thus harming the environment and depriving others of their fair share of water for irrigation. Poor and middle farmers are losing heavily in this scenario.

In the case of Chennai, there are ways of dealing with the problem. The large number of traditional tanks around the city could offer a solution—de-silting these would go a long way in recharging groundwater. The question again is: who will pay? The case study suggests that instead of investing in projects to bring water from across long distances, the state should invest in recharging groundwater in peri-urban areas. It is still not clear how increased recharge will be shared equitably among all stakeholders, but there is at least some ongoing dialogue on the issue.

The rural–urban conflict over water is emerging as a serious and potentially dangerous conflict with serious consequences for large populations. Such conflicts are particularly aggravated during drought years. Another case is that of Goa, although the scale of the conflict is small. Farmers in Goa sell water at the rate of Rs 40 per m³, a more profitable venture than using the water for rice cultivation. In Sangolda, two farmers dug deeper and installed heavy-duty pumps to extract water to sell to the city. Those deprived protested and the conflict escalated from the community level to reach the courts. Scientific studies proved that over-exploitation of the groundwater was unfair, and the court intervened to stop water extraction. But, in spite of the court order and the Goa Groundwater Regulation Act 2002, water continues to be extracted. This case study emphasises that although we need regulation and court orders, it is imperative to have appropriate institutions to enforce them.

A Matter of Trust

Majuli, in the river Brahmaputra in Assam, is one of the biggest river islands in the world. Its area is about 875 km² and it is home to more than 160,000 people. The island is under threat due to massive erosion: its area was about 1,245 km² in 1950, but it has lost almost one-third of it to the river in the past five decades. While it is not known to what extent human activities and faulty flood protection have contributed to this rapid erosion, the state of mind of the local people is fragile. There is already a history of displacements in northeast India due to large projects without proper compensation. The point of conflict in Majuli at present is a proposed bridge that will change the course of the river. The local population is displeased with the proposal, fearing that the bridge will further threaten the island and its fertile land. The situation is exacerbated by lack of comprehensive feasibility studies, ecological studies and an understanding of why the island is eroding. There are fears that this project will involve displacement, loss of land and livelihood. The main issue in Majuli is a deep mistrust of infrastructural projects.

Conclusion

Conflicts over water are likely to intensify at every level. Lack of understanding of ecosystems and large investments with insufficient data are further enhancing the conflict. The cases clearly demonstrate the complexities of water conflicts due to contending uses in the country.

These ten cases illustrate three salient points:

1. Structural interventions to natural systems might or might not increase water quantity but can certainly lead to conflict because of the changes they introduce. It is important that these be implemented with all relevant studies carried out in advance. It is equally essential that people be informed about these projects. Dialogue with the people is crucial if we are to avoid some costly mistakes.

2. Improving water resources through rainwater harvesting, check dams, de-silting of dams, etc. are good measures, but it is also necessary to reach agreements over who gets what from the improved water resources. Often the poorest of the poor and the ecosystems involved lose out heavily when it comes to receiving the benefits of improvements.

3. The rural–urban aspect of water sharing is complex. Court orders and legislation can only lay down the law and are limited when proper enforcing institutions are absent. Therefore, the country needs to evolve a framework to regulate how urban areas will tap rural resources. We need to work out how equity is to be built into the system while we strive to meet the growing demands of urban areas. Finally, we need to build trust between communities, and initiate dialogue before rushing to build bridges across rivers. Crossing turbulent rivers will be a lot more difficult when they are stirred by the neglected feelings and sentiments of the people.

Choking the Largest Wetlands in South India: The Thanneermukkom Bund

V. K. Ravi Varma Thampuran

Kuttanadu Rises from the Sea

Once upon a time, all of Kuttanadu was part of a shallow coastal area of the Arabian Sea that extended right up to its eastern border. That is also why the subsoil in Kuttanadu often contains sediments of marine origin and typical marine mollusc shells. It is presumed that it was separated from the sea by a narrow strip of land when the delta formed by the rivers Pampa, Achenkoil, Manimala and Meenachil joined with low-lying areas in and around Vembanadu lake. Even today vast areas below 0.5 to 3 m mean sea level (MSL) remain waterlogged with fresh water during monsoon and saline water in summer. The waterspread area consists of Vembanadu lake, the river systems, thousands of canals, channels and other waterways covering a total area of about 186 km².

The 1,152 km² Kuttanadu is spread over ten *taluks* in the three revenue districts of Alappuzha, Pathanamthitta and Kottayam in Kerala and has a population of approximately 1,500,000. The temperature ranges from 21°C to 36°C and humidity is high throughout the year. Average annual rainfall is approximately 3,000 mm, of which 83 per cent is received during the two distinct rainy seasons: the southwest monsoon from June to August and the northeast monsoon from October to November.

Considered the rice bowl of Kerala, Kuttanadu comprises 1,129 *padasekharams* (paddy field clusters) with individual sizes ranging from 4 ha to 1,000 ha and a total area of over 55,000 ha. Before every season water is pumped out and agriculture is practiced in the *padasekharams*. The government provides pumping subsidy to promote paddy cultivation. In addition, agriculture is also practised in thirty-two Kayal lands[1] that vary in size from 30 to 1,000 ha. These lands began to be reclaimed from Vembanadu lake about sixty years ago and it was thirty years before they were fully reclaimed. The annual rice production in Kuttanadu is around 2 lakh tons. Approximately 600,000 farmers, 200,000 agricultural labourers and 21,000 fishermen live in Kuttanadu. Sadly, today people are migrating to other areas for their livelihood and Kuttanadu is being faced with water pollution, disease and environmental hazards.

Multiplicity of Conflicts in Kuttanadu

The major natural resource in Kuttanadu is water and everyone living here connects with it in some way or the other. Unfortunately, its unscientific management has led to many conflicts and the major ones are summarised here:

Farmers versus Fishermen

Farmers, especially Kayal land cultivators, are the force behind the construction of Thanneermukkom bund (discussed later in detail) to block salt water intrusion. Fishermen have been against the bund from the very beginning because they require salt water intrusion for a better catch. The conflict between these two major communities is one of the crucial aspects of the Thanneermukkom bund issue.

Tourism versus Environment

Backwater tourism is at its peak in Kerala, and Kuttanadu is now an internationally popular destination. As a result, pressure on the environment is also increasing day by day. Almost all the boats, the major components of backwater tourism, empty their entire waste, including plastic, directly into the Vembanadu lake. Toilets on the houseboats directly let out waste into the lake. Many resorts have mushroomed, and their construction has damaged and destroyed mangrove tracts.

Fig. 1. *The rice bowl: Kuttanadu during the paddy season.*

Public Health versus Development

So-called development has increased pollution and also changed the geography of the lake. There has been a rise in infectious diseases like cholera, diarrhoea, jaundice and Japanese encephalitis. Asthma and skin diseases have also become common.

Development versus Natural Ecosystems

Natural ecosystems are ruined because the approach to development has been unscientific.

Fresh Water Needs versus Natural Flushing Needs

The salt water intrusion and tidal ebbs and flows are natural flushing processes for Kuttanadu. The bund blocks them both. A section of the people and the administration still believe that the bund keeps the Kuttanadu water fresh, even though the water is highly polluted. In their eyes, salt seems to be the only pollutant!

Fig. 2. *The land of inland transport system is fast being replaced by road transport. Countryside channels are becoming non-navigable due to uncontrolled growth of water hyacinth after the completion of the Thanneermukkom bund.*

Traditional Agricultural Practices versus Modern Green Revolution Practices

The Green Revolution brought in uncontrolled use of chemical fertilisers and pesticides which, in fact, increased pests and diseases because the natural enemies of many disease-producing insects also perished due to the overuse of chemicals.

Farmers versus Agricultural Labourers

Formerly, both farmers and labourers were involved with agriculture. It was their life. But continuous political intervention has destroyed the mutual understanding between these two groups, making farming an unprofitable venture.

Natural Inland Water Bodies versus the Sea

Lakes can act as a buffer between the rivers and the sea. If Vembanadu lake is damaged, the sea will intrude upstream into all the four rivers and create many environmental problems.

Tamil Nadu's Water Demand versus Kuttanadu's Environment

The central government is considering Tamil Nadu's proposal to link Vaippar to the rivers Pampa and Achenkoil in Kerala. If the project is sanctioned, the inflow to Kuttanadu will be reduced drastically and its unique ecosystem will be adversely affected.

Farmers, Fisherfolk and the Thanneermukkom Bund

The first component of the Kuttanadu development scheme was the Thottappally spillway that was completed in 1955. The spillway was aimed at flushing extra floodwater from the Pampa, Manimala and Achenkoil into the sea before it could flood Kuttanadu. A 9 km long, 366 m wide channel was planned and the spillway was placed at the tail end just before the channel opened out into the sea. Due to inefficient implementation, the channel was never widened properly and the spillway has never served its purpose. Moreover, due to lack of proper maintenance, salt water intrusion through the spillway has now become a permanent concern for the farmers.

Thanneermukkom bund is the second component of the Kuttanadu development scheme. Originally a regulator was planned at Thanneermukkom where the Vembanadu lake is at its narrowest width of around

Fig. 3. *Natural migration in peril: Kuttanadu, a well-known destination for migrating and native avian species faces decline in bird count due to environmental degradation of the wetlands.*

1,250 m. The lake is linked to the Arabian Sea at its tail end and its association with saline water means that salt water intrusion during high tide is a permanent problem for farmers. Earlier, people used to carry out rice farming by constructing *orumuttus* (temporary bunds to prevent salt water intrusion). The Thanneermukkom regulator was planned by the government to avoid these recurring expenses. The original plan was to construct ninety-three iron shutter, across the lake. The obstruction from Kochi port authorities delayed its construction. Thirty-one shutters each were installed at the western and the eastern ends. The middle portion remained incomplete.

In 1975, the farmers of Kuttanadu started an agitation to press for completion of the bund. Thousands of farmers reached the incomplete bund in country boats and worked free of cost for almost seventeen days to construct a permanent earthen bund in the middle of the lake and joined the western and eastern sides of the iron barrier. Thus the earthen bund replaced the originally planned thirty-one shutters.

The 1,250 m long bund is now a permanent wall across the lake. Though it helps agriculture to a certain extent, it also checks sea water and affects fisheries, giving rise to conflicts between farmers and fishermen. The fish catch has decreased after the bund was completed. The fishermen could do little when the farmers built the earthen wall. But it is alleged that they now regularly keep the iron shutters open by inserting wood and stone wedges. The general agreement is to close the bund fully by 15 December every year and open it by 15 March the following year. Fishermen protest if the bund is closed for more than three months. (Together with environmentalists they are now demanding that the bund be kept open for a full year on an experimental basis.) Having experienced the adverse impact of the bund, a section of the farmers is also now supporting this idea. Two years ago, the then state minister for irrigation, T.M. Jacob, had announced that the bund would indeed be kept open. However, an expert committee advised against this move. Fishermen are now questioning the reliability of the report.

As mentioned earlier, the tides provided a natural cleaning and flushing system for Kuttanadu that the bund now prevents, leading to numerous health problems for the people in Kuttanadu. For instance, the outbreak of epizootic ulcerative syndrome in 1991 exhausted the fisheries to a large extent. In 1996–97 the epidemic of Japanese encephalitis in Kerala had its origin in Kuttanadu. At least twenty-five people succumbed to the disease in Kuttanadu alone. The bund has also created siltation, and the depth of the Vembanadu lake—the largest wetland ecosystem in south India that was declared a Ramsar site in 2002—has decreased by 5 m.

Before the bund was built, the annual fish catch from Vembanadu lake was 14,000 to 17,000 tons. After the completion of the bund, the northern portion of the lake directly linked to the sea yielded a fish catch of 7,200 tons between 1989 and 1990, whereas the southern portion that was cut off from the sea yielded only 500 tons. Fishermen living on the southern side of the bund have lost their livelihood and some have even turned to construction labour. A large number of agricultural labourers have also shifted to other work.

Farmers are giving up paddy cultivation because it is no longer profitable. Paddy cultivation has fallen from 55,000 ha to 39,200 ha. There are many factors behind this decline. Farming has become more expensive due to increase in pests and diseases—a side effect of the Thanneermukkom bund. The bund, constructed against all the norms of a wetland ecosystem, has become a matter of concern for all. The question is, who will bell the cat?

Tourism, Roads and Aggressive Farming

The third component of the Kuttanadu development scheme is the 24 km long road that connects Changanacherry to Alappuzha. Constructed in the late 1950s, it passes through the middle of Kuttanadu. Thereafter, innumerable roads and bunds have been constructed across the low-lying waterspread area creating blocks in water flow. Previously country boats and motorboats were the only mode of transport in this area. The shift from water to road transport has led not only to the decline of water transport but also the neglect of the numerous countryside water channels. Increased pollution and flooding are the consequences.

Backwater tourism is the new phenomenon in Kuttanadu. All the 300 houseboats being operated in and around Kuttanadu dump their waste, including human excreta, into the lake and other water bodies. The numerous resorts are also adding to the pollution, and encroaching on the lake and other water bodies.

Aggressive farming started after the construction of the Thottappally spillway and the Thanneermukkom bund resulting in a dramatic spurt in the use of chemical fertilisers and pesticides. Whereas the national standard for the use of fertilisers is 75 to 100 kg/ha, the usage in Kuttanadu is 400 kg/ha. An estimated 22,000 tons of chemical fertilisers and 500 tons of pesticides are used in Kuttanadu.

The e-coli content in Kuttanadu water is ten times more than is permissible. The e-coli contamination is a result of the Sabarimala pilgrimage and the lifestyle of the people of Kuttanadu. At least 90 per cent of the latrines of Kuttanadu let out their waste into the nearest water bodies. The river Pampa that flows from the Sabarimala hills to Kuttanadu also brings a huge quantity of human waste.

Suggestions to Protect the Kuttanadu Ecosystem

1. Thanneermukkom bund should be completely demolished, or maintained only for road traffic.
2. The use of fertilisers and pesticides should be minimised and organic farming popularised.
3. Tourism sector should be properly educated and upgraded to follow eco-friendly practices.
4. The water transport network should be revived at least for cargo movement since it is a cheaper option.
5. The Pampa–Achenkoil–Vaippar river-linking plan should be abandoned.

Notes

*All figures by author.
1. Kayal lands are paddy fields reclaimed from the kayal, or lake, also known as backwaters. In other words, it is a stretch of lake converted to paddy fields by proper bunding and then dewatering. Each kayal land is located within the lake, surrounded by water. A kayal land can extend up to more than 1,000 acres and lies 2–6 m below sea level.

Biodiversity Versus Irrigation: The Case of Keoladeo National Park

Malavika Chauhan

Background

KEOLADEO NATIONAL PARK IS SITUATED IN EASTERN RAJASTHAN ON THE EDGE OF THE Gangetic plains, 2 km southeast of Bharatpur town and 50 km west of Agra (27°07'–27°12'N and 77°29'–77°33'E). The park, known locally as 'Ghana', is a mosaic of dry grasslands, woodlands, swamps and wetlands spread over 29 km². About 900 ha are divided into small, seasonally inundated reservoirs by a series of bunds and dykes. Bharatpur experiences climatic extremes—hot, dry summers and freezing cold winters, with temperatures ranging from 0 to 2°C in winter to above 48°C during summer. The three-month monsoon is usually very wet and is followed by a post-monsoon spell. The mean annual precipitation is 662 mm, with rainfall on an average of thirty-six days in a year.

Designated a Ramsar site in 1981 and a World Heritage site in 1985, Keoladeo was famous as a wintering site for a sub-group of the western population of the Siberian crane. Though this species is now locally extinct, extensive habitat management over the past century has resulted in exceptionally high biodiversity including over 370 species of avifauna. For a few years the park had a resident tigress (she died in May 2005) who highlighted its importance as a solitary natural habitat in a vast agricultural landscape.

Three-centuries' old records describe a depression south of Bharatpur town that supported a thick forest subject to monsoonal flooding. To protect the town, Suraj Mal, the then ruler of Bharatpur state, constructed Ajan bund sometime between 1726 and 1763, 1 km from Keoladeo, to

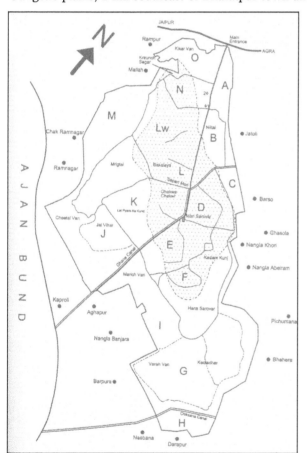

Map 1. *Keoladeo National Park showing Ajan bund to the south, and the two canals which enter the park.*

restrain destructive floodwaters and store water for the lean season. The bund continues to supply irrigation water to adjoining regions. During the 1850–90s, inspired by British game reserves, Prince Harbhanji of Morvi state, Gujarat, during his term as administrator of Bharatpur developed the depression into a duck shooting reserve by building bunds and dykes to create a series of reservoirs. Sustained by rich organic material, invertebrates, and millions of fish fingerlings that enter with the onrush of water during the monsoons, the area rapidly developed into a highly productive system of freshwater marshes attracting large populations of migratory waterfowl. The area was flooded for the first time in 1901 from Ajan bund via Ghana canal. Water was diverted into the reservoirs through a system of canals, gates and sluices. Ajan bund itself was fed by canals from two rivers, the now dry Banganga and the ephemeral Gambhir. Lord Curzon inaugurated the reserve with an organised duck shoot in 1902.

Game hunting was, however, only one of the reasons for the creation of Keoladeo; others included the provision of grazing facilities for village cattle and habitat for feral cows. This created a large and successful community of herders in surrounding villages who used the wetland area as buffalo pasture and the terrestrial area as dry pasture for cows. With the creation of the National Park, however, usage of the park's biomass was discontinued and the herding community gradually disappeared. Most shifted to other occupations, including agriculture and tourism-related business. Milch animals are now stall-fed while dry, old or diseased cows are dumped in the park, creating a large population of wild cows. Buffaloes, on the other hand, are precious and are never left in the park. Fodder, however, continues to be collected by many families.

A History of Conflicts

Keoladeo is unique in that it is a man-made, biodiversity-rich zone in a predominantly arid and highly populated rural landscape. In pre-Independence India the area was a common property resource used by local herder communities, but with Independence the first expressions of discontent surfaced, fuelled by a need for arable land and irrigation water. Since then the park has faced several threats, and survived, with its administration and interested groups continuing to learn how to manage high biodiversity under difficult conditions. Keoladeo is surrounded by twenty-one villages (total population approximately 15,000) and the adjoining Bharatpur town (population approximately 150,017). Under the circumstances it is inevitable that the issue of seasonal water requirement for the park and that of irrigation in the surrounding rural landscape becomes a contentious one; and has, in fact, been a long-standing reason for discontent and conflict in the region.

Keoladeo's Water Requirement

The park's basic ecological water requirement is estimated at 9.5 Mm³ (or 350.8 mcft) [Vijayan 1991]. However, to maintain the ecosystem at an intermediate phase of disturbance necessary to maintain high biodiversity in the wetlands, occasional high floods are required. It is now understood that water requirement stands at 15 Mm³ (*http://www.unep-wcmc.org/sites/wh/keoladeo.html*), with a minimum of 9.5 Mm³ and occasional high floods. In normal rainfall years the park receives at least 8.15 Mm³ of water from the Panchna dam. Additions to this are through an increase in the Gambhir's flow downstream of the Panchna from catchment runoff, and from direct precipitation onto the park area. Waters from the Panchna catchment are also rich in organics and organisms essential for the continued survival of the park. However, this in itself is not enough in the long run.

The usage of groundwater for the park is a purely temporary measure to allow resident species to survive the lean season, and is recognised as such by park managers. The groundwater is saline and cannot sustain the park. Politicians speak to the media about needing funds for developing groundwater resources for the park but their desire to provide a solution is not backed by biological facts, or by the existing hydrology and water quality of the region.

The Conflict

There were originally two principal sources of water to the park and the surrounding agricultural district of Bharatpur—the rivers Gambhir and Banganga—which fed water to Ajan bund. Canals from the bund distribute water in the surrounding region. However, the Banganga dried up several years ago as a direct result of catchment deforestation and water diversions leaving the Gambhir as the sole water supplier.

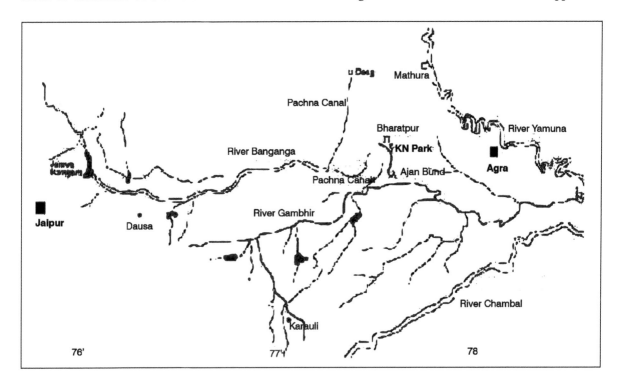

Map 2. *Map of the region surrounding Keoladeo National Park, showing the canal network and the Karauli region.*

In 1991, the Panchna dam was constructed on the Gambhir in Karauli *tehsil*, district Karauli, Rajasthan, to mitigate high floods and fulfil the irrigation needs of the local farming community. The dam is approximately 100 km from Keoladeo. Panchna (*panch*, meaning five) was constructed at the confluence of five rivers. Live storage of the reservoir is 52.65 Mm³; intercepted catchment area is 31 km²; average monsoon rainfall is 724 mm. The dam itself is earthen, with Full Reservoir Level (FRL) at 258.62 m.

There are two dimensions to the issue of conflict over water for Keoladeo National Park. One has been highlighted in the media, while the other has been a quieter but more persistent issue within Bharatpur district:

1. The first and older dimension has been conflict over the water in Ajan bund. Every year, water allocation for the park vs. that for local farmers is an issue of contention. The dispute often escalates and forest fires occur in the park with unnatural regularity during the dry season. Every year, the park administration has to lobby at the State Irrigation Department for their quota of water from Ajan bund.

2. The second dimension surfaced in the post-monsoon season of 2004; this had to do with the demands of upstream agriculturalists. These are farmers in the command area of the Panchna dam, which irrigates

Fig. 1. *The Irrigation Department's board at the Pachana reservoir.*

Fig. 2. *Panchna Dam, on the reservoir side.*

about thirty-five villages over a gross command area of 11,172 ha. There is no major difference in the demands of the two groups of farmers. The authorities need to work out a long-term solution that takes a holistic approach to providing for farmers in both regions.

The latest conflict developed in August 2004, after a third consecutive year of low rainfall. The agitation was precipitated by a decision of the Rajasthan state government committee headed by Chief Minister Vasundhara Raje to release 8.15 Mm^3 water from the Panchna reservoir to Keoladeo, which at the time had a storage of 35.7 Mm^3. The park had until then received a minimal 0.5 Mm^3 of water.

In September 2004, farmers from the command area of Panchna dam protested the release of water for Keoladeo. Men and women pitched tents on the dry riverbed below the dam and dared the irrigation department to release water. The Vasundhara Raje government reversed its previous decision, condemning Keoladeo to a dry year. The agitation was, according to some articles in the media, politically motivated. An article in *Outlook* (*http://www.outlookindia.com/pti_news.asp?id=248925*) said that Bhawani Singh Rajawat, Parliamentary Secretary of ministerial rank, alleged that farmers from Karauli (which falls in the Lok Sabha constituency of Union Minister of State for Environment, Namo Narain Meena) had been instigated by local Congress leaders. He also said that though the park was important, the rights of the local farmers needed to be safeguarded, particularly during drought situations. According to Atar Singh, an advocate from Karauli who led the farmers' agitation against water release for Keoladeo under the banner of the Panchna Pani Bachao Sangharsh Samiti, 'The dam was constructed specifically for irrigation purposes with World Bank assistance. It is named Panchna Bandh Sinchai Pariyojana (Panchna Dam Irrigation Project).' (*http://www/outlookindia.com/pti_news.asp?id=248925*).

The reversal of the committee's original decision kicked off a spate of pro-park protests and media articles. Organisations and individuals petitioned, agitated, sat on *dharnas*, led rallies and held prayers. These included the Tourism and Wildlife Society of India (TWSI) that petitioned the courts, the Ghana Keoladeo Natural History Society and the 'Protect Keoladeo National Park' committee. Rajasthan legislators formed a green lobby group for the conservation of environment and wildlife in the state to exert pressure on the government to release water from Panchna. 'We shall form a group of like-minded MLAs from all political parties to give voice to environmental issues like saving of Keoladeo National Park', Congress legislator C.S. Baid told reporters (*http://www/hindu.com/2005/02/20 stories/2005022003190500.htm*). Baid and his party colleague Harimohan Sharma said they would try to make policymakers understand the need to protect fragile ecosystems like the Keoladeo National Park. The General Secretary of Rajasthan Chamber of Commerce, K.L. Jain said the infrastructural resources of the Chamber would be at the disposal of the Tourism and Wildlife Society of India to ensure the protection of the sanctuary. TWSI has already launched a multi-pronged campaign demanding water, and the matter is also pending with the Central Empowered Committee of the Supreme Court.

The Opposing Stands, the Stakeholders

One of the best birding sites in Asia, Keoladeo attracts more than 100,000 visitors annually who range from serious bird watchers to school children and tourists—45,000 of whom are foreigners. In addition the location of the park within the 'Golden Triangle' en route from Delhi to Agra allows even low budget tourists to stop over. Local stakeholders include inhabitants of the eleven surrounding villages and Bharatpur town—rickshaw pullers, guides, tour operators, tourists, park staff, the royal family of Bharatpur, local businesses, shops and hospitals, and others who are linked to park-related markets.

However, apart from some hoteliers, most of Bharatpur's local stakeholders, though aware of the situation, are not seriously worried. One reason for this may be the fact that Keoladeo has seen consecutive years of water scarcity before and the biodiversity has always revived, though with subtle changes like the proliferation of weeds. A second reason may be that the tourism sector has not yet been seriously hit. However, if the park is subject to long-term water shortage, it will lose both biodiversity and tourism, jeopardising the Rs. 500 crore it earns annually, apart from losing genetic diversity— something that cannot be easily valued.

Current Status

Following numerous complaints and after being approached by TWSI, the Central Empowered Committee held its first hearing on January 31, 2005. The Rajasthan state government officials expressed their inability to release water due to irrigation commitments in Karauli district. They mooted the possibility of providing water from the Chambal river through a pipeline planned to bring drinking water to Bharatpur district. Submissions made by the Rajasthan Irrigation and Public Health Engineering Department on behalf of the state government were opposed because they did not represent the government—the presentation did not carry the signature of the chief minister who was abroad.

The next CEC hearing was scheduled for February 21, 2005, when the State Irrigation Department was to appear before it with 'alternative plans' for supplying water to the sanctuary. On March 10, a three-member special bench passed a notice subsequent to the submission of a twenty-page report by the CEC. The court directed the government of Rajasthan to release water to the park from Panchana dam ('Vasundhara Government Refues Water to Keoladeo Park'–*http://www.indianexpress.com/res/web/ple/full_story.php?content_id=64100*, 'Water Crisis Threatens National Park'–*http://www.deccanherald.com/deccanherald/feb062005/n13.asp*, 'Keoladeo—Wetland Turns Woodland'–*http://www.deccanherald.com/deccanherald/feb182005/n15.asp*).

Fortunately monsoons have been good and as of July 25, 2005, Keoladeo had already received more than 8 Mm³ of water. Though this makes the ongoing case temporarily redundant, there needs to be a policy that makes it mandatory for a certain amount of water to be set aside for the park, particularly during dry years.

Scope for Dialogue

The issue is one of a rapidly decreasing water supply amidst a growing number of users who require larger quantities with every passing year. The need is to increase quantities of available water. V.S. Vijayan noted, 'It is unlikely that the park will be able to continue to draw the required minimum of 9.5 Mm³ water from the Ajan Bund reservoir, and therefore a suggestion has been made to bring water from River Chambal' (Vijayan, 1991). The state government has since begun a multi-crore scheme to supply drinking water from the Chambal to the Bharatpur region. According to media reports quoting state government officials, this pipeline is also meant to supply water to the park. However, it is clear (see box) that the water supplied through this scheme will be treated as drinking water and will be made available to 930 villages in Bharatpur district.

If used for the park, the chemically treated water, rid of organic matter, will condemn millions of fish fry, invertebrates and young amphibian to oblivion.

The Chambal–Bharatpur Pipeline: Chemically treated water to be supplied

A Press Release of February 17, 2003, on the Essar website is titled—'Essar awarded the Rs.137 crore Chambal Dholpur Bharatpur Water Transmission Project'. The release states that:

'Essar Projects Ltd. has been awarded the prestigious Chambal-Dholpur-Bharatpur water supply project valued at Rs 137 crore by the Rajasthan Government's Public Health Engineering Department. The project will supply treated water to the Bharatpur area with the Chambal river water as its source. The construction contract is worth Rs 128.35 crore while the Operations and Maintenance (O&M) contract is for Rs 8.65 crore. The time-frame for the project is 30 months for construction and the O&M contract is for 5 years. The project will cover 69 villages in Dholpur district and 930 villages in Bharatpur district.'

Source: http://www.essar.com/constructions/pra/feb_17_2003.htm

---

A more realistic solution is to revive the Banganga river and increase catchment afforestation and water movement in the Gambhir basin. Intensive watershed management and floodplain revival programmes in both catchments, particularly in the stretch downstream of the Panchna reservoir, may be the solution. This, coupled with decrease in demand from the Panchna reservoir due to incoming water from the Chambal may be enough to allow the long-term survival of park biodiversity. Any such project would require work on community and riparian lands (both private and common) and therefore necessarily involve the state government, the forest department and NGOs with the capability to mobilise the local population.

Meanwhile it is imperative that people be educated on the park's global importance and its unique ecology. Knowledge has to come before any strategy can be successfully implemented.

Note

*All maps and figures by author.

References

Chauhan, M. and B. Gopal. 2001. 'Biodiversity and Management of Keoladeo National Park (India)—A Wetland of International Importance' in B. Gopal, W. J. Junk and J. A. Davis (eds), *Biodiversity in Wetlands: Assessment, Function and Conservation*, Volume II: 217-56. Leiden: Backhuys Publishers.

Chopra, K., M. Chauhan, S. Sharma, and N. Sangeeta. 1997. 'Economic Valuation of Biodiversity: A Case Study of Keoladeo National Park, Bharatpur'. Parts I And II, Unpublished Report, Capacity 21 Project. Mimeographed. New Delhi: Institute of Economic Growth.

Vijayan, V. S. 1991. *Keoladeo National Park Ecology Study: Final Report 1980–90*, Bombay: Bombay Natural History Society.

www.deccanherald.com/deccanherald/feb062005/n13.asp

www.deccanherald.com/deccanherald/feb182005/n15.asp

www.indianexpress.com/full_story.php?content_id=64100

http://mumbai.indymedia.org/en/2005/02/210058.shtml

http://timesofindia.indiatimes.com/articleshow/997096.cms

http://whc.unesco.org/pg.cfm?cid=31&id_site=340

Politics, Water and Forests in the Himalayas: Crisis in the Gagas River Basin

Anita Paul and Kalyan Paul

The Forest That Was...

THE HIGH MOUNTAIN WALLS OF DRONAGIRI, PANDOKHOLI AND BHATKOT (2,200 TO 3,000 M above sea level) form the natural northern boundary of the Gagas river basin and to the south are the 2,000 m high Ranikhet–Kalika ranges. The basin is spread over 1,500 km² in Almora district, Uttaranchal. The headwaters of Gagas flow through a thick canopy of broad-leaved oak forests on these high ridges. The Gagas is a perennial river. It flows southwards for about 10 km, then swings west to enter the valley at an elevation of 1,300 m above sea level, and then flows from the east to the west irrigating vast stretches of crop lands until it meets the Ramganga river at Bhikiasein.

The Gagas river basin is drained by over a dozen significant streams, one of which, the Khirau *gadhera* (the local word for stream) is the focus of this case study. The Khirau *gadhera* originates in the western aspect of Dronagiri and the headwaters are located in the Ukhalekh reserve forest and the Bijapur *van panchayat* areas. The *gadhera* flows for about 10 km till it joins the Gagas near Tipola. About 2,500 households live in the Khirau catchment, of which 2,000 households comprising 12,700 people in 14 *gram panchayats* live on both banks of the *gadhera*. Additionally, 522 households comprising 2,600 people live in the old town of Dwarahat, situated near the headwaters of the *gadhera*.

According to the 2001 Census, of the total population of 15,300 about 4,500, i.e. roughly 30 per cent, belong to scheduled castes (about 4,000 in the *panchayats* and 450 in Dwarahat).

The Great Water Hunt

The ecological status of the headwaters region is believed to be rather poor. The forest has been cleared of broad-leaved trees (converted to charcoal) and has been replaced by pine, which has an adverse effect on the soil and moisture regime. The villagers have been struggling to meet their domestic energy requirements (provided by firewood), not just for their homes but also for livestock. Increasing population and fodder requirements (there are approximately 10,000 cattle and 2,000 goats in the area) have put enormous pressure on the commons. Under the double burden of community needs and the revenue demands of the state, the forests in the Khirau *gadhera* have more or less ceased to exist, leading to a severe deterioration in the quality of life of the people, and also for the wildlife in the region.

Thus, despite an annual precipitation of 1,200 to 1,400 mm, primary and secondary water resources have dwindled to the extent that even drinking water has become a precious commodity. In fact, the Khirau *gadhera* and the entire Gagas river basin show up distinctly on satellite photographs as a region with thin vegetative cover and high soil erosion.

Fig. 1. *The cleared forest where broad-leaved trees have been replaced by pine.*

The state of resources in this forest region is best discussed in terms of what has happened to the different kinds of water resources. The primary water resources of the village are the *naulas*, the secondary water resources are *raulies* and *gadheras,* and the tertiary resource is the Gagas. Traditionally the villagers utilised *naulas* for drinking water and *raulies/gadheras* for minor irrigation. The Gagas river was used primarily by the villagers who lived along the valley.

The first to disappear along with the dying forest were the *naulas*. For the villages on the banks of Khirau *gadhera*, the current position of this primary water resource is as follows:

	Village	*Naulas*	Functional	Dry
1.	Bijapur	5	5	–
2.	Ghaglori	2	2	–
3.	Dairi	7	7	–
4.	Dharamgaon	6	4	2
5.	Malli Mirai	30	9	21
6.	Bunga	6	1	5
7.	Chatina	9	3	6
8.	Basera	6	1	5
9.	Chatgulla	9	2	7
10.	Kuna	6	1	5
11.	Dhaniari	4	1	3
12.	Malli Kiroli	9	2	7
13.	Salna	4	2	2
14.	Asgoli	21	5	16
	Total	**124**	**45**	**79**

Source: Pan-Himalayan Grassroots Development Foundation, Almora.

Only about 30 per cent of the *naulas* are functional while the rest have dried up because the fragile catchment areas adjoining the villages have been stripped of vegetal cover. This has led to insignificant rainwater percolation and poor recharge of the subsurface water veins that are traditionally trapped by the *naulas*.

Owing to this chronic water problem, the state created two line departments: one to plan and implement water supply schemes and the other to operate and maintain them. Since the mid-1990s these departments have spent large amounts of public funds to build and operate water supply schemes based on the Khirau *gadhera*, and this has continued even after the creation of Uttaranchal.

A total of twelve gravity-flow pipeline schemes depend on the surface-water of Khirau to supply water to twenty-two villages, of which only eleven are located within the Khirau *gadhera* basin. Only 50 per cent of these schemes are functional and that too only during the monsoon and post-monsoon months. Water shortage persists during the rest of the year. In this scenario, residents of the Khirau basin have become resentful of the schemes that take water to villages outside the catchment.

The location of Dwarahat town at the headwaters of Khirau has created its own share of complications. Political power rests in the hands of the town dwellers who have got the state to set up a pumped water supply scheme for them. This allows the 2,600 residents to draw significant quantities of both surface- and subsurface water from Khirau; a similar number of people in three large villages outside the Khirau catchment also benefit from this programme.

Two villages at the headwater zone, Bijapur and Ghaglori 'sacrificed' the main water source that supplied irrigation water in order to meet drinking water needs

Fig. 2. *Traditional water source (naula) being guarded.*

of Dwarahat town; the traditional *kuhls* (irrigation channels) are lying defunct. Yet, neither they nor the town dwellers or indeed the other three villages manage to get enough water because the town has also had a 'floating population' of about 500 people in the last ten to fifteen years thanks to the massive boom in the construction industry.

By appealing to the politically connected people in Dwarahat, Salna, an adjacent village, also managed to get water in the late 1990s after a brief agitation. The state arranged for a pumped water supply to reach the village along with six other villages. Equity issues have never been adequately addressed; water is supplied through 100 stand-posts, of which seventy-three are in Salna while the remaining twenty-seven serve the other six villages. Salna controls the scheme, further evident from the fact that, of its seventy-three stand-posts, fifty-five are private and only twenty-three are public.

Further down the Khirau, the problem gets much worse, with the fertile irrigated fields of Basera having lost traditional rights over the surface flow of the *gadhera*. The *kuhls* in Basera are dry and defunct, yet the state line department is busy spending funds to 'repair' them.

Meanwhile the disputes continue to mushroom. Dwarahat town generates massive quantities of garbage and sewage, which is drained into the Khirau; since the town is located at the headwaters the pollution affects all the downstream villages.

Also, the influence of the town dwellers has led the state to invest in another grandiose scheme to pump water from the Ramganga river 20 km away, at an elevation of 700 metres below Dwarahat.

So there is now fresh conflict brewing outside the Gagas river basin. The people around Chaukhutia town on the banks of the Ramganga resent the idea of the state pumping water from their source for distant Dwarahat town. The scheme had to be abandoned for a while after a flare-up at Chaukhutia. Although construction has now resumed, there are fresh demands from the residents of Chaukhutia.

Meanwhile, another conflict has been simmering in the Khirau catchment for the past fifteen years as a result of an engineering college on the outskirts of Dwarahat. Local politicians demanded the creation of a college in the name of development, and to provide opportunities for the children of the area. A huge amount of money was spent to employ security to keep the concrete college campus, bare of even the shade of a tree, safe from the agitators of Dwarahat town. The problem was their use of a perennial spring at the base of the campus, a significant primary source of water that was now denied to the Khirau catchment residents. The state therefore planned an ambitious water supply scheme for the college based on the distant Gagas river. This too has become a sore point with the villagers.

Clamouring Voices

A 'circle of reason' has supposedly been created by the formation in 2000 of the Khirau Ghati Sangharsh Vikas Samiti by people from eight villages within the Khirau catchment, and another eight outside with water supply schemes based on the Khirau. Their argument is that the *gadhera* is slowly dying and the village-level *naulas* are already dead. So if the state can invest funds for a pumping scheme for students at the engineering college from distant Gagas river, then a similar scheme ought to be executed for them too. Considering that the government has okayed the pumping of water from distant Ramganga to Dwarahat, perhaps the aspirations of the residents of Khirau are not so unreasonable.

Highs and Lows

During the last four years, the Khirau Sangharsh Samiti has staged massive *dharnas* and even *gheraoed* the offices of the two line departments. More than 2,000 villagers participated and a large percentage of them were women. In 2002 when the District Magistrate (DM) of Almora attempted to intervene during the course of an agitation the villagers almost burnt the vehicle of the Chief Development Officer who had accompanied the DM. It was only after the DM gave an undertaking to set up a project for pumping water from the Gagas that the protestors relented. Needless to say, the other villages located along the Gagas river were quick to express their unwillingness to allow any such plan to come into effect.

Discontent is already simmering. The Sangharsh Samiti is affronted that their scheme has not been executed while the one for Dwarahat is progressing at great speed. However, all is quiet in Khirau for the time being as the political leadership has assured them that their turn will come as soon as the Dwarahat project is completed.

Understanding the Terrain

It is indeed a matter of shame that the villages do not have access to safe drinking water. But it is even worse that the state does not realise the nature of the problem or the reasons for water shortage, not only in Khirau *gadhera*, but in all the streams in the Gagas river basin.

The political leadership at the local, state and the national level seems to have misplaced confidence in technology as the remover of obstacles. Therefore, plans continue to be drawn up by barely qualified engineers in the line departments. They work in tandem with private contractors who hardly make the grade. At the end of every such exercise public funds literally go down the drain.

The state also has a team of foresters who are supposed to be the custodians of natural resources but have little understanding about sustaining the bounties of nature. The connection between trees and water, soil and moisture conservation, precipitation and hydrological recharge are issues that require comprehension but do not seem to interest anyone, even though the drying *naulas* are a clear indication that a similar fate awaits the *gadheras* and finally, the river.

It is important that a platform be created to share traditional and scientific knowledge with the key players—the villagers, the department staff and the politicians. The Pan Himalayan Grassroots Development Foundation has been actively engaged in working with the communities in the Gagas river basin and elsewhere in the central and western Himalaya for a little over a decade. Field experiences clearly reveal an urgent need to employ appropriate strategies for the restoration of languishing streams and river basins for sustainable conflict resolution.

This involves a two-fold plan: one, introduction of appropriate technology applications cutting across the sectors of drinking water, sanitation, renewable energy, rainwater harvesting and catchment area treatment, which will improve the quality of life for the people. Two, taking measures to encourage aforestation on a large scale to save the denuded hill slopes and restore the fragile hydrological balance in the region.

Keynesian economics may be meaningful in sustaining the stock market but holds little water when it comes to sustainable development of human societies. We are really not dead. Our children survive us.

Note

*Figures 1 and 2 from Pan-Himalayan Grassroots Development Foundation, Almora.

The Ithai Barrage in Manipur:
A Lake in Trouble

Mihir Kumar Maitra

Manipur River Basin

THE STATE OF MANIPUR SPANS 22,327 KM². IT IS PREDOMINANTLY HILLY EXCEPT FOR THE flat Manipur valley that occupies nearly 10 per cent of the state. Imphal, the capital, is located in the northern part of the valley. Compared with the sparsely-peopled hills, the valley is highly populated. Annual rainfall varies widely, ranging from 600 mm to above 1,600 mm, with an average of 1,400 mm.

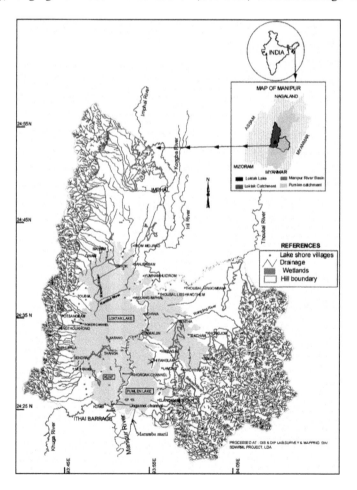

Map 1. *Location of Loktak Lake.*

Manipur valley is drained by the Manipur river system. The Imphal river rises in the north and flows past the capital and drains southwards. The Imphal is joined on its left bank by Iril, Thoubal and Sekmai rivers and is thereafter called the Manipur river. Further downstream, Khuga joins the Manipur on its south bank at Ithai and the waters finally merge with the Chindwin in Myanmar. The catchment area of the Manipur river is estimated at 3,820 km². The river system thus carries a huge discharge with only one outlet in the southern periphery of the valley.

Pretty Picture

A number of shallow floodplain lakes or *pats* exist on both sides of the Manipur river. The most prominent among them is the Loktak lake on the right bank, which is connected to the Manipur through two main channels. These channels act both as inlet and outlet depending upon the water level of the river. Loktak therefore acts as a buffer storage reservoir.

The other notable lakes on the left bank are Waithou *pat*, Ikop *pat*, Kharung *pat*, Khoidum *pat* and Phumlen *pat*. These *pats*—commonly known as the Phumlen system—are connected with Manipur river through three east-west running natural channels and one 12 km long south-flowing stream called Maramba Maril which links the Phumlen system to the Manipur river further south of Ithai.

The Barrage as a Solution for the Shrinking Lake

The Loktak lake has a a dense and widespread aquatic and terrestrial vegetation known locally as *phumdi*. A 40 km² compact block of *Phumdi* in the southeast forms the Keibul Lamjao National Park (KLNP); it

Fig. 1. *Loktak Multipurpose Project.*

is the natural habitat for the Sanghai deer, a highly endangered species. The Loktak lake serves a wide range of wetland functions by providing fish, food, fuel and fodder to the local population in fifty-five lake-shore villages. Over the years, eutrophication, siltation and encroachment gradually began to shrink the lake. (By 1983, the dry season size was about 55 km² with an average depth of only 1.19 m.) The demand for urgent measures to save Loktak lake began to build up.

The solution came in the form of the Loktak Multipurpose Project (LMP). It was conceived in 1967 by the Ministry of Irrigation and Power, constructed by the National Hydroelectric Power Corporation (NHPC) and commissioned in 1984. The project involved the building of the Ithai barrage at the confluence of the Khuga and Manipur rivers. The objectives were: generation of hydropower, rejuvenation of the Loktak lake, flood control and irrigation. This would entail storage of river water in the Loktak lake through backflow and diversion of water to produce 105 MW hydropower through three thirty-five MW generators. The project was designed to maintain the water level in Loktak lake between 768.5 m (maximum) and 766.2 m (minimum). The project report estimated that flood control measures would reclaim 22,941 ha of land around the lake.

Flood-Prone

Out of 225,000 ha of cultivable land in the state, nearly 100,000 ha are located along the Manipur river. Since the construction of the Ithai barrage, the agricultural scenario in the state changed drastically as a large area along the river became flood-prone. About 83,450 ha of agricultural land involving a population of 130,000 is estimated to have been affected partially while around 20,000 ha were completely submerged (Kabui, 1993). As per remote sensing imagery, the seasonally affected lands around the lake that were permanently flooded total approximately 6,350 ha.

Many farmers gradually converted their lands for intensive pisciculture, some encroached upon the Loktak lake for capture-fishing using *phumdi* traps (*athaphum*), and others took up petty trading or migrated to Imphal for unskilled jobs. No systematic assessment of the flood-affected areas and consequent socio-economic changes has been attempted so far, perhaps because of the sensitive nature of the problem.

A Dying Ecosystem

Although Loktak lake is now larger in size, the process of degradation has been accelerated due to the barrage that hampers the natural flow of water, which used to carry with it sufficient nutrition, silt and *phumdi*. Recent years have seen the rapid proliferation of *phumdi*, and there has been a reduction in both the quantity and diversity of fish population because they no longer have access to their natural breeding grounds. At least four species of fish are known to have disappeared permanently (Kabui, 1993). The region has paid a heavy price, both environmentally as well as on the socio-economic front due to the loss of biodiversity (Wetlands India South Asia, 2005).

Highs and Lows

Community Initiatives

In the early years the flood-affected people made representations demanding alternative land or financial compensation. In July 1985, MLAs from fifteen assembly constituencies formed the Loktak Flood Control

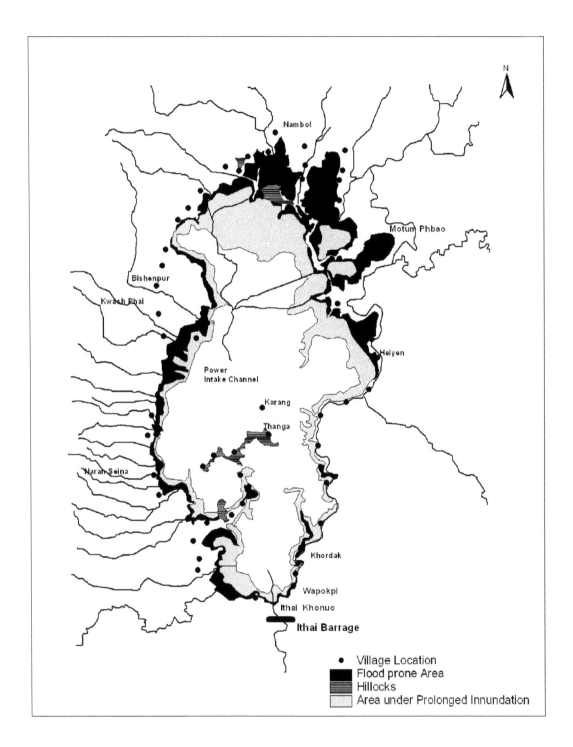

Map 2

Demand Committee. A series of representations were made to the Chief Minister, Governor and Prime Minister to initiate measures to halt the continuing inundation of paddy fields. In October 1991, social activists filed a public interest litigation and academics too did their bit by observing Loktak Day in an effort to create awareness among the general public. Other small local forums were also formed.

Of these the most active community initiative was the All Manipur Ithai Barrage People's Organisation. They held a state-level workshop in 1993 and their recommendations centred around conducting special surveys in the flood-affected areas, carrying out scientific studies, preparing an integrated plan, lowering the water level in the lake, and improving the socio-economic condition of the people.

State Government Initiatives

Since the state government could not do much in terms of providing alternative land or financial compensation, it maintained a low profile. But the Loktak Development Authority (LDA), an arm of the Irrigation Department, was established in 1987 to address the problems associated with Loktak lake.

However, the non-availability of reliable information and data was an impediment to serious intervention. Therefore, a six-year project called Sustainable Development and Water Resource Management of Loktak Lake (SDWRML) was launched in 1998 by the LDA in collaboration with Wetlands International South Asia (WISA) with funding support from India Canada Environment Facility (ICEF). The project emphasised the collection of basic data and scientific information as well as educating local communities to take up alternative income-generating activities. The most important outcome of this project is a water management plan for the entire Manipur river basin, which is useful in flood forecasting, settling water-related disputes and preparing an optimum barrage gate operation schedule.

NHPC Initiatives

The NHPC is aware of the problems of flooding but maintains a studied silence on the subject. Specific technical information is shared with the LDA. But though local officials attend select meetings and workshops they are unable to contribute in any useful manner. All policy and operational decisions on this project rest with the corporate office in New Delhi; in any case their mandate is to look into power generation and not local issues like flooding and fishing. The local NHPC authorities do recognise the importance of the LDA's role in preventing flooding in the valley. In order to maintain the water level in the lake at 768.5 m, the barrage gates are kept closed for most of the year. Gates are opened in consultation with the LDA only when the valley is threatened with large-scale flooding due to heavy rainfall in the catchments.

Two Sides of the Divide

The state receives 12 per cent of the total power produced by the multi-purpose project free of cost, and purchases an additional 18 per cent to meet its needs. However, the Manipur government is not happy with this arrangement since their loss on account of floods and environmental damage is more than what they gain. The government will be happier with at least 30 per cent share of the profit from this venture; this way they can afford to spend some amount on improving the environment. Besides, the irrigation component in the project never took off and remains defunct till date.

One estimate puts annual crop loss due to flooding at 3,475,920 quintals of paddy per year. Others say that crop loss from paddy alone (from 83,450 ha) is Rs. 350 crore annually. If the loss due to vegetables and pulses is added to this already huge total, the impact of flooding on agriculture alone would work out to more than 400 crore annually. To this one must add the environmental and ecological damage to the region. On these grounds the flood-affected people have demanded that Ithai barrage be demolished. Wetland experts feel that it was a grave error to convert productive wetlands into a storage reservoir without paying heed to flood control. Even at a glance, it is evident that Manipur's loss is greater than its gain.

Negotiate for a Better Future

Peaceful and prolonged negotiations alone can bring about meaningful change in the valley. To begin with the LDA needs to form a forum comprising representatives from those affected by submergence as well as the state government to spearhead the talks with the NHPC. The forum needs to come to an understanding on the problems it needs to tackle before engaging in negotiations with the NHPC. It should also be in a position to suggest solutions.

The NHPC has brought out an information-sharing brochure on this project with the opening statement: 'National Hydroelectric Power Corporation takes pride in being instrumental in providing the first major infrastructural input aimed at transforming the economy of a State'. The locals however beg to differ; they feel that the barrage is the cause of their poor economic condition. The NHPC should therefore prioritise sharing information that sheds light on the positive aspects of the project by emphasising how much power is generated and how it has resulted in transforming the state's economy.

During the ICEF-funded project, the LDA carried out an intensive hydrological study in Manipur valley. It opined that even if the water level in the lake were reduced seasonally by 0.5 m, the power generation would not be affected; there is a total head difference of 2.3 m between the design water level and the power channel intake. In fact, the marginal reduction would help farmers grow seasonal crops in a large part of flood-affected agricultural land. The NHPC needs to examine the efficacy of this proposal.

The LDA's former chief engineer and project director, Col. N. P. K. Nambiar, said in his keynote address at a workshop held at Imphal in 1991: 'Ithai barrage has played a vital beneficial role in the field of power generation and irrigation. The criticism comes mainly from the flood control aspect, for which the responsibility lies with the man who operates the barrage and not the barrage itself.' However, the NHPC alone has the authority to operate the gate and with the exception of opening it in order to release excess water during the rainy season, the gates are kept closed for the rest of the year.

Perhaps the NHPC can consider handing over this responsibility to the LDA. Also, the optimum gate operation schedule worked out by the LDA should be jointly evaluated and adopted by the NHPC. If flood control is prioritised the gates should be kept open for most of the year and closed only during the lean period.

Meanwhile the LDA should initiate flood mitigation measures of its own. These would include providing cross-drainage, clearing the connecting channels, and enlarging the Maramba Maril channel to drain the Phumlen system.

Local communities should be involved in disposing *phumdis* during the rainy season. The gates should be kept open for this purpose. Future water management plans in the valley must not consider Loktak alone but include the entire Manipur river basin with due emphasis on the Phumlen system and Chakpi region, which is a flood-affected area. The flood-affected people should understand that the LDA cannot be blamed for errors committed by others. The State government must encourage the authorities to work with the people so that they can implement community-endorsed solutions.

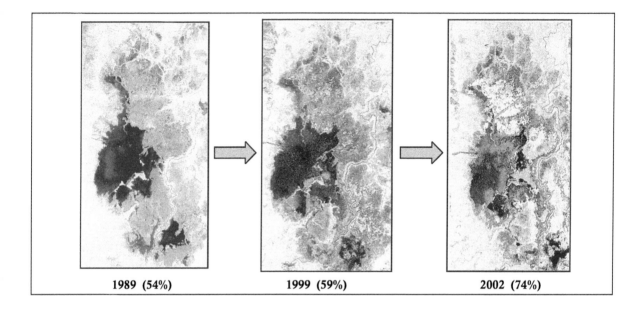

1989 (54%)   1999 (59%)   2002 (74%)

Fig. 2. *Prolific growth of Phumdis.*

Note

*Map 1 and figures from Loktak Development Authority, Manipur.

References

Kabui, Gangmumei (ed.). 1993. *Ithai Barrage: A Boon or Scourge for Manipur.* Published proceedings of the conference organized by the All Manipur Ithai Barrage People's Organisation, Imphal.

Singh, H. Tombi and R.K. Shyamananda Singh. 1994. *Ramsar Sites of India—Loktak Lake, Manipur.* New Delhi: World Wide Fund for Nature.

Wetlands International South Asia. 2005. *Conservation and Management of Loktak and Associated Wetlands Integrating Manipur River Basin*—Report submitted to the Planning Commission, Government of India, New Delhi.

Youth Volunteers Union (Undated). *Development of Phumlen and Khoidum Lake—An Integrated Plan.* Thoubal Manipur: Public Library-cum-Information Centre. The Youth Volunteers Union is a NGO which prepared a proposal for the construction of a 22.7 km long earthen embankment with culverts, sluices, etc., in consultation with the affected people, to protect a part of the flood-affected area. The proposal contained a description of the problems, interactions with some government officials, a proposed solution and a budget.

Rural Needs or Tourism:
The Use of Groundwater in Goa

Sujeetkumar M. Dongre and Govind S. Poteker

Goa: The Smallest State in the Union

Sandwiched between Maharashtra in the north, Karnataka in the east and south, and the Arabian Sea in the west, Goa is the smallest state in India, with an area of 3,702 sq km and a coastline that is 105 km long. Topographically, Goa may be divided into three main regions: the coastal region, the midlands and the mountainous Western Ghats that rise to a little more than 1,000 m above mean sea level (MSL). Along the coast, there are some coastal highlands that form small plateau-topped hills that rise to a height of between 30 to 100 m above MSL, separated by vast expanses of coastal floodplain. The most commonly found surface rock in Goa is laterite, a rock that can be broadly classified into hard massive laterite, which forms the top surface layer; gravely laterite; followed by semi-permeable clayey laterite. Porous laterite, which is fractured at many places, is an excellent rock formation that stores water.

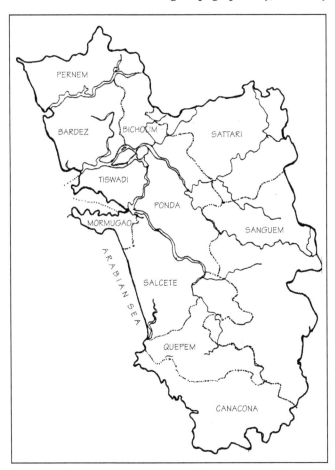

Map 1. *Location of water crisis areas in Goa.*

High Rainfall, but Summer Scarcity

Located as it is on the western coast, Goa receives the southwest monsoon. Rainfall increases as we move from the coastline up to the Ghats and the average precipitation is 2,500 to 3,500 mm. Hydrologically, in the coastal as well as low-lying flat agricultural areas sandy and loamy formations with moderate to good primary porosity and permeability are the main aquifers with a shallow water table. The fractured and weathered crystalline rocks located below the lateritic clay sequence form a semi-confined to confined aquifer. Though Goa receives more precipitation than other parts of the country, crises arise due to the unequal distribution of rainfall. About 90 per cent of the total rainfall occurs in a short period between June and September.

The rainfall is heavy but 60 per cent of it is washed away as surface runoff, as there is little time for the water to percolate into the ground. So the major water source in the off-season is the water stored in the vegetated Western Ghats. It drains down along the two major watersheds of the Mandovi and Zuari rivers. Three major water reservoirs—the Selaulim, Opa and Anjunem reservoirs—supply water to most parts of Goa. For domestic use and irrigation, traditionally the people in the villages depended on well water, tanks and artificial freshwater lakes. The rural development agency/block development office subsidises wells and power for farmers.

Tourism, Construction, Industries: Development and Water Demand

Over the last three decades the boom in tourism, construction and industries on the coastal belt and plateaus has dramatically increased the demand for water. One of the major consumers is the hotel industry. Though the government supplies water to the hotels, they require additional supply to maintain swimming pools, gardens, etc. There are hardly any measures to control water use by customers. The extra demand is being met by extracting water from wells in adjoining villages. This is supplied in tankers for a payment of only 250 rupees for 6,000 litres. In Saligao and Sangolda in Bardez *taluk* of north Goa, this water is mainly extracted from wells owned by five farmers in the first village and two farmers in the second.

Water Sales are More Profitable than Agriculture

In these villages selling water has become more profitable than growing crops. This has disturbed the village economy, ecology and traditional agricultural practices. Farmers who sell water have stopped growing winter crops. Today hardly any village along the Goa coastline depends on agriculture and it is considered only a seasonal occupation. The impact of these activities has been felt in a lowering of the water table. This is what alerted the local people in Saligao and Sangolda.

The Sangolda Story

The first agitation in Sangolda against the over-extraction of groundwater for commercial use began in the early 1980s. Two farmers belonging to Bella Vista Vaddo and Mae de Deus Vaddo in Sangolda had started selling water illegally for construction activity. Heavy-duty pumps were used to extract 150,000 to 180,000 lpd of water from two wells, both of which were at that time at a lower level than the rest of the wells in the Vaddo.

The villagers of Sangolda submitted a petition that carried signatures of forty locals to the *sarpanch* during the committee meeting, protesting the sale of water for commercial purposes. Although the *sarpanch* assured the petitioners that he would put a stop to this activity, the sale of water continued. The villagers discussed the concept of a water table with the two farmers and requested them not to over-extract water from the wells as this had led to a lowering of the water level in the adjoining wells. This too had no effect. The villagers then tried to convince the truck drivers not to buy water from these wells, but to no avail. Left with no option the villagers had to complain to higher authorities. They met different departmental heads—the local *panchayat* secretary, the Block District Officer, the *mamlatdar*, PWD engineers, officials of the Irrigation Department and even the Collector. Although the officials intervened in the matter, the sale of illegal water continued.

At present, over-extraction of the groundwater results in a loss of 21.6 million litres of wholesome drinking water per year from the Sangolda water table. Assuming that an average family uses 200 litres per day, 450 families are being deprived of their share and the effect of this practice is strongly felt during the drought months of March, April and May. In October 1987, a detailed study was carried out by Professor F.B. Antao, then Head of the geology department, Dhempe College of Arts and Science. In his report, 'Geo-Hydrological Consideration for the Sangolda Village Area', he elaborated how water was being extracted from the two wells through 100 to 150 mm flexible pipes and transported in large tankers of 6,000 litres capacity for supply to various building sites.

The result of such large-scale withdrawal of groundwater is a lowering of the water table. The maximum impact is felt when the water table reaches its lowest seasonal level in April and May. In October 1987, the water level in village wells located at a distance of about 150 m to the east of the road was 3.5 to 8.5 m below ground level, but it soon dropped to 7.25 m to 12.13 m. When levels fall to such an extent, wells generally go dry in summer. Thus, except for one or two, all other wells in the area now go without water in April and May.

This study was used to file a case against the water sellers in the civil court at Mapusa and in its judgment on 28 April, 1998 the Court ordered the two farmers to stop illegal pumping of water from their private wells. The judgment order reads: 'The defendants, their agents, servants, family members and/or labourers claiming through the defenders are hereby restrained from selling water from the suit wells to any water tankers or in any way pumping out water from the suit wells into the tankers and defendants shall use water only for the purpose of agriculture and domestic needs.'

. . . and the Saligao Story

Saligao, a neighbouring village, also faces a similar problem due to a spurt in hotels and construction activity. In the vicinity is the Pilern Industrial Estate. According to a local activist, the first major extraction of groundwater from the wells in Saligao started in the late 1970s for the construction of the Taj Hotel at Fort Aguada. Since then water has been extracted on a large scale from wells located in the fields at lower levels and is sold to hotels along the coast. This has resulted in decreasing water levels in domestic wells, especially in Don Vaddo, part of Sonnarbhat and Mollembhat.

This illegal extraction of water for commercial purposes was brought to the notice of the *sarpanch* of Saligao village *panchayat* by the villagers during the *gram sabha* in 2002. The *gram sabha* passed a resolution to constitute a committee attached to the Saligao Civic and Consumer Cell (SCCC) under the chairmanship of Sebastian D'Cruz. Members surveyed the wells in Don Vaddo and Sonnarbhat, Cotula, Mollembhat and Muddawadi and Tabra Vaddo area of Saligao village. The report said:

1. The water crises are partly caused by large-scale groundwater extraction by private well owners, who sell water to hotels, industries, construction sites and INS Mandovi at Verem.
2. The extraction of groundwater for sale outside the village has risen sharply in the last five years, to a level of up to 435,000 litres per day. Groundwater levels are falling, and wells for household use and irrigation are running dry in some parts of Saligao, while other areas of the village are unaffected.

3. More than half of the twenty-eight household wells surveyed in Don Vaddo and Sonnarbhatt went dry during April–May 2001, although monsoon was normal.

The amount of groundwater extracted varies in different months. November and December are the peak months because it is the tourist season. The findings were shared with Dr A.G. Chachadi, lecturer in Goa University and an expert on groundwater. His six years of research in the area provided excellent information on groundwater occurrence, movement and availability in the north Goa coastal aquifers.

The quantity of groundwater that is being pumped out from only four wells—estimated by the committee to be about 435 m³ per day—is actually a very large quantity if one were to take into consideration the retaining capacity of the shallow sandy aquifer. Extensive pumping draws in groundwater from a very large area, which has led to a dramatic fall in the water table in the surrounding areas. This is the main reason why even shallow wells around Saligao are going dry despite good monsoons.

Lessons

The rapid growth in tourism and construction activity in the absence of adequate municipal water supply has led to reckless increase in commercial exploitation of groundwater with utter disregard for safe yield limits of the aquifers. In spite of the court order to stop over-extraction of groundwater and the Goa Ground Water Regulation Act 2002, this exploitation of water continues.

In the villages, well water is used for domestic as well as irrigation purposes around the year. When water began to be sold to outsiders, there was a shift from cultivation to the sale of water and this had an adverse impact on the traditional village economy. At one time water was available at a depth of about one metre, but today one has to dig at least three metres. The impact has been felt not just by the economy but also by the ecology. Since the villages are close to the coast there is every likelihood that salt water intrusion may occur and cause irreversible damage to the groundwater.

The water level can be raised by rainwater harvesting, by constructing percolation tanks atop hills, and building check dams in the valley regions to prevent surface runoff during the monsoon.

The villagers in Saligao have started rainwater harvesting on the hilltops with the help of NSS units of Green Rosary Higher Secondary School, Dona Paula. Additionally, it is important to implement the Goa Ground Water Regulation Act and heed the court order so that the commercial extraction of groundwater comes to an end.

Note

*Map 1 from Chachadi (2002).

References

Adyalkar, P.G. 1985. A Collection of Papers Submitted for the Seminar held at Panaji, Earth Resource for Goa's Development, Hyderabad.

Alvares, Claude (ed.). 2002. *Fish, Curry and Rice*. Goa: Goa Foundation.

Antao, F.B., 1987. 'Geo-hydrological Consideration for the Sangolda Village Area'. Technical Report. Panaji: Dhempe College of Arts and Science.

Chachadi, A.G. 2002. Technical Report of Ground Water Management of Coastal Areas. Lisbon: LNEC.

Mascarenhas, Mario. 'A Brief Historical Overview of the Illegal Sale of Water in Sangolda'. Unpublished paper submitted to the Saligao Panchayat and the Saligao Civic Consumer Cell.

Official Gazette, Government of Goa, Series 1 (43), January 29, 2002, Panaji.

'Saligao Water Crises and Commercial Groundwater Extraction'. First Report of the Saligao Civic Consumer Cell, May 2002.

Social Undercurrents in a Gujarat Village: Irrigation for the Rich Versus Drinking Water for the Poor

Anjal Prakash and R. K. Sama

Where Every Drop Counts

$\mathbf{F}$OR MORE THAN A DECADE GUJARAT HAS BEEN AT THE FOREFRONT OF DEBATES CONCERNING water scarcity and a declining groundwater table. Water for drinking and irrigation is of critical concern in the state today, particularly in areas where groundwater recharge is low and rainfall is scanty and erratic. Consecutive droughts between 1999 and 2001 have aggravated the problem.[1] Vadali in Chotila *taluk* of Surendranagar district is one of the numerous drought-stricken villages in the state. It is 35 km south of Chotila town and inhabited by 354 families. According to the 2001 Census, the village has a population of 1,517. It is dominated by Ahirs (around 120 families) followed by Kolis (100 families) while the rest are from Dalit, Suthar, Bawa and Bharwad castes. Ahirs are a landed class that owns around 90 per cent of the total irrigated land in the village.

The villagers depend on groundwater for both irrigation and drinking water. There are around 125 open wells for irrigation. The area is characterised by hard rock and therefore not suited for deep tubewells. The depth of open wells ranges from 12 to 24 m. Drinking water needs are met through deep wells that have been dug in the centre of the village. Water is accessed

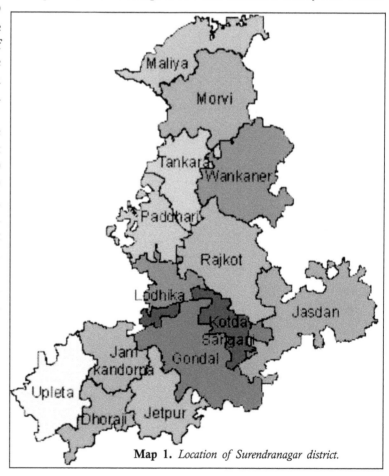

Map 1. *Location of Surendranagar district.*

Fig. 1. *The routine scene of people waiting in queue for water, irrespective of age.*

through pumps and is collected in overhead tanks from where it is distributed through stand-posts to different social groups.

Over the past four or five years, the village has been facing a scarcity of drinking water. The present drinking water well dries up in early March and is recharged only during the monsoon, i.e. late June or early July.[2] During the time when water is not available in the wells, the village is supplied water through tankers by the Gujarat Water Supply and Sewerage Board (GWSSB). However, tanker supply is resumed only between late April and early May. For the remaining months the villagers manage to meet their drinking water needs through farm wells. In the year 2003–04, only 20 to 30 per cent of the total farm wells had water that was available for around thirty minutes to one hour every day.

In order to ensure drinking water security, GWSSB constructed two check dams on the two small rivulets that flow through the village. However, these measures were not sufficient and provided little relief for the villagers. The village was covered under the community-managed development of water supply and sanitation programme of the Water and Sanitation Management Organisation (WASMO)[3] with active support from the Aga Khan Rural Support Programme, India (AKRSP [I]). The village was helped to draw up an action plan to manage drinking water and sanitation activities. One of the components of the plan was to de-silt the check dams built by GWSSB; these are close to the tubewells and helped recharge the aquifer. The plan was to dig a drinking water well near the dam and supply water to the village through a pipeline.

According to plan, the check dam was de-silted in early 2004. The subsequent monsoon filled the check dam and recharged twelve farm wells in the vicinity. The villagers said that the water actually overflowed from the wells, something that they had not seen in the last five years. Looking at the water availability, the farmers around the dam sowed winter crops such as cotton and groundnut, and were able to reap benefits. However, in late 2004 while monitoring the progress of the Pani Samiti by the Coordination Monitoring and Support Unit (CMSU) of WASMO, Surendranagar and AKRSP (I), the topic of constructing a drinking water well in the vicinity of the check dam came up. It reflected the village action plan. However, having seen the bountiful harvest that the check dams had enabled, the farmers in the area opposed the drinking water well. They feared that their farm wells would have to share the aquifer and that they would get less water for irrigation. These farmers belong to the dominant Ahir caste that has a stronghold over the functioning of the *panchayat* and Pani Samiti (Table 1).

Table 1: Stakeholder Analysis Matrix

Stakeholder	How they are affected by the scarcity of drinking water	Motivation to participate in addressing the problem of scarcity	Relationship with other stakeholders
Pani Samiti/ *panchayat*	Some members who are from resource poor groups are affected but a majority are rich farmers who are not directly affected	As a representative group they are motivated to address the problem	Partnership with some but in conflict with others as they try to force decisions that favour their interests
Rich Farmers/ Well owners	Not directly affected	Not motivated to address drinking water problems because they benefit from the tanker water supply programme of the government	In conflict with the landless and Dalit and in partnership with *panchayat*, Pani Samiti AKRSP (I) and WASMO
Landless/Near-landless/Dalit/ Dalit women	Directly affected/ dependent on well owners for drinking water during times of scarcity	Motivation is high because they suffer the most	In covert conflict with rich farmers, in partnership with *panchayat*, Pani Samiti, AKRSP (I) and WASMO
AKRSP (I)	Not affected directly	Motivation is high since they are agents of change	In partnership with all stakeholders
WASMO	Not affected directly	Motivation is high since they are agents of change	In partnership with all stakeholders

A pertinent question is, why is there opposition if the drinking water is meant for everybody in the village? Does drinking water scarcity not affect them all in a similar way? The answer lies in understanding the social structure of the village. As mentioned before, Ahirs dominate the village both economically and socially. They are mostly landed households who own a significant number of farm wells. When the water in the community well (which is the source of drinking water) dries up, landed Ahirs do not face any problem. Around twenty to thirty farm wells continue to hold some water; while it is not sufficient for irrigation, it is enough to meet their drinking water needs. During periods of scarcity, these families stay on the farm so that they have enough water for themselves and for their livestock; those families who stay back in the village cart water from the field. The water problem affects those who do not have land or farm wells such as the Kolis and the Dalits. In the bad months the women from these families have to depend on the upper caste for drinking water. Manjuben—a Dalit woman from the Vadali village—says, 'during summer we face acute problem of drinking water. The Ahirs have their own farm wells and they either carry water from those wells or stay there during scarcity months so they do not face the problem. We do not have wells and so have to depend on those who have.'[4]

The dependency makes lower caste women vulnerable and prone to harassment. As one Dalit woman puts it, 'We cannot afford to disobey upper caste families as we are dependent on them for employment on their farms and other needs including drinking water when our village well runs dry. The GWSSB's tanker supply is uncertain and unreliable.'[5]

The Ahirs on the other hand also benefit from the tanker programme of GWSSB. As soon as the village is declared scarcity-affected, the GWSSB is obliged to make drinking water available through tankers. The supply of water is done through contractors who fetch water from the Ahir farm wells and distribute it in the nearby villages. Most contractors are from the richer classes and benefit from the job. This makes them less interested in drinking water projects that aim to resolve water scarcity since they directly gain from the scarcity situation—both economically and socially.

The village dynamics forced AKRSP (I) and WASMO to seek an alternative. A number of meetings were held with various stakeholders—farmers, Pani Samiti and *panchayat* members, and women from various social groups. The farmers in the vicinity of the newly de-silted check dam refused to cooperate

Fig. 2. *Women fetching water at the village wells.*

and threatened dire consequences if the well was dug at that site. Dalits and Kolis voiced their opinion in secret but never raised any issues in public meetings out of fear for their jobs. It was clear that a compromise formula that was socially and technically feasible needed to be worked out. An alternative site was selected by the Pani Samiti that is close to the present drinking water well in the village. Both AKRSP (I) and WASMO are involved in the project; they have understood that conflict can lead to possible cancellation or a change of plans.[6] De-silting another check dam on the rivulet might help recharge the new location. A field visit in March 2005 assured us that the new well is indeed being constructed and is expected to be 24 m deep.

The selection of a new construction site for the drinking water well minimised the conflict but also revealed the internal dynamics of the community. It highlighted the fact that power structure and social and economic hierarchy go hand in hand and unless the issue of resource inequity is tackled through policy and advocacy, the real issue will not be solved. Community-based programmes under innovative institutions such as WASMO can help speed up the service delivery systems and minimise the corruption that was prevalent earlier; however, bringing about social change within eighteen months of a project cycle is too much to expect.

Notes

*Table 1 from field-notes of Vadali village and discussions with major stakeholders; map 1 and figures from CMSU, Surendranagar, WASMO.
1. During 1999–2000, the government declared 8,666 villages (from a total of 18,637 villages) scarcity-hit in the wake of monsoon failure. Total scarcity was declared in 6,675 villages while 1,991 villages had semi-scarcity conditions and 7,467 villages faced severe shortage of drinking water. According to the government, the deficient monsoon resulted in 29 to 31 per cent decline in crop production in Saurashtra, Kutch and north Gujarat districts. The estimated figures of production showed a fall of 45 per cent in pearl millet, 83 per cent in sorghum, 72 per cent in groundnut and 41 per cent in *moong*. The total crop loss was estimated at Rs. 4.59 crore (GoG 2000). The problem continued in 2002 when 13 out of 25 districts received less than normal rainfall. In total, 5,144 villages in these districts were declared *scarcity/semi-scarcity hit*. The production loss for the kharif season was estimated to be 23 per cent amounting to Rs. 1.87 crore while in the Rabi season the loss was of the order of Rs. 96.9 lakh (GoG 2004).
2. Gujarat has only one rainfall period between June–July and September–October. Rainfall ranges between 1,000 and 2,000 mm in the southern rocky highland to between 250 and 400 mm in Kutch. The distribution of rainfall determines the water regime in the state. Around 70 per cent of Gujarat's total geographic area falls in the arid or semi-arid zone and is drought-prone (Patel 1997).
3. WASMO is an autonomous organisation created by the government of Gujarat to promote, facilitate and empower village panchayats and rural communities to manage local water resources, and have their own water supply system and environmental sanitation facilities. It empowers the village community through Pani Samitis and they plan, approve, implement, operate and maintain their own water supply systems, manage water resources and ensure safe and reliable drinking water supply throughout the year. WASMO works with NGOs and Pani Samitis by providing financial and technical support (*www.wasmo.org*).
4. Interview conducted during field visit at Vadali village.
5. Ibid.
6. This is a difficult situation for outsiders trying to bring about change. The nature of both the external implementing agencies is based on consensus building and partnership. The project cycle is around eighteen months and as one of the officials puts it, one cannot expect social change to come about in such a short

period of time and within the framework of partnership. This is in the context of the feudal structure of the village in question and the relationship that the poor have with rich Ahir farmers. Without an alternative, Dalits cannot afford to be at loggerheads with upper caste farmers who are part of their social and economic security system. A Koli labourer puts it succinctly: 'when we need money we go to Ahir farmers, when we need work we go to them, how do you expect us to disobey them?'

References

'Assessment of Economy in Brief: Socio-economic Review of Gujarat State 2003–04'. Gandhinagar: Directorate of Economics and Statistics, Government of Gujarat.

Patel, P.P. 1997. *Ecoregions of Gujarat*. Vadodara: Gujarat Ecology Commission.

Whose is the Chilika?
Fishing in Troubled Waters

R.S. Deshpande and Satyasiba Bedamatta

Chilika: A Ramsar Site

WATER CONFLICTS ARE NOT ALWAYS RELATED TO WATER SHARING ISSUES BUT CAN ALSO arise out of differences in harvesting the product of water bodies, even though that may not involve water use per se. Largely, the cause of the dispute is related to the customary rights of traditional communities. Chilika lagoon, the largest brackish water wetland system, is a Ramsar site—a wetland of international importance. More than 200,000 people from the local communities depend on it for their livelihood. The waterspread in the Chilika varies from 1,165 sq km in the monsoon to 906 sq km during summer.

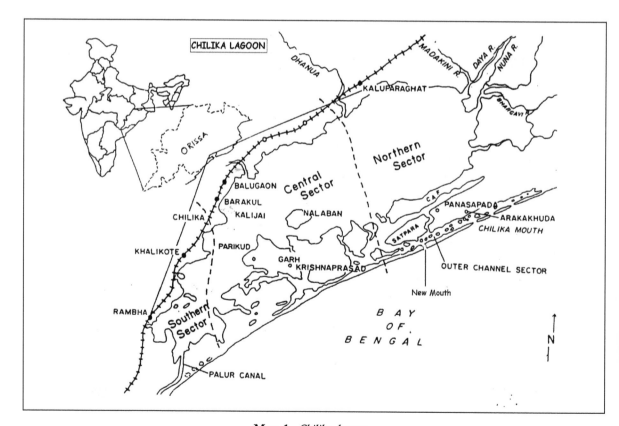

Map 1. *Chilika lagoon.*

The lagoon can be broadly divided into four ecological zones based on salinity and depth. These are the southern zone, the central zone, the northern zone, and the outer channel. In the recent past the Chilika Development Authority (CDA) has been taking keen interest in the ecological restoration of the lagoon. It is the first site in Asia to be deleted from the Montreux Record that lists threatened Ramsar sites.

The satellite image of the lagoon shows varying depths and different aqua-ecological regions indicating variations in the availability of fish, and when this is superimposed on the property rights of the local fishermen one understands the basic economic reason behind the stress.

Threatening Traditional Occupations

The traditional fishermen of Chilika were given fishing rights or *pattas* by King Parikud, and the British recognised these rights. The non-fishing sections among the Khandayata communities largely handled fish trading. Earlier the caste structure determined the mode of fishing. Now, however, most communiteis practise the *khanda* mode of fishing. Almost 63.8 per cent of *khanda* nets are used in an average household. The changes imply that diversified fishing methods developed by different communities in response to varying situations are dying out; the fisherfolk have switched to the mechanical system irrespective of caste, community and regional or ecological requirements. The reasons for the change are not hard to pinpoint. One is the increase in prawn fishing (it fetches high market value), and the other is the *mahajan* (the usurer) factor (*mahajans* provide easy credit but make it mandatory to use *khanda* nets). There are 132 fishing villages and quite a few non-fishing communities/new fishing villages that directly depend on the Chilika for their livelihood.

Fig. 1. *Chilika lagoon: A satellite image.*

Fishing groups called Sairats used to obtain exclusive rights from Chulimunda *zamindars* and they had evolved a complex system of marking out territory in the lake. However, after the abolition of *zamindari* in 1953, the demarcated traditional fishing areas were leased out to cooperatives of local fishermen. In 1959, based on earlier experience, the government of Orissa

Fig. 2. *Change in fishing practices.*

identified four fishing grounds for the effective management of Sairats. These were the Bahani Fishing Ground, the Jano Fishing Ground, the Prawn Fishing Ground and the Dian Fishing Ground. The Central Fishermen Cooperative Marketing Society (CFCMS) was formed around this time (Das, 1993). As prawn became increasingly remunerative, outside interests began to enter the Chilika. The leasing system broke down completely under commercial pressure and thus spawned the seeds of a simmering conflict. In 1991, the Orissa government outlined a new policy that auctioned leases to the highest bidder, handing over the advantage to moneyed people, most of whom were outsiders. The CFCMS challenged the government order in the High Court. These changes in leasing policies have resulted in encroachment on centuries' old fishing rights of traditional fishermen by 'contract fishermen' (Kadekodi and Gulati, 1999). The court ordered that prawn culture, except through pen method (enclosure), not be permitted in the lake and directed the government to make changes to safeguard the interests of the traditional fisherfolk. But in the absence of any corrective policy measures in this respect, the conflict continued and the local people were marginalised by powerful outsiders (Das, 1993).

The situation was further complicated by the entry of refugees from Bangladesh who soon took up moneylending. The end result was the degradation of the lagoon and excessive commercialisation. More non-fisherfolk entered the Chilika system and took up fishing and shrimp culture. Over the years, the conflict between and within communities has cast its shadow on issues involving credit, labour, product and property rights markets.

Entry of Other Castes

Caste plays a pivotal role in the conflicts in the Chilika region, not so much due to the usual disputes related to caste hierarchy but because of occupational and economic interests in fishing. Fishing and agriculture are the two primary sources of livelihood for these people. The fisherfolk belong to lower castes and most of them are either landless or possess tiny landholdings. Non-fisherfolk on the other hand belong to higher castes (many are migrants) and are engaged in agriculture and other non-fishing occupations. The fisherfolk have to travel to their prescribed fishing grounds even if it is far from their village, and thus, there is always the temptation to transgress into others' areas. Also, many non-fisherfolk are now engaged in fishing, particularly after they discovered that there is a lucrative market for prawn. This was one of the reasons for conflict.

Shrinking Resources

The dual dispute involves, first, the contentious issue of the snatching of traditional fishing grounds—made possible by the new policy—and second, that of extending prawn culture into fishing zones. The other problem has to do with caste and communities. The main cause for the conflict is ill-defined property rights and the leased area which

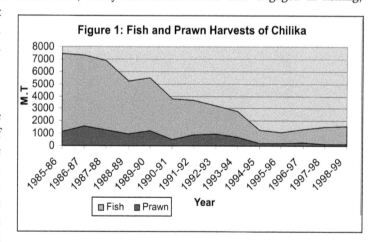

Fig. 3. *Fish and prawn harvests of Chilika.*

was further exacerbated as a result of changes in occupational structure wherein increasingly more households (including migrants) got into fishing or prawn culture, resulting in a decline in produce. The dictates of the market forces made things worse as did the new lease policy of the state government. Caste and religious divisions added fuel to the fire.

Fishing as a profession has always carried a stigma—it has been considered a lowly occupation, spurned by the higher castes. Profit motives have now brought the upper castes into this arena but socially they continue to maintain their distance and largely concentrate on marketing. The low-caste fisherfolk still live in conditions of abject poverty with little education and a large number of them are in debt. Marketing and the introduction of prawn, controlled by the non-fishing upper castes, and the leasing policy, have aggravated these evils. Thus a four-fold pattern slowly emerges: first, the already shrinking fishing grounds were further destroyed by the introduction of prawn culture; second, the non-fishing communities entered this traditional domain and soon transformed into moneylenders/fisherfolk and the competition became fierce; third, these groups snatched away traditional fishing grounds through their lending activities and consolidated their control on the resource and the fishing community with the help of the new leasing policy; and last, migrants from across the border also followed the same route to quick success. The contrast between a traditional fishing household that reels under dreadful poverty and the newly ensconced migrant who flaunts his money and exercises control over the former could not be more stark. In addition, the opening of an artificial mouth and the de-silting of the lead channel not only rejuvenated the lagoon's ecosystem but also made fishing more lucrative: the average annual income increased by about 30 to 50,000 (http://www.chilika.com/restore.htm). The traditional fishing villages near the mouth protested against the second opening but the monetary benefits brought them into the mainstream.

A Small Victory

Initially the fishermen agitated against the lease policy as well as the consequent environmental degradation. During early 1991, there were severe small conflicts/agitations in and around the lake that largely concerned the new lease policy. Then the High Court ordered that the 'capture-to-culture' ratio be maintained at 60:40 and permitted non-fishing communities to enter the business of prawn culture. In the early 1990s, the Tata Aquatic Farm project was being implemented by the state and this led to a series of protests because it posed a threat to the livelihood of traditional fishermen. Finally, the company withdrew from the region.

In March 1999, the fishermen of Chilika began the 'Do or Die Movement' demanding total prohibition of shrimp culture in the lake. There were eight other demands as well and the Chilika Matchhyajibi Mahasangha gave the Chief Minister an ultimatum to fulfil them by April 15, 1999, particularly the one that dealt with the demolition of all shrimp *gheris* (enclosures) in the lake (Samal, 2002). However, the government took no action and on April 24, the fishermen themselves began to destroy the shrimp *gheris* spread over 600 ha across the lake. On May 27, they also blocked the national highway for five hours. On May 29, the movement entered the next stage when, after a public meeting at Ghodadauda field, Sorana village, eleven shrimp *gheris* spread over 400 ha were destroyed. The fishermen were arrested at Sorana amidst protests and a few of them blocked the police vehicle. This led to police firing that claimed four lives. The government finally suspended lease for shrimp culture in Chilika in December 2001. Though it is now illegal, unauthorised shrimp culture continues in the lake and the fishermen have been agitating against the mafia that controls it.

A Question of Survival

The view from the other side of the fence is also quite interesting. The non-fishing community had earlier been engaged in salt making and agriculture. But after the abolition of salt production by the British these communities became dependent on agriculture, which became increasingly non-remunerative. On the other hand, prawn culture was a lucrative option and they soon took to this new-found occupation. Some of them also became moneylenders and traders. Non-fishermen who live inside the lagoon argue that even though they are not fisherfolk by caste they have to undertake fishing since they live on the island but do not have enough agricultural land (sometimes the land they have is also saline), and no irrigation facilities. This forces them to take up prawn culture, and they continue with it in spite of the ban and the district administration's periodic destruction of the *gheris*. They also have their own association, Chilika Ana-matchhyajibi Mahasangha (Chilika Non-fishermen Association), and non-fishermen cooperative societies.

Fig. 4. *Prawn gheris.*

The fishing community would like a complete ban on the entry of non-fishing communities into their traditional occupation and prawn culture. But non-fisherfolk who have settled in the region are now economically and politically powerful, and it will be difficult to implement any ban on their activities. This is also not a viable solution given their political and financial power. One idea would be to re-allocate fishing rights to these communities in a manner that favours the poorer group. Restricting rather than banning prawn culture and maintaining the capture-to-culture ratio will also go a long way in managing the conflict peacefully.

The Waves Subside

It is interesting that the situation has undergone considerable changes, both economic as well as ecological, after the state's intervention in September 2000. On the commercial side, fish landing (catch) has gradually improved and touched an all-time high of 14,000 tons (from capture fishery) in 2003–04. This has resulted in a significant increase in the per capita income of the local community. Chilika is an important example of the open use of common property resource. The changes introduced here led to commercialisation and stronger markets. On the face of it, it seems that the unrest and agitation in the local communities has now simmered down to a large extent as a result of increase in fish production after the restoration of the lagoon's ecosystem.

Encroachments on the lake continue despite the ban on shrimp culture. The fishermen reacted by organising an agitation in December 2001. In February 2003, the district administration started destroying prawn *gheris* inside the Chilika lake. The majority of members of the fact-finding committee on Chilika fisheries felt that fishing by erecting large scale *gheris* was the major source of water pollution besides posing a serious challenge to aquatic life in the lake (Sahoo, 2003). On April 29, 2003, there was a group clash in Chilika between two neighbouring villages over access to fishing rights in which two fisherfolk died and sixty were injured. Basanta Kumar Nayak, the Vice-President of Chilika Fishermen's Association, told the district administration that it was responsible for the clash and demanded that the district administration and the state government take strict action against the culprits and formulate a clear Chilika lease policy to prevent such social tension (*Dharitri*, 2003). A draft legislation of December 2003 fixed the ratio of fishing rights as 70:30 favouring fisherfolk. Yet, a large number of fisherfolk households opposed this draft legislation. One hopes that an acceptable draft will soon emerge.

Notes

*Map 1 and figures from *www.chilika.com*

References

Chatrath, K. J. S. 1992. 'The Challenge of Chilika' in K J S. Chatrath (ed.), *Wetlands of India*, pp. 135-70. New Delhi: Ashish Publishing House.

Das, G. S. 1993. The Report of the Fact Finding Committee on Chilika Fisheries, Submitted to the High Court of Orissa, August 16.

Dharitri (Oriya daily newspaper), May 3, 2003.

Kadekodi, Gopal K. and S.C. Gulati. 1999. 'Root Causes of Biodiversity Losses in Chilika Lake: Reflections on Socio-Economic Magnitudes'. Report submitted to WWF-India, New Delhi. Centre for Multi-Disciplinary Development Research, Dharwar and Institute of Economic Growth, Delhi.

Mitra, G. N. 1946. Development of the Chilika Lake. Cuttack: Orissa Government Press.

Mohanty, S. K. 2002. 'Fisheries Biodiversity of Chilika Lagoon', *Chilika Newsletter*. 3: 11–12.

Sahoo, A. K. 2003. 'Fishing in Chilika Lake? Not Possible Anymore', *The Asian Age*, February 20.

Samal, Kishor C. 2002. 'Shrimp Culture in Chilika Lake: Case of Occupational Displacement of Fishermen', *Economic and Political Weekly*. 37(18), May 4–10.

Shankar, U. 1992. 'Chilika: A Lake in Limbo', *Down to Earth*, August 31, pp. 25-31.

www.ids.ac.uk/ids/civsoc/final/india/ind9.doc

www.chilika.com

Bridge over the Brahmaputra: Unleashing Nature's Fury

Chandan Mahanta and Anjana Mahanta

Islands in the Stream

THE MEGA TWO-TIER BOGIBEEL BRIDGE IS THE FOURTH ONE TO BE CONSTRUCTED UNDER THE National Rail Vikas Yojana and also the fourth to be built on the river Brahmaputra in India. The bridge is meant to connect Dibrugarh in the south to Lakhimpur in the north; the rail link will join Chaulkhowa station and Sisibargaon–Siripani on opposite banks. The construction is expected to be completed by 2009. The bridge will fulfil a long-felt need of the people of Upper Assam and Arunachal Pradesh. The hitch is the Majuli Island downstream of the proposed bridge.

Majuli used to be the largest inhabited river island in the world—its area is 875 km^2 and population is 160,000. It is located in the mid-reaches of the Brahmaputra in Assam, about 630 km upstream of the Assam–Bangladesh border, and in the broadest part of the Brahmaputra valley, where the river divides the

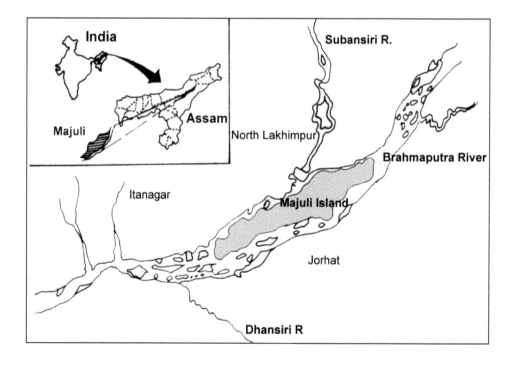

Map 1. *Location of Majuli in Assam.*

erstwhile district of Sibsagar and Lakhimpur; the island is tucked into an isolated block with the Mikir hills on one side and Dafla hills on the other. Neither of the ranges is too far from the river. At a few places the hills actually intrude on the river but for the most part the Brahmaputra flows between sandy banks that are subject to its many changes in course (Goswami, 2001).

The physiography of Majuli is characterised by an extremely dynamic flow system; the ecology is unique and there is a constant threat of flood and erosion. The island forms a significant part of the vast alluvial floodplain of the Brahmaputra, dominated by a profusion of depositional landforms including sandbars, bed-forms, and abandoned channels, not to mention vast lengths of wetland. The danger from floods and erosion has intensified significantly after the great Assam earthquake of 1950 that measured 8.7 on the Richter scale. It triggered extensive sediment transport, accelerated the rate of erosion, caused the riverbed to rise, and is responsible for frequent channel changes. The threat from erosion has been particularly severe in recent times and the size of the island has shrunk from 1,246 km² in 1950 to 875 km² at present.

Turbulent Waters

Water has always been associated with progress; the world's greatest civilisations have sprung up on the banks of rivers. But in this instance, life-giving water has jeopardised life and development in Majuli, even as the island is vying for recognition as a world heritage site. Flood management here is carried out entirely by government agencies. Due to a sustained loss of fertile land to the river, there is a simmering tension between the local people and these agencies about the merit of the structural measures taken for erosion control until now. People's apprehensions have been fuelled by worries over the construction of the Bogibeel bridge less than 100 km upstream. Therefore, instead of welcoming the bridge as essential infrastructure that will lead to better connectivity for the people of the island, there is fear that it will escalate erosion. There has been growing concern over the lack of special environmental impact assessment (EIA) including hydraulic modelling or a sound environmental management plan for mitigating the adverse impact of the bridge prior to its construction. It is this issue that has emerged as a major cause of conflict. The feasibility study carried out by the railways was reportedly restricted to the structural safety of the bridge and did not cover its impact on Majuli island or other vulnerable areas in Dhemaji and Lakhimpur districts. The Majuli Suraksha Samiti, a local NGO, asserted that the design of the bridge is based on Parkar Stability Analysis and this system for a stable river course is applicable only in the case of streamlining watercourses within the guarded reach below the bridge. Pointing to the devastation already caused by erosion, the people believe that it was not the forces of nature that were responsible for the loss of landmass as much as erroneous decisions, such as, closing natural floodways like the Kherkatia Suti and the Tuni river; construction of multiple dykes; building of roads that crisscross the island with scant regard for floodwater storage, etc. There is growing discontent over the indifference of the project authority with the result that the issue is snowballing into a major conflict involving the entire population of the island.

A Question of Compensation

The conflict over the issue of erosion is at an early stage. There have been protests by the local population and NGOs like the Majuli Suraksha Parishad. The people have not yet taken the matter to court, apparently for lack of sufficient outside support and hard scientific evidence. The government too has not perhaps gauged the gravity of the problem, and until now no attempt has been made to undertake a detailed technical survey either to corroborate or disprove the stand taken by the people. In a related development, villagers were evicted near the bridge construction site without adequate compensation. This triggered a year-long agitation by the Mishing Students Union. However, even that dispute has been dormant due to the extremely slow pace of construction. Ironically, erosion and floods at the project site forced railway authorities to halt construction work twice in one year.

It appears as if the construction of another rail-cum-road bridge was of more importance to the government than the people of the island. The state administration did propose to set up a strategic planning system but no concrete steps have been taken in that direction so far.

Unfortunately, although the issue has been covered in some local and regional language dailies it has not received the kind of media attention it deserves. While there are no activist movements at this time, things might take an ugly turn in the future. The Majulis are essentially a peace-loving community and have remained remarkably restrained until now in spite of the lack of political support.

The problem with the Mishing tribe arose after the railways had completed the rail approach through a 23.6 km stretch that covered nineteen villages in three *mouzas*. A total of nineteen cases of land acquisition were framed for acquiring about 1,300 *bighas* of land from these villages. But the Mishing Students Union began agitations from February 2002, and the eviction drive has not yet been executed. The railway authorities had released around Rs. 5 crore to district officials to initiate the compensation procedure. The Assam government fixed the value of land at Rs. 70,000 per *bigha* but there was a discrepancy in calculations; the amount fixed for the tribal areas was only Rs. 17,000. Also, since the terms were not too clear and did not specify whether compensation would be paid in case of accelerated bank erosion, it will be difficult to link erosion with the construction of the bridge.

The strong opposition from the All Assam Mishing Students Union (AAMSU), locally known as Takam Mishing Poring Kebang (TMPK), and the local people has put a question mark on the timely completion of the 1,700 crore Bogibeel project. The union insists that it will not stop protesting until its demands are met.

The PWD is yet to survey the Kulajan to Kaba stretch of the land allotted for construction of a service road connected with the project. The villagers have been demanding a proper study and the TMPK has submitted a memorandum to the general manager, Northeast Frontier Railways, stating that the land belongs to poor farmers.

This is a project rife with controversies. Not only has there been a lack of a comprehensive feasibility study on the part of the authorities, but the issue also involves displacement and loss of land and livelihood. The lack of transparency and public participation in decision-making has made it a target of wrath.

Progress versus Traditional Lifestyles

If the conflict persists and all the stakeholders maintain their rigid stand, the traditional Mishing society with its hierarchical social structure and distinct cultural traits will be greatly affected. Loss of land will not merely destroy a traditional resource base and the economy of the region, it will also impact severely on the social fabric of the community. Land, kinship, power centres, wealth and religion are all interrelated. Sudden and far-reaching change in even one of these components will affect the other aspects as well and thereby hurt the entire culture. Changes like this may lead to stress and adaptation problems for the people, and eventually sow seeds of discontent. On the other hand, if the idea of the bridge is abandoned and shifted to another location, the local population will most certainly lose out on valuable opportunities for economic development.

Tough Choices

The vociferous protest by the Majuli Suraksha Samiti against the construction of the Bogibeel bridge is based on earlier unpleasant experiences when, during and after the construction of three bridges over the Brahmaputra, there was enhanced braiding of the river course leading to unprecedented floods and erosion in villages immediately downstream from the site (Palasbari, Gumi downstream of Saraighat Bridge, Morigaon, Nagaon region downstream of Koliabhomora Bridge and Dakshin Salmara, Pancharatna and Mancachar areas downstream from Naranarayan Setu).

What is required at the moment is a dialogue between the authorities and the people. The railways can begin by conducting a realistic environmental impact assessment based on a study of the physical hydraulic and sediment transport models. The report should involve an expert team of ecologists, environmentalists, and technical people, and offer a sound environment management plan to mitigate the problem of erosion. Before the conflict gets polarised, the government should come forward with a problem-solving, cooperative approach that involves collaboration—merging resources to seek solutions that address everyone's interests and are mutually beneficial. An impartial and compassionate attitude on the part of the agencies can help initiate a dialogue, and together both parties can isolate areas of common interest such as the development of the economy through better facilities. By recognising and addressing the underlying issues of the conflict, the government will gain the confidence of the people. Stakeholders too should accordingly change their attitude of defiance to break the impasse. They need to perceive the threat and recognise the urgent need for intervention.

Protests against structural interventions on the unpredictable Brahmaputra are not new. The Bogibeel case merely reflects growing doubts about the viability of constructing such bridges on an already capricious river. The perception is that they cause more harm than good in the floodplains: hence the public fear over projects like the Tipaimukh, Pagladia and Subansiri dams.

Note

*Map 1 from Goswami (2001).

References

Das, A.K. 2003. 'The Brahmaputra's Changing Ecology', *The Ecologist Asia*, 2(1).

Das, R. 2004. 'Chased from Classroom to Cowshed—Erosion by Brahmaputra Forces School in Bogibeel to Change Location', *The Telegraph*, Kolkata, June 4.

Goswami, D.C. 2001. 'Geomorphology of Majuli' in K.C. Kalita (ed.), *Majul*, pp. 22–35. Kolkata: Cambridge India.

Goswami, D.C. 2003. 'The Brahmaputra River, India', *The Ecologist Asia*, 2(1).

Patowary, A. 2005. 'Affected People to Launch Stir against Railways', *The Assam Tribune*, February 1.

Staff Reporter. 'Union Minister's Concern over Erosion at Palasbari', *The Assam Tribune*, October 2.

'Concern over Delay in NF Railway Projects', *The Assam Chronicle*, January 11, 2003.

'A Report on the Progress on Bogibeel Rail-cum-Road Bridge Project in Dibrugarh, Assam', *http://www.dibrugarh.nic.in*, January 1.

Madhukar, V.K. and P.K. Gupta. 2004. 'A Comprehensive Account of the Construction of the Bogibeel Bridge over the River Brahmaputra, Assam', *Times Journal of Construction and Design*, *www.etcconstructionanddesign.com*, August.

'Erosion Posing Great Threat to South Kamrup', *www.axom.faithweb.com*, August 22, 2000.

Unequal Power, Unequal Contracts and Unexplained Resistance: The Case of the Peri-Urban Areas of Chennai

S. Janakarajan

Genesis

RAPID INDUSTRIAL GROWTH AND EXPANSION OF POPULATION DUE TO FACTORS SUCH AS scarcity of land for urban use, pollution, lack of adequate drinking water and sanitation, degradation of coastal ecology, and sea water intrusion puts mega cities like Chennai under severe stress. In such situations, metropolitan cities reduce stress by encroaching upon peri-urban areas. But this often results in conflict of interests in the use of natural resources. Peri-urban areas are neither urban nor rural because they lose their traditional agrarian identity without acquiring an urban identity; they lose their traditional livelihood options without getting benefits like urban employment, and do not have the infrastructure that an average city dweller requires.

Chennai's Persistent Drinking Water Crisis

Over the past three or four decades, successive governments in Tamil Nadu have spent more than Rs. 4,000 crore on water supply to the city. But the problem persists, and per capita water supply in Chennai at 76 litres per day still remains the lowest as compared to the major cities in India (Joel Ruet, et al., 2002), not to speak of uncertainty and unreliability. But this is only in exceptionally good years. In bad years, hardly any water flows through the pipes; instead, it is distributed through tanker-trucks in a haphazard fashion.

Chennai does not have access to a perennial river and has to depend primarily on three major irrigation tanks and one small reservoir across a river that floods only for a few days during the monsoon. All these sources together supply about 300 mld (million litres per day) in a good year. In the last two decades, during the dry season, these have had to be supplemented by other sources such as wells in the peri-urban areas, contributing about 175 mld. The current water needs of the city and its urban agglomeration are almost double, of the order of 750 mld, and, it is estimated that by 2011, at 100 lpcd (litres per capita per day), the city would require about 660 mld for an estimated population of 6,600,000.. For the rest of the Madras urban agglomeration, an estimated 300 mld would be required for its 300,000 population. If the estimated industrial requirement in 2011 is also added, then the total requirement of the city and its urban agglomeration would be of the order of 1,210 mld. This is only a conservative estimate. But the current supply from surface sources is nowhere near what is needed. Is there then a way out for Chennai?

The Conflict: Urban versus Peri-Urban

Cities spread and subsume peri-urban areas in much the same way as a snake eats a frog: initially, the frog screams loudly, but the volume gradually reduces and finally all protest ceases. How long this will take is determined by many socio-economic, political and environmental factors. In the process, a few people benefit but the majority in peri-urban areas suffer. As long as urban development and available institutional mechanisms absorb the displaced peri-urban population, there is little to worry about; but when it does not, it is a serious problem. In other words if its 'spillover effects' are minimal, mutually beneficial urban expansion may be sustainable; however, if they are forced upon, or are negative and exploitative or can lead to the destruction of existing livelihood options, then urban expansion not only becomes unsustainable but also a contested terrain, where conflicts surface.

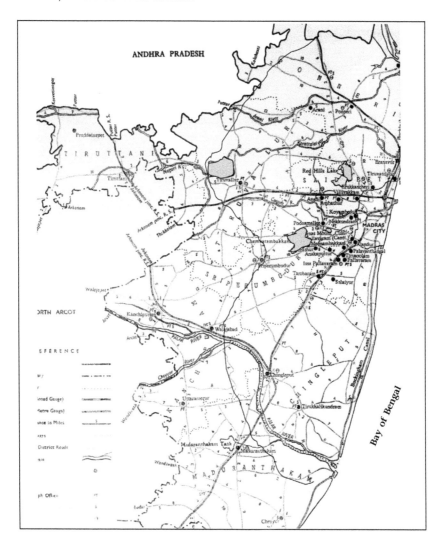

Map 1. *River basins in the Chennai region.*

Reasons for Conflict in the Peri-Urban Areas of Chennai

- Urban stress is transferred to peri-urban areas when urban populations migrate and settle there.
- Industries relocate to peri-urban regions due to the availability of better land and water.
- Land in the peri-urban areas is bought for urban use, resulting in dramatic changes in land use pattern.
- Increasing urban activities in the peri-urban areas lead to pollution and degradation of natural resources.
- Water is used more and more for non-agricultural purposes, and the water that had been earmarked for farming is forced to compete with supply to urban areas.
- Changing land use leads to fall in agricultural employment in peri-urban areas, weakens agriculture and causes serious livelihood problems.
- The village commons—land and water bodies—are either encroached upon or suffer neglect.
- While the need for infrastructure grows, the institutional vacuum in peri-urban areas leads to inadequate infrastructure development; a neither here nor there situation.
- Women who lose agricultural employment are the worst hit among the peri-urban population.

Conflict in Velliyur Village

What happened in early 2005 in Velliyur village might be taken as the strongest manifestation of this conflict so far. For over ten years, groundwater in the village has been pumped from irrigation wells and sold. Recently groundwater levels dropped steeply due to a three-year-long drought. Many irrigation wells failed and agriculture was badly affected. Cropped area dropped considerably, and landless labourers and poor farmers started migrating to the city in search of employment. In early 2005, the village issued an ultimatum to the Chennai Metrowater Board to stop pumping groundwater; water sellers from the village were also warned not to sell water but Metrowater did not take the warning seriously. In the first week of February, about 400 farmers came together and demolished all Metrowater structures and pumping was immediately suspended. The District Collector visited the spot and forty-four people were arrested and remanded to judicial custody for fifteen days and later released on bail. They are now fighting the case in court. The struggle has spread to other villages from where groundwater is being pumped by Metrowater. There is animosity between farmers who sell water and the others in the villages, a situation that is being exploited by Metrowater.

Clashing Viewpoints

The various groups involved in the conflict approach the issues very differently. Water sellers who have an abundant groundwater supply benefit a great deal by selling water to Metrowater for urban use and can even afford to abandon agriculture.

However, such water sale affects groundwater availability in the adjoining wells and the resultant fall in the groundwater table adversely affects the livelihood of a majority of the villagers. Often bores installed along the river exhaust even the riverbed aquifer, reducing surface as well as groundwater in the adjoining villages. The village commons are encroached upon, reducing access for the poor. In each one of these developments the majority lose while only a few make significant gains. In effect, urban users exploit peri-urban natural resources, and transfer scarcity and pollution to the peri-urban areas.

Metrowater maintains that groundwater is pumped only from areas where groundwater is available in abundance and where farmers are ready to sell. It argues that there is no compulsion involved, farmers

Fig. 1. *Pipes transfer groundwater from various wells to a sump in Karanai village.*

Fig. 2. *Groundwater being pumped to a tanker-truck for sale in Chennai–Nalur village near Chennai.*

sell water voluntarily and for profit. The Board does not take into consideration the serious long-term implications for peri-urban areas—depletion of groundwater, sea water intrusion and the destruction of farmers' livelihoods.

Urban dwellers are more educated, politically active, organised and influential than their rural counterparts, and they are also increasing in number. Political power in India is likely to shift more and more towards urban areas, making it increasingly difficult for farmers to resist water diversion from rural to urban areas (Moench and Janakarajan, 2003).

Scope for Dialogue

There is a case for organising a dialogue between peri-urban farmers, the Metrowater Board, researchers, activists and other sections of society, despite the prevailing atmosphere of intense competition and bitter conflicts, and the difficulties of bringing together the various groups for dialogue and coordinated action. These reasons include:

- Though urbanisation is an inevitable process, peri-urban areas should not suffer as a result. There is a need to explore ways in which urbanisation can be managed and shaped to the advantage of both populations.
- The Chennai Metrowater Board has tried all kinds of mega projects, yet the city's water needs are met only marginally and on an ad hoc basis. A permanent solution is nowhere in sight.
- The peri-urban population has stood up to the government and resisted pumping of groundwater from its villages.
- A large number of irrigation tanks in these villages (around 3,600 in adjoining districts) are either unutilised, or under-utilised or abandoned. Rehabilitation of these tanks with the help of local farmers will help agriculture, and also recharge groundwater. Surplus water can be supplied to the city, thus facilitating peaceful co-existence of urban and peri-urban populations.
- The dialogue process can pressurise Metrowater to introduce waste-water recycling, reduce wastage by arresting leakages, improve the water supply system, and take up rainwater harvesting wherever possible, particularly by utilising the large number of tanks.
- The urban population needs to be educated about water-saving devices and practices in bathrooms and kitchens.

The Ongoing Dialogue Initiative

The dialogue initiative began in December 2004, and three committee meetings have been held so far. The agenda is being discussed with stakeholders while at the same time research is also being carried out, particularly into poverty and livelihood analysis, water budgeting, mapping of water bodies in peri-urban areas using Geographic Information System (GIS)—which helps to map all quantitative and hydraulic information—and by analysing Metrowater laws. Research and the continuing dialogue process might provide the scope to find a lasting solution. It will be very difficult to set a deadline for arriving at a solution to the problem, but the stakeholders' committee is hopeful of reaching a conclusion soon.

Note

*Map 1 and figures by author.

References

Joel Ruet, Saravanan and Marie-Helene Zerah. 2002. 'The Water and Sanitation Scenario in Indian Metropolitan Cities: Resources and Management in Delhi, Calcutta, Chennai, Mumbai'. CSH Occasional Paper No.6. Delhi: French Institute.

Moench, Marcus and S. Janakarajan. 2003. 'Are Wells a Potential Threat to Farmers' Well-being? The Case of Deteriorating Groundwater Irrigation in Tamil Nadu', Working Paper No. 174. Chennai: Madras Institute of Development Studies.

Rural Livelihoods, Urban Needs:
Diversion of Water from the Ganga Canal

Binayak Das

Farmers Gather in Protest

In AUGUST 2002, ABOUT 5,000 FARMERS FROM MURADNAGAR AND ADJACENT REGIONS GATHERED in Bhanera village to protest against diversion of water from the Upper Ganga canal to provide drinking water for Delhi. The problem is further compounded by the fact that for the first time in India's water management, a multinational, Degremont, is sourcing this water for their Sonia Vihar plant in Delhi. The area immediately affected by the conflict lies in northeast Delhi around Bhanera village in Muradnagar district of Uttar Pradesh.

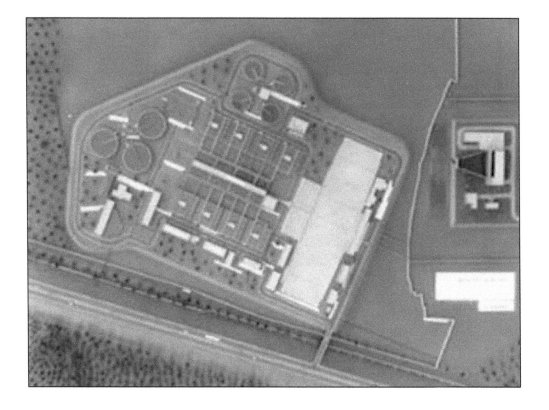

Fig. 1. *The Sonia Vihar plant.*

The Sonia Vihar Plant

Delhi, the capital city of India, currently faces a shortage of 2,000 mld (million litres per day) of water and it is expected that with an increasing population, the demand will overshoot supply by more than half. At present, Delhi procures its water from various sources—the Yamuna, groundwater, Bhakra storage, and from the Ganga. In a bid to augment water supply, the Delhi Jal Board (DJB) has contracted the building, operation and management of its treatment plant in Sonia Vihar to the French multinational Ondeo Degremont. The 200 crore plant is to be supplied with water from the Tehri dam through the Western Ganga canal.

The water is brought to the plant from a point near Muradnagar on the Western Ganga canal through a 30 km long, 3.5 m diameter pipeline at a cost of Rs. 111.31 crore. The Upper Ganga canal has been lined to prevent seepage. In January 2005, the plant with a design capacity of 635 mld began to operate on a trial basis. Delhi already receives 100 mgd (million gallons per day) of water from the Western Ganga canal for its Bhagirathi treatment plant through a pipeline from Muradnagar.

Built in 1847, the Upper Ganga canal is one of the oldest irrigation canals in the country. The 304 km long canal extends from Haridwar to Kanpur and carries about 300 cumec of water and is the main source of irrigation for a 924,000 ha tract which covers Haridwar, Roorkee, Saharanpur, Muzaffarnagar, Meerut, Ghaziabad, Gautam Budhnagar, Bulandshahr, Aligarh, Mathura, Hathras, Mainpuri and Etawah. It is considered one of the most fertile agricultural tracts in the country, and the principal crops include wheat, rice (basmati and coarse varieties), sugarcane, maize, potato, and gram. Farmers in this region have been using the water for more than 150 years. Besides the canal, tubewells are used as an important source of irrigation.

Over the last few years, water availability in the region has been decreasing. The groundwater table has plummeted by about 6 m due to overdraft. In western UP, the failure of monsoons in 2002 and the subsequent drought have had an adverse effect; even the Yamuna has gone dry. Since the overall water availability in the region has, for a variety of reasons, decreased, the Upper Ganga canal has become even more important as a source of canal irrigation and also for recharging the aquifers.

Farmers' Apprehensions

The farmers in the villages from both sides of the Upper Ganga canal are apprehensive about the diversion of water to Delhi. According to them, this move will affect their crop yield in the lean season. Studies conducted by the New Delhi-based Research Foundation for Science, Technology and Ecology corroborated that the transfer of 635 mld of water from the Upper Ganga canal will lead to a net annual loss of 1.4 MT (metric ton) of crop harvest. The farmers are also concerned over the lining of the canal as it will prevent groundwater recharge and so deprive the land of the much-needed groundwater for irrigation. Farmers have also been asked to refrain from digging wells in their field and this too has worried them.

When Degremont began construction of the Sonia Vihar plant pipeline, discontent spread in the twelve villages on both sides of the Upper Ganga canal from Haridwar to Muradnagar. On August 9, 2002 about 5,000 farmers representing various interests gathered at Sonia Vihar and protested against the ongoing project and demanded that the lining of the canal be stopped, and that Degremont discontinue all work on the water diversion project. They also demanded that Degremont quit India.

Fig. 2. *The Sonia Vihar project is designed to procure water through a 30 km long, 3.5 m diameter pipeline from the Western Ganga canal near Muradnagar.*

The Different Protesting Groups

The rally was the culmination of a 300 km mobilisation drive by farmers and inhabitants of Garhwal in Uttaranchal protesting against the construction of the Tehri dam. The event was launched with a rally in Haridwar. Several religious leaders supported it and activists declared that the Ganga was not for sale. The participants belonged to three different groups and had various issues on their agenda. The first group comprised the people from the Tehri region who are opposed to the dam because it is seismically unsafe, because it will submerge the old town of Tehri, and will lead to the displacement of 100,000 people from the town. The second group comprised anti-privatisation activists and other citizens who were opposed to the handing over of the Ganga to a private firm, and a multinational at that. They took the lead in demanding the cancellation of the contract with Degremont.

The third group were farmers from the Muradnagar region. They were more worried about their share of the irrigation water being taken away, and were not concerned with the nationality of the entity behind this. They expressed full sympathy with Delhi's need for drinking water, but also emphasised that decisions taken behind closed doors had resulted in the loss of irrigation water for them.

The farmers received the support of organisations like Navdanya Foundation and of eminent individuals like Vandana Shiva and Rajendra Singh, who travelled to Muradnagar to support their cause.

Those Defending the Diversion

Degremont obviously expected the government to provide water for the plant. Though no written assurances had been given, the failure of the DJB to get water for the initial run of the plant has not gone down well with the firm, which has told the DJB to bear the resultant losses. Talks are on between the firm and DJB, and there have been discussions about what is to be done if such a situation arises again. However, neither the DJB nor Degremont has so far revealed whether assurances and counter assurances have been given.

As things stand, it is estimated that after channelisation of the water for the Sonia Vihar plant, only 30 per cent of the earlier amount will be left over for irrigation in Muradnagar. The government claims that the water for farmers from Upper Ganga canal will not be reduced because eventually Tehri is expected to supply the water that is diverted to Delhi. According to them, the canal will have more water and will also benefit the farmers. However, there is no guarantee that this will happen.

Chronology of events

December 2000	Commencement of work on Sonia Vihar plant
August 2002	Protest by farmers in Bhanera
January 2005	Sonia Vihar plant starts trial run

Current Status

The Sonia Vihar plant started functioning on a trial basis from January 2005 and is presently drawing water from the Western Ganga canal near Muradnagar. Local farmers have not yet voiced their concerns but there are murmurs of dissent and things can flare up during the lean season, or when the plant starts operating at full capacity. Despite an agreement between Delhi and UP that the plant will receive water from the Ganga canal, water has not been released due to poor rains and political pressure. For the time being, the farmers' apprehensions have been allayed, but the issue remains alive.

Scope for Dialogue

The most important factor in this conflict is Delhi's ever-increasing demand and continual dependence on other states to quench its thirst. The problem that has arisen with the farmers of Muradnagar can in future raise its head elsewhere too. The biggest hurdle in resolving the problem is Delhi's refusal to acknowledge that a conflict exists. The government of UP, with whom the contract to procure water for Delhi has been signed, is wary of hurting the farmers' interests but has not shown any inclination to seek a long-term solution.

The conflict is further aggravated by the fact that a multinational has been authorised to draw water from the canal. The issue has the potential of flaring into a UP versus Delhi conflict. While there are guidelines for the procuring of water by the private firm, there are no guarantees that farmers will be provided adequate water before the rest is diverted to Delhi, or indeed that their needs will be prioritised during the lean season.

In summer, Delhi too will require more water, and since the private firm has been entrusted with the responsibility of providing water for Delhi, the farmers fear that this will be done at their expense. It is important that the governments of Delhi and UP come together and formulate common guidelines for water allocation to both the parties. Given the possibility of reduction in supply of water for irrigation, farmers may have to adopt new technologies for saving water and grow crops that require less water. The government has to monitor Degremont and put in place a transparent mechanism of regulation and monitoring to ensure that the firm follows prescribed norms and does not overdraw water from the canal. However, this will serve as a solution only in the short run. It is imperative to develop a common framework for such allocation conflicts. Stop-gap measures will only postpone the conflict.

Table 1: Estimated Crop Loss due to Diversion

Crop	Yearly production loss (in T)
Wheat	331,055
Rice (Basmati)	55,115
Rice (coarse)	92,710
Potato	965,729
Total	**1,444,609**

Note

*Table 1 from Shiva *et al.* (2003). 'The Impact of the River Linking Project', Jal Swaraj, Water Sovereignty Series No. 2. Navdanya Foundation, New Delhi; figures by author.

References

Shiva, Vandana and Jalees Kunwar. 2003. *Ganga: Common Heritage or Corporate Commodity?*. New Delhi: Navdanya/ RESTE.

Sharma, Sudhirendar. 2004. 'Delhi: Demanding More from the Ganga', *Disputes over the Ganga*. Kathmandu: Panos– South Asia.

PART 2

Equity, Access and Allocations:
A Review

Suhas Paranjape and K.J. Joy

ALL THE TEN CASES INCLUDED IN THIS SECTION DEAL WITH EQUITY IN ACCESS AND allocation. Of course, equity in its wider sense is a very broad concept and many of the other sections in this book may also be said to be dealing with equity. For example, the section on contending uses is, in one sense, an issue of equity; what forms an equitable allocation between different kinds of uses. Here 'equitable' is used in the sense of 'fair'; what constitutes a fair allocation between the different kinds of uses. In this sense, many of the case studies in the different sections also incorporate a strong sense of what is and what is not equitable. Second, as has been said in the introduction, sequencing and grouping in a publication does not give you the same kind of flexibility that a web publication does, and it becomes necessary to identify one of the dominant aspects in order to group cases together. In this sense, what characterises the ten cases in this group is that they are all broadly about issues related to access and allocation in the context of a single use. Thus, whether it is irrigation or drinking water, the issue is, who gets how much and why. Having said that, however, there are other issues that are interwoven with it, and those are what make the cases all the more interesting.

New and Old Entrants

What happens when a new project comes up in a basin in which there are already established water rights? What happens when, for many reasons, reservoir storage begins to fall short of whatever was the assumed design capacity? Who gains, who loses? Is there some kind of parity between those who lose? Every project planned, constructed and implemented over the years has to find answers to the question—who gets how much. And how that issue is resolved, or left unresolved, has a bearing on how far we have come in making equity a central concern inbuilt in the project design.

There are two case studies in this section that deal with the issue of holders of prior rights and new rights. The first relates to the old and new *ayacutdars* (irrigators) in the Lower Bhavani project (LBP). The old *ayacutdars* had rights to eleven months of irrigation for 20,000 ha, presumably to protect prior rights. The new *ayacutdars* were promised water for one crop for 400,000 ha. This worked fine, whenever the LBP filled to capacity according to the design schedule. It did do so in its early years, but recently, storage has fallen short of design capacity year after year. The issue is how the shortages will be shared. The old *ayacutdars* insist that they be assured of their eleven-month irrigation irrespective of the water available in the dam. The new *ayacutdars* insist that water should be supplied to everyone for at least one crop and only what remains should be made available to the old *ayacutdars* for the next crop.

The issue is how to share shortages. Projects specify allocation of design storage. No protocol is specified for when there may be a shortfall or a surplus. First, it is important to recognise that water shortages are here to stay, especially when measured against designed availability. Upstream developments, massive dispersed groundwater withdrawals as well as diversion to other uses have reduced inflows into most dams. The extent and pattern of reduction may vary considerably, but this is a possibility that needs to be taken into account and protocols for sharing of shortages need to be made part of the project design. Second, it is important to evolve a consensus on the principles for equitable sharing of shortages. Unfortunately, the thinking is in terms of ad-hoc fire-fighting local measures and little attention is paid to the necessary long-term measures.

Then there is the case of the decline of the *phad* system in the Panzara valley in Maharashtra. The *phad* system is a traditional irrigation system based on diversion. The designated block of land that could be irrigated is called a *phad*. At least all farming caste households in a village had access to some land in the *phad,* and thereby to irrigation. Incidentally, it is also important to realise that such an arrangement was made possible by thinking in terms of bringing everyone's land to the water, instead of bringing water to everyone's land! The *phad* or the irrigated block was determined first, and then everyone, at some given historical point, was given land in the *phad*. Though not a solution we can easily adopt, it goes to show that suitably modifying social arrangements is often the cheapest and the optimum solution.

The *phad* system farmers were not as lucky as the Bhavani basin's old *ayacutdars*. They received many promises but little protection. The smooth functioning of the *phad* depended on the availability of water for irrigation. As projects proliferated in the Panzara valley, the water available in the diversion channels of the *phad*s decreased, and *phad* after *phad* fell apart. Only a few *phad*s are now functioning in the valley and they too have undergone substantial changes in their practices. The study also shows that the old *phad*s used much more water than would have been their share on the basis of a per capita approach.

Could the *phad*s have been preserved? In retrospect, one would say that the *phad*s could only have survived by spreading, not in their original form, which was too closely tied to the social milieu of the past, but by evolving a set of social arrangements incorporating the learnings from the *phad*s. The Tembu case study is about just that kind of an attempt.

Equity, Affordability and Pooling of Costs

The Tembu Lift Irrigation scheme (TLIS) lifts water directly from the Krishna and makes it available to the severely drought-prone, rain shadow region in the eastern portion of the Krishna basin in Maharashtra. The scheme is also one of the most controversial, mainly because it lifts the water through more than 300 m to irrigate the Atpadi *taluk* at the tail end of the planned command.

The potential beneficiaries, especially those in Atpadi *taluk*, are organised under the banner of the Shetmajoor Kashtakari Shetkari Sanghatana (SKSS). They have demanded that their allocation be distributed equitably and have shown that it is sufficient to supply 3,000 m^3 of water to *every* household in the *taluk*. To understand the significance of this, we need to compare it with the conventional method in which the canal and the sub-channels run by gravity define the command area that will receive water. This splits farmers in a village(s) into those who receive water and those who do not, and second, those who happen to have more land in the command get more water than those who happen to have less.

In contrast, first, the SKSS scheme provides access to *every* household in the *taluk*, thereby taking care of locational inequities and second, delinks the supply of water from the amount of land held, by providing minimum water to every household irrespective of its holding. Obviously, restructuring the scheme has a cost but it is not very high compared to the present total cost of the project. The SKSS has a telling argument in this respect: that equity does not come free, it involves cost; and accepting equity is accepting those costs while not accepting those costs amounts to denying equity.

The greater and more controversial cost, however, is not the relatively small cost of equitable restructuring. It is the much larger annual, recurring energy and financial cost of lifting water over more than 300 m. At first sight, this appears to be an issue of affordability and not equity. The issue is better brought out by a parallel example. Let us take the issue of transport and transaction costs of supplying essential foodgrain to *adivasi* villages living in difficult and remote areas. Assume that the cost is double that of supply to the plains. Should we then treat supplying these *adivasi* areas with essential foodgrain as unaffordable, and second, should we then charge them double for it? To do so would amount to penalising the *adivasis* for their disadvantaged status.

The proper way is to pool the costs of supplying foodgrain to all areas and charge a common rate so that locational inequities are ironed out. Should we not look at water in the same manner? It may take little energy to serve some areas but much more to serve other locationally disadvantaged areas with the minimum water required to stabilise farming livelihoods. Should such areas then be similarly penalised for the locational and meteorological disadvantage that they face?

This is not a simple question. The problem appears to be naively simple if we reduce it to the 'affordability' of large lifts for isolated areas, since it masks the element of equity. It is important that the SKSS has brought it out and made it part of the discourse, and hence has been able to develop a more nuanced and multi-pronged approach to the problem as part of the stakeholders' dialogue it has participated in. Among other things, it envisages: (*a*) use of a portion of water for environmental regeneration and recharge; (*b*) integrating Krishna water with local water harvesting; (*c*) a progressive reduction in the Krishna water requirement and so, lifting costs and energy; (*d*) reserving a portion of water for energy crop to meet the financial costs of the lift as well as cover the cost of energy in kind.

Inter-basin Transfers: Upscaling Conflicts

The old method of delineating gravity commands was more comfortable since there was no great scope for variation. But once we delink project command areas from gravity commands, the question is wide open. What principles should govern how far the water should be carried and how high should it be lifted? The issue of equity in access and allocation has to be faced squarely and responsibly. While TLIS represents one side of the coin, the Nar–Par–Damanganga diversion represents the other.

For many of us who think in terms of technological fixes, including, with all due respect, our scientist President as well, water often becomes an abstract entity to be manipulated at will, delinked from the ecosystems it is embedded in. All that is needed to solve the problem of any area is to bring water from however afar it may seem to be available. After all, bringing water from a few thousands of kilometres away is but a small achievement for modern science. Such supply-augmenting solutions are politically convenient, populist measures that prevent us from facing the much more difficult task of improving our own water use pattern and making it more environment-friendly and equitable. Every such technological fix that is suggested virtually creates new conflicts and stakeholders. And naturally, it requires politics to manage these conflicts. Water fixes have become a virtual political industry.

Take for example the idea of diverting the Nar–Par–Damanganga west-flowing rivers to bring water to the Tapi and Godavari basins. Leave aside, for the time being, the erroneous conception that water flowing out to the sea through the rivers is 'wasted'. The point is that without this idea farmers in the Tapi and Godavari basins would not view themselves as stakeholders in the Nar–Par–Damanganga waters. But now they do, and not only that, they view themselves as antagonistic stakeholders. And as the study shows, the *adivasis* in the middle region shoulder all the disadvantages.

Actually, all supply-side solutions as well as many simplistic demand-side solutions disregard an important equity issue. They welcome water-saving technologies, but are silent on what is to be done with the saved water. At present water-saving technologies serve only to double and triple the irrigated areas of the farmers who *already* have access to irrigation. If water is saved through such technologies, and that too with State subsidies, should that saved water not go to those who do *not* presently have access to water? If such equitable measures are not built in, even demand-side solutions will not prevent the emergence of water 'shortages'. It is only after this is done on a priority basis that we can talk about inter-basin transfers. In fact, if provision of a small amount of water from the west-flowing rivers were to become an incentive to sustainable and equitable use in the recipient areas, one may keep all reservations aside and accede to the demand. Unfortunately, the Tapi–Godavari basin areas that the Nar–Par–Damanganga water is being diverted to are far from fulfilling these conditions.

Tail-enders and Head-reachers

Two case studies deal with issues of allocations within a project. The Palkhed case from Maharashtra shows blatant double standards with respect to the irrigated area added later. Projects seem to start as small and modest in size and then grow bigger by populist additions that continue even after a project's completion, putting allocation norms under severe strain. In the Palkhed case, it is difficult to see a rationale for the extreme variation in the quota from about 1,500 m³/ha for the Left Bank Canal (LBC) to more than 6,000 m³/ha for the Right Bank Canal (RBC). In fact, we find that the Palkhed left bank canal is charged with *all* the losses of the project. The discrimination is fertile ground for a conflict just waiting to happen.

The other case is that of the Indira Gandhi canal (IGC) in Rajasthan. The IGC is in a mess and the case study brings out only the tip of the iceberg. Two kinds of problems are overlaid here. The first is that of transient rights becoming established rights. A similar example is found on the Krishna banks in south Maharashtra. Many of the projects planned on the Krishna would have taken decades to be completed, so farmers were allowed temporary permits to lift water from the river. Lifts mushroomed and now, as major diversions on the river approach completion, the so-called temporary rights have become a major problem. The second problem is that of unforeseen shortages, in this case, not only natural shortages, but those brought about by the Punjab government as well. How the issues are handled, and how transparently, is also a problem in both cases.

Well Owners and Others

Two cases, the Pondicherry case of Keezhaparikkalpet as well as the north Gujarat case of Mehsana district, relate to the division between those who own wells and those who do not, although from totally different perspectives and in different contexts.

In the north Gujarat case, the conflict is latent and simmering and centres on the ever deepening of the wells and the difference in water charge between those who are part of the groups that own wells and those who buy water from them. Non-members are charged in terms of the share of crop that amounts to almost double of what members are charged. While non-members resent the double standards, they are too dependent on the water they buy to raise their voices against this system.

The Pondicherry case is different and offers valuable lessons in how a well-managed process of tank rehabilitation can lead to a gradual change in power relations in a village, and make for a more equitable outcome. Many farmers were dependent on the borewell owners in the village. Even after rehabilitation, the tank could only assure irrigation for one season and many farmers still depended on the borewell owners for the other seasons. Borewell owners even got some of the farmers to refuse tank water and oppose the Tank Association (TA). The solution found was a community borewell owned by the TA supplying the farmers with groundwater for the later seasons. Initially the borewell owners tried to stall the move, but soon they found themselves isolated and had to accede to the community borewell. It is to the credit of the agency working there that they thought of imaginative methods to isolate trouble-makers.

Women and Dalits

The water sector does not exist in a vacuum; it is part and parcel of the society that it inhabits. All the inequities that exist 'outside' the sector, nevertheless, pervade it as much as they do other sectors, no more and no less. The issues related to class are more easily discussed and tackled, because differences between the big and small farmers are visibly relevant. But other inequities often define exclusion-inclusion relations in such a way that relevant issues are kept out of the discourse.

One such issue is gender. What do women have to do with water, especially with its productive use? By definition they are out of irrigation. They are allowed some presence in drinking water, but here too, more as 'collectors'. The Jholapuri case in Gujarat is an example of how that mould can be broken. Utthan, an organisation that has been working with women of coastal Gujarat, sets an example of a stakeholders' dialogue process that includes women and their concerns. Though it still remains almost exclusively a drinking water concern, it does go some way towards gender participation and gender equity.

The other case of the Mangaon Dalits, unfortunately, does not have that happy an aspect. Caste oppression, and particularly that of the Dalits, is one of the ugliest features of Indian society. The Mangaon case illustrates how efforts to work around the problem rather than facing it head on only postpones the conflict that erupts once it finds favourable conditions. Baba Adhav, social reformer and activist from Maharashtra, has long advocated 'ek gaon ek panavatha' meaning, 'one village, one watering place', so that all castes are required to draw and share water from a common source, not only in theory but visibly so. Instead, what we have today is a politics of compromise under the name of accommodation.

This politics essentially provided each 'community' with a separate water source, and all it achieved was to push the conflict below the surface. It brought about a sort of ghettoisation, caste-isation of water sources. The fierceness of the underlying casteism came to the forefront once Dalit wells failed in the Konkan villages during a drought. Though finally a semblance of order was re-established, it was only after a better monsoon revived Dalit wells and re-established the ghettoisation of water sources that the conflict subsided. There is surely a lesson to be learnt here—that there really cannot be equity within the water sector if there is inequity within the society that surrounds it. It is interesting to note that Mahad is the site where Babasaheb Ambedkar led the historic Mahad *satyagraha* over the very same issue in the 1920s.

Missing Norms of Equitable Allocation

Allocation norms have evolved according to local situations, size and nature of project and historical socio-political relations. For example, an old project may first have come up as a drought relief measure taken up by the British more than a century ago. Initially, finding that the water had no takers the British may have offered liberal terms and assurances to the farmers at the head-reach. However, they did not offer such favourable terms to the later additions. The project area may have greatly stretched after independence on the basis of a hypothetical crop pattern, maximising the cost-benefit ratios. And with the large-scale advent of electric pumps, now an entirely different pattern may be in place, turning all project assumptions on their head. In such a situation, even attempting to pin down allocations is a legal minefield, leave alone trying to determine what is equitable.

However indiscriminate and chaotic, the populist extension of command areas has an implicit principle that water must be provided as widely as possible, an entirely acceptable principle. The unfortunate part is the ad-hoc manner in which it is applied without disturbing underlying relations and entrenched interests. There seem to be no proper underlying equity norms and principles that will provide us with a road map of where to go.

We need a better concept of a right or an entitlement to water. How much water should a person or a household be entitled to *as a right*? Here we need a livelihood-needs framework that sees assurance of minimum livelihood needs and the corresponding water requirement as an associated right. This assurance must, in turn, accommodate two kinds of assurances; in a physical sense, a sufficiently high degree of dependability, and in a social sense, availability at a sufficiently affordable price for the poorest sections. There are ways of combining assured and variable water in a manner that makes it possible to provide minimum water assurance very widely even while allowing sufficient incentive for enterprising farmers. It is important to note that *rights* need to be defined in terms of minimum livelihood needs of persons or households and not in terms of rights accruing to land owned in the command.

There is a need to combine two additional norms with these equity norms. The first is establishing synergy between small and large, and local and external sources so that they will expand resource availability and dependability. The second is to take into account minimum ecosystem needs and curbing wastage. Elaboration of these principles may not mean they could be applied tomorrow but it will help in more than one way. First, it will set a goal to work towards. Second, it will give us better road maps of how to go about it. And most importantly, it will allow us to share shortages and surpluses in a principled manner.

However, equity also means doing away with the obstacles that deny the disadvantaged sections of our society their rights. The water sector is a separate 'sector' only conceptually; it is as much part and parcel of the society that incorporates it. There is, therefore, a much greater need for proactive action, and to form bonds with movements and measures favouring equity in other sectors.

Water Users in the Bhavani River Basin in Tamil Nadu: Conflict among New and Old *Ayacutdars*[†]

A. Rajagopal and N. Jayakumar

The Lower Bhavani Project

Bhavani is an important tributary of the Kaveri in its mid-reach in Tamil Nadu. It originates in the Silent Valley forest in Kerala and flows in a south-easterly direction for 217 km till it joins the Kaveri at a town named Bhavani. The total area of the Kaveri basin in the State is about 43,000 km², of which the Bhavani sub-basin constitutes roughly 5,400 km². The Kaveri basin which drains Karnataka, Pondicherry, Kerala and Tamil Nadu comprises about 82,000 km², of which the Bhavani river basin is 6,000 km². A major portion (87 per cent) of this area is in Tamil Nadu.

Lower Bhavani Project (LBP) is a major multi-purpose reservoir, mainly constructed for water storage and distribution to canal systems in the basin. The reservoir is also used for hydel power generation and fishing. Apart from this, anicuts like Kodiveri and Kalingarayan are used to divert water into different canal systems. These are old systems that have been in existence for several centuries. The upper part of the basin is not well developed and mostly depends upon wells and rain-fed agriculture.

Parched Land, Thirsty People

The river plays an important role in the economy of Coimbatore and Erode districts by providing water for drinking, agriculture and industry. Due to increase in population and unplanned expansion of the command area, and domestic and industrial water demand, the basin is already 'closing' and stressed. As a result, there is intense competition among water users, and a sizeable gap exists between the demand and supply in agriculture and domestic sectors.

Water shortage downstream is even worse due to a drought that has lasted several years. Out of about 1,700 Mm³ (million cubic metre) of normal water supply in the LBP dam, the actual realisation declined to 1,275 Mm³ in 2001, 793 Mm³ in 2002 and 368 Mm³ in 2003. There was already a conflict of interest between farmers in the valley, the original settlers and the new command farmers of LBP. Old command farmers are entitled to eleven months' water supply which they used for growing two or three paddy crops and annual crops such as sugarcane and banana, whereas the new *ayacut* (command area) farmers were only able to grow a single paddy or dry crop in a year.

As long as water supply in the dam was adequate, the conflict remained subdued. But supply was at an all time low in 2002 and water was not released to the new command area. This prompted the new *ayacutdars* to file a case against the State in the high court seeking water supply for at least one crop. Their contention was that water should be provided for the second crop in the old settlement only after meeting the requirements of the first crop in the new command as per the government order (G.O.) issued as early as 1963 (G.O. No. 2274 dated August 30, 1963). The court asked the Water Resources Organisation to arrive at a compromise formula for water sharing between the two areas. The department prepared a plan on the basis of the size of the command area—60 per cent of the available water was to be given to the new *ayacut* (for irrigation of 80,000 ha) and 40 per cent to the old *ayacut* (about 20,000 ha). However, the old settlers objected on the grounds that they were entitled to eleven months of uninterrupted water supply as per their riparian rights. The impasse prompted the local central ministers to bring the two sides to the negotiating table, but this attempt to seek a solution also failed. The court, in its interim order, has now told the State to take prior permission from the court to begin the system every season. Under the original regulation the canal opens on April 18 for the old settlement and August 15 for the new *ayacut*.

The expansion of irrigation, and hence demand, has mostly taken place in upstream areas (and to some extent in old *ayacut* too) through unauthorised tapping of river water by direct pumping. Apart from direct pumping, the other major issue is the unregulated exploitation of groundwater in the catchments, which the State is unable to control. This practice is actually encouraged by liberal institutional financing. Supply of free electricity by the State has also contributed to the growth of this problem. Downstream farmers did take the issue to court and even won a favourable judgment but the ineffective bureaucracy has been unable to implement the court's orders.

Highs and Lows

Water conflicts played a major role in the fourteenth general elections held in 2004. There was a flurry of negotiations between politicians and farmers' associations in the basin.

The water-sharing problem between the old and new settlers had been compounded by growing demand from industry and the domestic sector. The drought in 2004 had added to the situation and there was a severe shortage of water in the dam. As election time approached, the ruling party's ministers sought to build up their vote banks by trying to get the farmers to compromise. The effort failed and in the meanwhile, the drought worsened and a new dimension was added to the dispute. The water shortage in the region forced the farmers in the Kaveri delta to clamour for water. This group constitutes a sizeable electorate and since it was potentially more beneficial for the ruling party to get their support, water from the dam was released to the Kaveri delta, which was totally against the norms for operating the reservoir. This angered the Bhavani basin farmers who chose to nominate an independent candidate and named him 'Water Candidate', a first in the electoral history of India. However, this attempt, too, failed due to ideological differences among various agrarian groups and finally it was a candidate from the main opposition party in the state who won the farmers' vote.

This has helped the new command farmers to take up water issues with the irrigation bureaucracy and to get more water from the reservoir easily as the politician went on to become a minister in the central government. The new command farmers' association has also taken the old settlers to court with help from the minister.

Stubborn Stakeholders

While the case is now pending in court, the water situation remains grim, and the domestic water consumers—especially the middle class—have resorted to purchasing water for drinking purposes. Farmers affected by

Fig. 1. *Pollution of water by effluents.*

Fig. 2. *Drinking water scarcity in the basin: women waiting with empty vessels.*

pollution have sought legal remedies and have got some of the polluting textile and chemical units shut down. These are the spontaneous actions of different stakeholders, with each group working in isolation to resolve the issue. Unfortunately, it seems that farmers have more faith in the legal system than in other efforts.

The basin water management situation is precarious due to uncoordinated action and counteraction by various stakeholders. The situation is likely to get critical in future as the demand from the non-agricultural sectors continue to grow rapidly. Under these circumstances, there is a need for an integrated approach and a mechanism for coordinated development of water resources in the basin with the participation of all those concerned, especially the State. This could be undertaken by an external agency that needs to get all the parties—farmers, industrialists, domestic users and the State—to come together and establish a forum where solutions are sought through dialogue.

Multi-Stakeholders' Dialogue Meeting

A stakeholders' meeting was held on February 21, 2005, to get ideas and feedback from farmers, NGOs, government departments, industrialists, social activists and academicians. They agreed to discuss these issues further and negotiate their way out of the tough situation.

Notes

[†] The study is part of a major research project funded by SIDA–Research Links programme and coordinated by Professor Jan Lundquist, Department of Water and Environment Studies, Linköpings universitet, Sweden.
*Figure 1 by authors; figure 2 by Mats Lannerstad, Linköpings universitet, Sweden.

Rehabilitating the Keezhparikkalpet in Pondicherry: Bore-well Owners Give in to Farmers

T.P. Raghunath and R. Vasanthan

Description of the Area

KEEZHPARIKKALPET IS A VILLAGE LOCATED 20 KM SOUTH OF PONDICHERRY ON THE NORTHERN bank of river Ponnaiar facing Cuddalore town in Tamil Nadu. The east coast road that connects Chennai to Kanyakumari runs along the eastern boundary of the village, which is about 6 km west of the Bay of Bengal. The soil is a mix of clay and loam; the average rainfall is 1,100 mm, of which a major part falls during the northeast monsoon between September and November. The Keezhparikkalpet tank is one of a cluster of system tanks fed by the Ponnair river with a catchment spread over a 15 km radius.

The village has three aquifer zones—the topmost at 12 to 18 m depth, the second at 36 to 48 m depth and the third at 150 to 200 m depth. Almost free power supply for farmers, green revolution and a three-season paddy have together resulted in the over-exploitation of groundwater, and because it is near the sea, saline water has intruded into the top and middle aquifers. Only big farmers can afford deep borewells (each requires an investment of about Rs. 2 lakh) to tap the deepest aquifer that has good quality water. Those who have their own borewells have become water sellers: the returns are up to Rs. 7,500 per ha per season. Twenty-four hours' power supply has meant that a single 200 m deep borewell can irrigate up to 8 ha. Small and marginal farmers who do not own deep borewells depend on the deep borewell owners of such wells for their irrigation requirements.

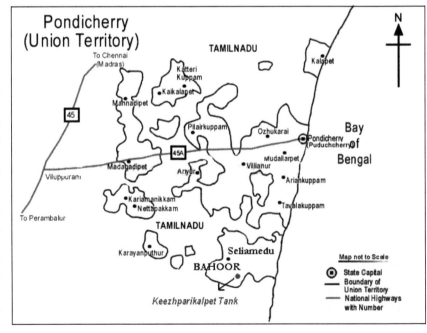

Map 1. *Location of Keezhparikkalpet in Pondicherry.*

Table 1: Composition of farmers in the tank *ayacut* area

Class of farmer	Holding (ha)	Per cent of farmers
Marginal farmers	Up to 1 ha	35.37
Small farmers	1 to 1.99 ha	26.83
Medium farmers	2 to 5 ha	25.60
Large farmers	More than 5 ha	12.20

Source: *Centre for Ecology and Rural Development (CERD), Pondicherry.*

Big Farmer Borewell Owners and Small Farmer Water Buyers

The Tank Rehabilitation Project, Pondicherry (TRPP) was introduced in Keezhparikkalpet village as a pilot project in July 1999. The big farmers who owned deep borewells thought that if the tank was rehabilitated, it would adversely affect their income. The small and marginal farmers, on the other hand, saw an advantage because once the tank was rehabilitated and able to provide water they would be able to save the huge sums they had been paying as water charges.

Fig. 1. *Keezhparikkalpet tank.*

It is important to understand the relationship between the three paddy seasons since much of the conflict centres around it. Samba is the main season because it gets the northeast monsoon showers. Navarai and Sornavari are the other two seasons. They have overlapping periods. Post-rehabilitation tank irrigation would be available for three-fourths of the Samba and one-fourth of the Navarai season.

An NGO called the Centre for Ecology and Rural Development (CERD) was entrusted with the task of looking after the TRPP in the village. The Tank Association (TA) was formed in December 1999 with major participation from small and marginal farmers, landless Dalits and women. After completion of the participatory planning phase and collection of funds from the interested parties, the TA began work on rehabilitation in July 2000 and most of it was completed by the end of that year. In the very next season there was good rain in the local catchments and the tank filled up. The conflict between the borewell owners and small farmers deepened when the latter started using the tank water for irrigation and did not buy water for the main Samba season. The owners forced the tank users to pay water fees even though most of them did not need water from the borewells. They were able to force their hand because the small farmers depended on them for the other two seasons when tank water was not available. Every borewell owner had his or her own service area that provided for a group of small farmers who depended on him or her for

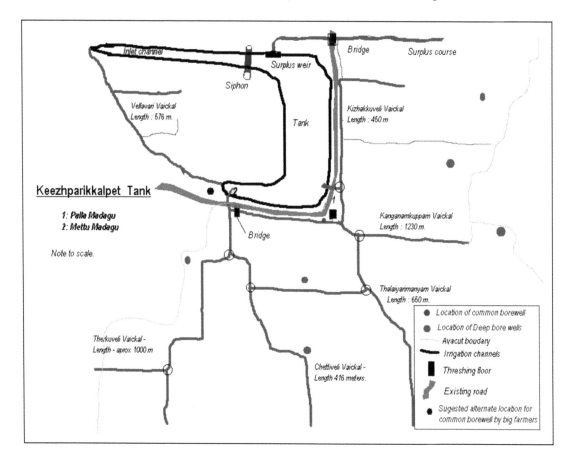

Fig. 2. *Tank borewell details.*

irrigation water. Even if a farmer approached other borewell owners, the water would have to pass through the previous water supplier's fields, because of the way the irrigation system had evolved. He obviously would not allow that. Moreover, in most such cases the borewell farmers supported each other. This created a major conflict between the small agriculturalists and the borewell owners. After negotiations the borewell owners agreed to reduce the water rent from six to five bags of paddy.

The small and marginal farmers represented by the secretary of the TA then approached the Department of Agriculture for sanction of a community borewell to be installed by the government free of cost and maintained and operated by the farmers with a minimum water charge. This well would be used during the times the tank was dry. The Project Management Unit (PMU) of the TRPP supported the proposal. This was the first TA to be sanctioned a community borewell.

The conflict intensified when the community borewell was sanctioned for the Tank Association. The immediate problem was the location of the borewell. The big farmers and borewell owners wanted the well to be sunk on private land owned by a rich landlord. The intention was to retain control over the borewell. The small and marginal farmers, on the other hand, wanted it located on government land to ensure easy access. The conflict spilled over into the realm of the TA, which was split in two. Some political parties also joined hands with the landlords in an attempt to halt the installation of the borewell.

Current Status

The Keezhparikkalpet tank currently supports paddy cultivation in the *ayacut* during Samba and part of the Navarai season. Because the tank has had water for three consecutive years and also due to the extensive irrigation from the tank in the *ayacut*, the water level and quality in the shallow aquifers in the *ayacut* has improved and many shallow borewells have become functional again. The community borewell has been installed and the government is in the process of handing over its operation and maintenance to the TA.

Highest and Lowest Points

The conflict was at its peak after the sanction of the community borewell when the location was being decided and the farming community and the TA split in two. To resolve the conflict on behalf of the small and marginal farmers, the TA approached the PMU and the Department of Agriculture with a memorandum that provided details of the landholding pattern in the *ayacut* along with the location map, and a list of potential beneficiaries of the community borewell. Subsequent to this, a meeting between the two conflicting groups was organised by CERD on October 29, 2001. Other participants were PMU staff, government officials and consultants.

The meeting began in a tense atmosphere. The PMU staff explained the risk that the government was taking in sanctioning the community borewell with special permission from the Lieutenant Governor. They stressed that since the village as a whole would benefit from government funds it was important for the government to understand its responsibility towards the project. This created a difficult situation for the big farmers in the village who understood that in case the proposal was rejected, they would be blamed for it, especially since the government had taken special efforts to get the community borewell sanctioned. Therefore they decided to support the small farmers and the borewell was finally located in the *poromboke* land (common land of the government) in the tank premises.

NGOs Have to Take the Lead in Opposing Borewell Owners

The big farmers never thought that the project would succeed. Since there was no elected *panchayat* and no recent history of any community action, they thought that it would be impossible to bring small farmers, the landless, and above all, the Dalits, under a common banner. They believed that political clout would influence the authorities into rejecting a democratic Tank Association. They thought that even if the project did come through, the rehabilitation work would be carried out like a usual government project without people's participation. They had even floated plans to pay for the entire scheme themselves or get it from contractors so as to bypass the TA. Once the project decided to involve all the stakeholders in the project and go ahead in a participatory manner, they tried to prevent marginal and small farmers from joining the TA. These tactics were defeated because the NGO insisted on a democratic process of election of representatives and prevented interference of vested interests in the project.

There was initially little opposition to the big farmers since the small and marginal farmers did not believe that an alternative was possible. There were no similar projects to learn from since Keezhparikkalpet itself was the pilot tank. The opposition had to be taken by the NGO, which had to do a lot of groundwork to create a healthy opposition to the power groups in the village. After the TA was formed, a new group of small and marginal farmers including women and Dalits gained strength and managed to confront the other group successfully, as in the case of the location of the community borewell.

Chronology of Events

September 2000	First village meeting about the project is conducted
October 2000	Stakeholders are identified
November 2000	*Ayacutdars'* meeting is conducted
December 2000	TA is formed
January – June 2001	Procedural works, trainings, planning meetings, estimate preparation is done
July 2001	Implementation of tank work begins
July 5, 2001	Government officials visit the tank
July 18, 2001	Meeting of the office bearers of the TA
July 20, 2001	Executive committee meeting of the TA held to discuss the issue
September 12, 2001	Meeting with Deputy Director of Agriculture regarding the demands of the farmers
October 29, 2001	Meeting between the small farmers and big farmers with TA EC members
March 2002	Government sanction for the community borewell

An Example is Set

The pilot tank of Keezhparikkalpet set an example in many ways. The eviction of encroachments by the TA through non-legal and community pressure, the planning and execution of all the works by the TA, and acquiring the ability to negotiate with the Public Works Department on technical grounds—many such aspects have set an example for other TAs to follow. This pilot simplified the task of community organisation in the later batches not only for CERD but also for other NGOs in the TRPP. Citing Keezhparikkalpet as an example, other tanks also approached the Department of Agriculture for community borewells, and succeeded in getting out of the clutches of the deep borewell owners.

Note

*Map 1 by authors; figures 1 and 2 by T.P. Raghunath.

References

1. Half-yearly report of the Keezhparikkalpet tank.
2. Profile of the Keezhparikkalpet tank.
3. Report of the community organiser for the month of October 2001.
The reports are available at

1. Centre for Ecology and Rural Development
    No. 46, II Street, P.R. Gardens
    Reddiarpalayam
    Pondicherry 605 010

2. Project Management Unit
    Tank Rehabilitation Project Pondicherry
    No. 3, Eveche Street
    Pondicherry 605 001

Diverting Nar–Par–Damanganga to Tapi–Godavari: Linking Rivers or Lurking Conflicts?

Datta Desai

THE PROPOSED NAR–PAR–TAPI–GODAVARI WATER TRANSFER PROJECT IS SITUATED IN northwestern Maharashtra and aims to divert water from two supposedly surplus basins of west-flowing rivers in Maharashtra (one, the North Konkan region comprising mainly the catchments of the Pinjal, Ulhas, and Vaitarana rivers and two, the Damanganga–Par basin including the catchment of Nar, Auranga and Ambika rivers) and divert it to solve the problems of two 'thirsty' catchments (the Tapi basin comprising

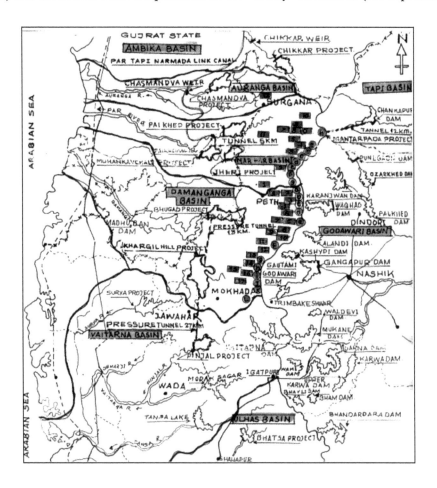

Map 1. *The diversion of the west-flowing rivers to the Tapi and Godavari basin.*

mainly the catchment of Girna, a major tributary of the Tapi, and the Upper Godavari basin up to the Jayakwadi–Paithan dam comprising the catchments of the Mula, Pravara, Darna, and other tributaries of the Godavari). The project includes almost a hundred schemes, linking various small and major rivers, existing dams, dams to be newly constructed and gravity and lift diversions at various points. The web it weaves involves Gujarat, Daman, and many districts and regions in Maharashtra: the districts of Thane (Konkan), Nashik, Jalgaon, Dhule (north Maharashtra), Ahmednagar (western Maharashtra) and Aurangabad (Marathwada), and Mumbai.

The main argument is that all the twenty-five *taluks* falling within the 22,349 sq km area of the Upper Godavari sub-basin are classified as drought-prone and, except for a small Western Ghat portion, the sub-basin receives less than 700 mm annual rainfall. (Nevertheless, about 35 per cent of the entire water yield in the sub-basin has been used to create an irrigation potential of 2,68,434 ha out of a cultivable 17,19,000 ha.[1] The basin faces an acute shortage of water because of the water apportioned to the projects under construction, the lift schemes, the requirements of the Jayakwadi dam, and the water diverted for non-irrigation use.) The Damanganga–Par sub-basin, a tribal belt with an area of 2,508 sq km, comprises eight *taluks* of Nashik and Thane districts and has an annual average rainfall of 2,198 mm. Hence, it is argued, the basin has surplus water since most of the water of the west-flowing rivers is allowed to flow into the Arabian Sea.[2] Nevertheless, it should be noted that four out of its eight *taluks* are classified as drought-prone.[3]

The Demand from the Tribal Belt in Nashik District

Before we proceed further, it is important to note two pre-existing competing water demands: that between the tribal belt and the sugar belt. Since 1960, the five *taluks*—Kalwan, Satana, Malegaon, Devla and Nandgaon—of Nashik district that come under the Tapi basin have been demanding water from the Tapi,[4] a demand that was also voiced in the State Legislative Assembly in 1974 by the then MLA, Mothabhau Bhamre. The 1997 Tapi Valley Development Corporation's (TVDC) plan to transfer 2300 Mm³ water from the Tapi to other areas, bypassing their demand, has become a political issue between Dhule–Nandurbar–Jalgaon districts and Nashik district.[5] Recently, at a farmers' rally held at Pimpalkothe–Baglan, Nashik district, they demanded 567 Mm³ water for these *taluks*, a demand that was also raised at a Secretariat meeting with the Irrigation Minister, and in 2001, with the Chief Minister and in the State Legislature by MLA Prashant Hire.[6]

The Upper Godavari sub-basin has been witnessing intense conflicts. Recently, the issue of releasing water from Mukne to Nandur Madhameshwar dam by an express canal pitched Ahmednagar–Nashik districts against Aurangabad district (Gangapur–Vaijapur *taluks*) and led to agitations including a *chakka jam* in late 2003.[7] The local farmers were also agitated that watershed development work was not permitted as 'it would reduce the inflow into the Dam'.[8] In 2002–03, two MPs and seven MLAs belonging to the major political parties demanded: diversion of 223 Mm³ water from Damanganga–Pinjal to Godavari[9] and to Mumbai (allowing 176 Mm³ water to be diverted from Vaitarana to Godavari);[10] inclusion of the five Nashik *taluks* under the Tapi Valley Development Scheme; lifting of water from Tapi at Prakasha in Nandurbar district; and diversion of 367 Mm³ water from Auranga–Nar–Par to Girna–Tapi basins.

The Maharashtra Pani Parishad Voices Sugar Belt Concerns

The Maharashtra Pani Parishad is a water forum formed by ex-minister Balasaheb Vikhe-Patil, a sugar baron from Ahmednagar district, ex-minister Ganpatrao Deshmukh and a group of retired officials from the Irrigation Department. The Parishad submitted a list of thirteen water-transfer schemes to the government of Maharashtra (GoM).[11] In January 2000, the GoM decided to study the possibility of diverting Nar–Par water to Upper Godavari including the 'drought-prone' sugar belt in the Mula–Pravara sub-basins in Ahmednagar district, a hotbed of water conflicts. The Parishad took up the matter once again in June 2003 with the Maharashtra Chief Minister and Suresh Prabhu, the then Chairman, Task Force on Interlinking of Rivers. The Parishad demanded diversion of 28 Mm^3 and lifting of 2,830 Mm^3 from west-flowing rivers to Bhandardara dam on Pravara river, and asked that the GoM should oppose the National Water Development Authority's (NWDA) scheme of transferring the Nar–Par water to Gujarat and act immediately to divert the water to Maharashtra. The Marathwada Janata Vikas Parishad also supports this demand as it expects Nar–Par water for drought-prone Gangapur–Vaijapur *taluks* and Aurangabad city.[12]

NWDA Interlinking Schemes

The NWDA schemes comprise two links:

The Damanganga–Pinjal Link

Transfer of surplus water from the Damanganga to the proposed Pinjal reservoir for meeting drinking water needs of Mumbai. The proposed Bhugad and Khargi Hill dams on the Damanganga will be connected to Pinjal by tunnels, and after reserving 91 Mm^3 for Madhuban reservoir in Gujarat and 290 Mm^3 for Gujarat's other future demands, 577 Mm^3 will be supplied to Mumbai.

The Par–Tapi–Narmada Link

The project suggests seven dams, three diversion weirs, a 5 km link canal, a 27 km feeder canal and a 224 km Tapi–Narmada link to transfer water from the Par, Auranga, Ambika, Purna and Tapi basins to the Narmada basin. It allocates 221 Mm^3 water for irrigation in Maharashtra, 487 Mm^3 in Gujarat, 1,350 Mm^3 surplus water for transfer to Tapi, and 1,554 Mm^3 water to be diverted from Tapi at Ukai dam—a total of 2,904 Mm^3 water to be used for developing command areas to the north of the Sardar Sarovar dam and in Saurashtra and Kutchh.[13]

The Government of Maharashtra's Stand and Proposed Scheme

The GoM wants the Damanganga–Pinjal Link to include diversion of 223 Mm^3 water from Damanganga to Godavari basin by gravity and lift, and 1,083 Mm^3 instead of 577 Mm^3 water to be diverted to Mumbai. Regarding the Par–Tapi–Narmada Link, the GoM argues that the Iyengar Committee and NWDA have allocated disproportionately more water to Gujarat from the Tapi basin and since Maharashtra contributes 82.8 per cent of the area of the Tapi basin and Gujarat 1.5 per cent, their shares of water should be 9,374 Mm^3 and 167 Mm^3 respectively. Moreover, the 2002 NWDA report states that the water yield in the Tapi basin has reduced and there is no surplus. Therefore, it argues, out of the 1,819 Mm^3 water from Nar-Par, 487 Mm^3 be allocated to Gujarat and the remaining 1,332 Mm^3 to Maharashtra (566 Mm^3 for local use, and the Godavari and Girna basins, and 766 Mm^3 for Tapi basin to be diverted at the Ukai dam, of which 231 Mm^3 upstream of Ukai will be used by Maharashtra in addition to 5,420 Mm^3 allocated by the Iyengar Committee).[14]

The GoM demands that Nar–Par–Tapi–Godavari project should be part of the NWDA interlinking plan and should include:

- diversion of 2,259 Mm^3 water from the west-flowing rivers Auranga, Damanganga, Nar–Par, Vaitarana, etc. to be allocated to Tapi basin (306 Mm^3), Godavari basin (1,637 Mm^3), and local use (317 Mm^3) through diversions by lift at 55 dam sites and by gravity at 27 dams;
- the Tapi allocation to be brought to Chanakapur dam (20 per cent reserved for domestic and industrial use and the remaining for irrigating 63,311 ha);
- of the Godavari allocation: 289 Mm^3 available at Punegaon to be carried by express canal to Kolhi dam in Aurangabad district (40 Mm^3 reserved for the future use of Aurangabad city, the rest to irrigate 55,172 ha in the drought-prone *taluks* of Nashik and Aurangabad); 512 Mm^3 to go to Upper Godavari sub-basin to irrigate 1,13,207 ha; 803 Mm^3 to be diverted to Darna, Mukne and Aalandi dams to irrigate 1,32,190 ha; 31 Mm^3 water to be diverted to Bhandardara dam to irrigate 6,938 ha in Pravara sub-basin.

The GoM estimates a total irrigation of about 4 lakh ha, power required about 700 MW, a benefit-cost ratio of around 0:61 and a total cost of Rs. 10,700 crore, submerging an estimated 4,140 ha of forest and 6,210 ha of private land.[15]

Opposition from the Tribal Catchment Areas

The opposition to this scheme comes from the tribal Surgana *taluk*, which is the crucial Western Ghat area where most of the diversions will take place. The main objections are: (*a*) 75 per cent of the land required for this project in Surgana *taluk* is forest land; (*b*) the eleven big dams to be built in Surgana *taluk* will submerge 60 per cent of the *taluk* area including villages; (*c*) 60 per cent of the population in Surgana *taluk* will be displaced; (*d*) Peth and Harsul (Trimbakeshwar) *taluks* will face a similar fate.

The suggested alternative is the construction of a series of percolation and small tanks on the *nalas*, minor and medium dams, KT weirs (which will submerge less forest) between elevations of 500 and 1,500 m above MSL. Harvested rainwater accumulating in such reservoirs between June and September can be continuously lifted and diverted to Tapi and Godavari basins and the last filling can be stored for local irrigation.[16]

Lurking Conflicts

The project is likely to sharpen existing conflicts and give rise to a host of new ones: local, regional, inter-state, upstream-downstream, catchment vs. command, head-reach vs. tail-enders, and micro-watersheds vs. irrigation projects.

- A conflict is likely to arise between Maharashtra, Gujarat and Daman over the sharing of west-flowing rivers. The feeling that Gujarat is already benefiting from large inter-State dam projects like the Sardar

Sarovar, Ukai and Madhuban, and that water rightly belonging to Maharashtra is being diverted to Gujarat at the behest of the NWDA, may lead to a backlash in Maharashtra.

- The Konkan region, especially the tribal belt of Thane, has very little water resource development and irrigation. If the water is diverted without taking into account the needs and due share of this region, it may accentuate the Konkan vs. western Maharashtra divide.
- The ecological aspects and likely damages have not been studied. While the Save Western Ghat Campaign is being carried out emphasising biodiversity protection, the proposed diversion schemes will submerge forests and destroy forest cover and biodiversity. Nor have the likely effects of the reduction of flow into the Arabian Sea on salinity ingress, fishery and soil deposits on the coast been studied.
- Tribals on both sides of the Western Ghats are being displaced and Peth–Surgana *taluks* sacrificed in the name of 'development' of the Upper Godavari basin, in the land of the 'ever-thirsty' and mighty sugar barons who commissioned private feasibility studies, and at whose behest the Chief Minister issued orders in June 2003 to the Irrigation Department to submit a project proposal for Nar–Par–Tapi–Godavari scheme within two months.[17] This will intensify the social divide and lead to severe conflict.
- This process is already accentuating conflicts between and within different regions: Peth-Surgana *taluks* vs. Malegoan *taluk* in Nashik district; Nashik district vs. other districts of north Maharashtra; Nashik–Nagar districts vs. Aurangabad districts; tribal areas in Konkan vs. Mumbai, etc. Marathwada, lying at the tail of the proposed schemes, also hopes to get Nar–Par water, but experience shows that the tail-enders are used only for mobilising support; they rarely receive water. Consequently, there is the possibility that it will add to the gathering frustration in Marathwada over increasing regional disparity.
- Presently, although all the twenty-five *taluks* in the Upper Godavari sub-basin are classified as drought-prone, they still boast of twenty sugar factories. If the pattern continues, additional water will be of little help and the vital issues of equitable distribution and regulation of the water-intensive cropping pattern will be ignored. The increasingly neo-liberal, market-driven policies will further monopolise water resources and intensify future conflicts.
- The entire scheme suffers from arbitrariness and conventional 'irrigationism'. It does not take into account the water needs of each river basin, conjunctive use of water or integrated water resource management. The implications of lifting water through a height of 400 to 500 m and consuming 700 MW electricity in a state ridden with power crises have been completely overlooked.

Possible Solutions

The problem is so complex that it needs a detailed critique if alternatives are to evolve. However, certain broad directions can be spelt out.

- While the proportionate share of Gujarat in the basins of Nar–Par, Ambika–Auranga, Damanganga, etc. needs to be worked out to avoid objection from Gujarat, the due share of Maharashtra on the basis of catchment area will have to be defended.
- Sub-basin wise planning of water resources with an integrated and 'bottom-up' approach at every level from micro-watershed, watershed to sub-basin is the first priority. Farmers are already demanding watershed development to be carried out. Samaj Parivartan Kendra in Ozar, Nashik district, has shown the way by combining watershed, groundwater and canal irrigation. The alternative suggested by MLA and *kisan sabha* leader Jiva Pandu Gavit from Surgana needs to be considered seriously.

- The myth that 'there is no more water available in Tapi–Godavari basins' is a socially constructed 'reality', a half-truth. The water used by industries, urban centres and irrigated agriculture in this region is not accounted for properly. Efficient water management (for example, by disciplining cultivation of water-intensive crops like banana and sugarcane, recycling of water in the urban-industrial sector) can release a considerable amount of water from these sectors. Also, there should be no increase in water-intensive crops and industries in future in these water-scarce basins.
- Equitable per capita distribution of water within Tapi and Godavari basins needs to be given top priority.

All the proposed projects and schemes need to be considered separately. Large dams and lift diversions should be treated as exceptional. Maharashtra can fully utilise its share of water through decentralised and dispersed water resource planning. Above all, the Thane–Nashik tribal belt must be protected and made part of the entire planning and designing process of any project involving it.

Notes

*   Map 1 from Brief note on Diversion of Water, undated, Irrigation Department, North Maharashtra, Nashik.
1.   Report of Maharashtra Water and Irrigation Commission (MWIC), Vol. II. P.19, 1999.
2.   Report of Coordination Committee on Diversion of West-Flowing Rivers (CCDWR) to Godavari-Tapi, September 2001. p.1
3.   Report of MWIC, Vol. II, p.451.
4.   Unpublished papers and press releases by ex-MLA Mothabhau Bhamre dated March 15 and September 30, 2003.
5.   Ibid
6.   Ibid
7.   Sakal. Marathi daily newspaper, November 2, 10, 11, 12 and 17, 2003.
8.   Ibid., November 13 and 14, 2003
9.   Ibid., September 28,2003
10. Ibid., September 19 and 20, 2003
11. *Maharashtra-Tutichya Khoryat Adhik Paani Uplabdha Karun Denyasathi Niyojan.* Maharashtra Pani Parishad (MPP) booklet, 2000.
12. Letter dated September 17, 2003 by the Deputy Secretary, Irrigation Department, Government of Maharashtra, to Managing directors of River Valley Development Corporations and Chief Engineers.
13. *Marathawada Vikasacha Jahirnama.* Aurangabad: Marathawada Janata Vikas Parishad (MJVP) booklet, 2004, and item 2 above.
14. Brief Note on Diversion of Water, Irrigation Department, North Maharashtra, Nashik.
15. Ibid., and *supra.*
16. Ibid.
17. Article by J.P.Gavit, *Jeevanmarg (JM).* Marathi weekly newspaper, October 19-25, 2003.

The Collapse of *Phad* System in the Tapi Basin: A River Strains to Meet Farmers' Needs

S.B. Sane and G.D. Joglekar

Once Upon a Time

THE PANZARA RIVER, A TRIBUTARY OF THE TAPI, RISES IN THE SAHYADRI HILLS AT 600 M ABOVE mean sea level at Warsa, and meets the Tapi river at Mudawad in Sindkheda *taluk* at an elevation of 125 m. The total length of the river is 136 km. It traverses Sakri, Dhule and Sindkheda *taluks* in Dhule district, Maharashtra. The total catchment area is 2,849 sq km.

A large number of ancient *bandharas* (diversion weirs) and canal systems exist on the Panzara river that have been operated and maintained through the centuries by those who benefited from them. The irrigation systems on these *bandharas* are called *phad* systems. It is not known exactly who constructed them and when, but local scholars believe that they date back to the fourteenth century. The earliest reference to these *bandharas* is found in Bombay Province gazettes, which mention that they are the legacy of the Farooki kings (1370–1600).

Map 1. *Location of Dhule district in Maharashtra.*

The Panzara river has been referred to as a perennial river in a 1964 Government of Maharashtra Project Report of the Panzara irrigation project; this was due to the adequate soil and forest cover the valley then had. The geology of the valley also favours infiltration of rainwater in the soil, resulting in good groundwater recharge that steadily regenerated the river round the year.

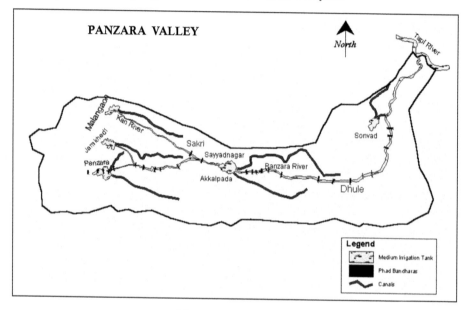

Map 2. *Panzara valley.*

Fig. 1. *A phad system Bandhara (Kokale/Gondas) in operation.*

The *Phad*s: A Unique Legacy

However, the situation has changed drastically in the last fifty to sixty years owing to various factors, and the river goes almost dry post-winter in the lower Panzara valley. This has created a piquant situation in the entire valley, especially in the riparian villages that have enjoyed bountiful crops over the past several centuries. The farmers in the valley are agitated and feel helpless because there are no laws to protect their riparian rights.

Phad System

Out of a total of forty-five *bandharas* on the Panzara before 1964, thirty were taken over that year by the Irrigation Department. Though all thirty of them are termed *phad* system *bandharas*, only in the twelve in Sakri *taluk* is the *phad* irrigation system practised as originally envisaged. *Phad* refers to a block of land where a single crop, usually irrigated, is grown. The command of a *bandhara* was usually divided into three or four such blocks called *phad*s; and each *phad* grew only one crop. A system of rotation ensured that a particular crop was grown on a *phad* only after a period of three to four years. The merits and features of the *phad* system are given in the Box below.

Main Features of the *Phad* System

1. One village, one service area.
2. Small compact command area and hence easy distribution of water.
3. Managed by a committee appointed by beneficiary farmers.
4. No interference from the government.
5. Employees and staff for irrigation management appointed by the committee.
6. Staff paid by the farmers directly in cash or kind or both, proportionate with the area irrigated.
7. Transparency in framing rules of operation, management and implementation.
8. Assured supply of water.
9. One crop in one block makes it easy to manage irrigation.
10. All concerned farmers have lands in each *phad* and hence they all benefit.
11. Rotation of crops keeps the soils healthy. Fertility is not adversely affected.
12. No storage is provided behind the *bandharas*. So, the series of *bandharas* ensures that excess water, if any, is immediately and automatically made available to the next *bandhara*.
13. Beneficiaries pay for the maintenance of the system, which is based on actual requirement and decided by the farmers' association.
14. The employees, being local people (and irrigators themselves, in some cases) carry out their work diligently and honestly.
15. Farmers come together for common social good and take collective decisions in a transparent manner, which is the greatest advantage of the system.

During the great famine of 1951–53, a medium sized irrigation project was constructed on the Panzara river at Sayyadnagar. It comprised a masonry pick-up weir with a right bank canal that had an irrigable command area of 4,800 ha. There were no storages constructed in the valley until 1972. As such, all these old *bandharas* on Panzara, including the new one at Sayyadnagar, were dependent on the perennial nature of the river.

Since 1972, a large number of irrigation projects (minor and medium) have come up in the sub-basin. A few are still under construction. Details of these are given in Table 1. These interventions have affected the flow of the river and the supply to the *bandharas*.

Table 1: Irrigation Projects in the Panzara Valley

Projects	Live storage (Mm³)	Command area (ha)	Status
Medium			
Panzara (Latipada)	35.63	7,777	Completed
Malangaon	11.33	2,185	Completed
Sonvad	14.36	3,450	Completed
Jamkhedi	12.34	4,130	Under construction
Total	**73.66**	**17,542**	
Minor – 48 Nos.	45.00	7,637	Completed
Percolation – 110 Nos.	14.50		Completed

Note: Besides these, a large number of wells (about 21,000) too have been dug in the valley.

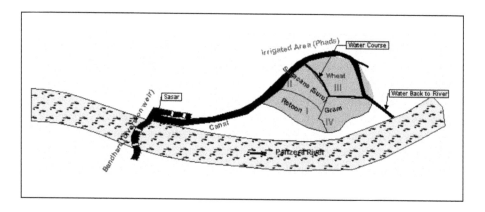

Fig. 2. *A Typical phad system Bandhara.*

As a result, the year-round availability of water in the river has been drastically reduced, affecting the irrigation of the *phad* system *bandharas*. In the twelve *bandharas* where the *phad* system is still in operation, storages are managed by the Irrigation Department (except the one on the Kan river, a tributary) and are supplied water from Latipada dam, which is part of the Panzara medium irrigation project. The *bandhara* on Kan river is supplied water from Malangaon medium irrigation project.

Not Enough to Go Around

The farmers on the *phad* system are deprived of their riparian rights because of the various irrigation projects in the valley. Farmers from some of the *phad* system *bandharas* say that they had opposed the construction of these new irrigation works right from the beginning and knew that they would suffer due to these interventions. The result is that only some of the thirty *bandharas* are operational at present, and even those supply water for just eight months (during the *kharif* and *rabi* seasons).

The main problem faced by the *phad* system farmers is the non-availability of water, not only in the post-monsoon season but also during the monsoon. Before 1947, the forty-five second class irrigation *bandharas* were the only irrigation projects apart from a few irrigation wells. Data from the 1964 project report of Panzara medium irrigation project shows that the crop-wise areas irrigated by these forty-five *bandharas* were:

Table 2: Crops in the *Phad* System Areas

Crop	Area
Wheat	1,373 ha
Paddy	740 ha
Sugarcane	626 ha
Other crops	258 ha
Total	**2997 ha**

These areas were situated along narrow strips parallel to the river throughout its entire length of 136 km. The estimated water consumption of these crops would be:

Table 3: Estimated Water Use by the *Phad* System Areas

	Crop	Area (ha)	Per ha requirement of water (mm)			Estimated water use (Mm³)			
			Kharif	*Rabi*	Hot Weather	*Kharif*	*Rabi*	Hot weather	Total
1.	Paddy	740	1000			7.40		—	7.40
2.	Wheat	1,373		600		—	8.24	—	8.24
3.	Sugarcane	626	500	1,000	1500	3.13	6.26	9.39	18.78
4.	Other crops	258	150	300		0.39	0.77	—	1.16
	Total	**2,997**				**10.92**	**15.27**	**9.39**	**35.58**

The average water yield available in the valley, as per the latest study by the water resources department, is estimated at 539 Mm³. One can therefore conclude that the water in the river was sufficient to meet the irrigation requirements of the forty-five *bandharas*, after providing for the domestic and drinking water needs of the population in the valley. The problem now is that except for the twelve *bandharas* in Sakri *taluk*, the rest are not even getting water for their *kayam bagayat* or the area of assured irrigation.

During deficit rainfall years, the District Collector reserves water from the medium irrigation projects for use by Dhule town and other villages in the valley. The allegation is that this amount is far higher than the actual requirement, about four to five times the net requirement of drinking water for Dhule city and other villages in the valley, even after accounting for en route losses, including unauthorised lifting from the canals and the river. This diversion makes water unavailable to the *phad* system farmers.

The problem is further aggravated by uncontrolled exploitation of groundwater throughout the valley. The latest figure for the number of wells in the valley is 21,000 and together they would account for about 210 Mm³ of water at an average draught of 1 ha-m (10,000 m³) per year per well; this could explain why there has been no post-monsoon flow in the Panzara river in recent years.

There is no dramatic conflict between those for and against; this is an instance of a dying traditional system coming face to face with 'progress'. The problem began in the 1960s and the system has deteriorated gradually since then. And yet, the old way had certain features which need to be preserved.

A related problem is that even prior to independence, the *phad* system farmers have been paying a water cess—*pankasar* in local parlance—along with land tax, to the Revenue Department. The Panzara river was notified in 1976, which means that permission from the Irrigation Department was made mandatory for all those who utilised the river water; and the farmers were charged a fee by the Irrigation Department for use of the *bandharas* taken over by them, in addition to the above mentioned water cess that was paid to the Revenue Department. While the *phad* system farmers are willing to pay the Irrigation Department, provided they are assured water for their fields, they demand that the Revenue Department stop collecting water tax. To this end they have approached various authorities since 1983:—the Settlement Commissioner, Director, Land Records, the Chief Engineer, the Irrigation Department, and even the Civil Court, Dhule. While there is an exchange of correspondence between the concerned parties, the court case awaits conclusion and no final solution is in sight yet.

Whether the ancient *phad* system should be revived or allowed to die a natural death is a matter for debate. Even if it were to be revived, the question is whether it will be possible to do so in the face of physical changes that have occurred in the valley, as well as the rising aspirations of large sections of the people in the valley and the consequent demand for irrigation. If the system is to be revived, greater discipline will have to be observed in ensuring that water earmarked exclusively for the *phad*s in the new irrigation projects actually reaches them.

One argument against the revival of the system is that it only covered 3,000 to 3,700 ha at most out of 2,84,900 ha, which makes it just 1.3 per cent of the total area of the valley. The number of families dependent on this ancient system are 2,500 to 3,000—a population of 12,500 to 15,000 out of 10.5 lakh.

Can the *Phad* System be Revived?

The decline of the ancient *phad* system post-1971–72 has hurt not only the people who made their living from it but the government as well. Several proposals for its revival have been mooted at the highest level.

Most of them consider reserving a certain quota in the reservoirs that have been proliferating along the river. The quota required by the farmers (provided it is the same cropping pattern that existed before 1964) would work out to 100 Mm^3, as shown in Table 4:

Table 4: Estimated Water Requirement for *Phad* System Revival

Crop	Per cent	Area (ha)	Quantity (Mm$^{3)}$)
Sugarcane	20	740	22.20
Wheat	50	1,850	11.10
Paddy	25	925	9.25
Others	5	185	0.74
Total	**100**	**3,700**	**53.29** (approx. 100.00 Mm^3)
Add transit and evaporation losses			53.29
Total quota			107.58

The issue of double taxation (irrigation cess paid to Revenue as well as Irrigation Departments) is important to the *phad* system farmers, but it does not present a difficulty. It is simply a matter of political will and it is unfortunate that the issue has not been sorted out over the last twenty-five years. This has probably happened because the amount of water cess (*pankasar*), which is only Rs. 25 per ha, is very small. Most of the *phad* farmers, except those of Raywat, Kadana and Mana *phad*s, have found it more convenient to pay the cess than get into difficulty with the officers of the Revenue Department. The farmers of these *phad*s had gone to civil court in 1983 and obtained a stay on paying the water cess to the Revenue Department. More concerted action along these lines is needed.

Meanwhile, on 14 January 2005, the Chief Engineer of the Tapi Irrigation Development Corporation in Jalgaon, in charge of the irrigation projects in Dhule district, also wrote to the Secretary of the Water Resources Department, Government of Maharashtra, asking that the *pankasar* cess charged by the Revenue Department be revoked. This should also help resolve the issue.

The more difficult part of the problem is providing the *phad*s with the 100 Mm^3 water that they require. A medium irrigation project at Akkalpada on the Panzara has been under construction for two decades. There is a proposal to reserve part of the water from this storage for the *phad* systems between Sayyadnagar and Dhule. Also, the Mukti irrigation tank that was previously serving the *phad* system below Dhule, and is presently used to supplement drinking water supply to the town, is likely to be freed when supply to Dhule is fully switched over to the Tapi river.

Note

*All tables based on Government of Maharashtra (April 1994) Panzara Project Report; maps 1 and 2 from SOPPECOM, Pune; figures by authors.

References

Datye, K.R. and R.K. Patil. 1987. *Farmer Managed Irrigation System Indian Experience*. Pune: Centre for Applied System Analysis in Development.

Government of Maharashtra. April 1964. Panzara Project Report.

Government of Maharashtra. 1974. Gazetteer of Dhule District.

Kulkarni, D.N. and R.K. Patil. 1984. *Water Management Factors—Water Management through Farmers' Organisations—Phad System: A Case Study*. Fort Collins: USA.

Second Maharashtra State Water and Irrigation Commission and Sinchan Sahayog, Dhule. 1997. Workshop on 'Planning of Water in Panzara River Basin and Modernisation of *Phad* System'.

Tembu Lift Irrigation in the Krishna River Basin: Conflict over Equitable Distribution of Water

Namrata Kavde-Datye

A Parched Land

THE TEMBU LIFT IRRIGATION SCHEME (TLIS) ON THE KRISHNA RIVER WAS INITIATED IN 1995. The scheme derived its name from Tembu village in Karad *taluk*, Satara district, Maharashtra, from where the water is to be lifted. The project will utilise 623 Mm^3 of water annually in five different stages through a 317-m high lift and irrigate about 79,600 ha of land in 173 villages in six *taluks*—Karad, Khanapur, Atpadi, Kavathemahankal, Tasgaon and Sangola from Satara, Sangli and Solapur districts.

Shetmajoor Kashtakari Shetkari Sanghatana (SKSS), a rural organisation, works on drought proofing and equitable water distribution in south Maharashtra. They brought together farmers hailing from the TLIS

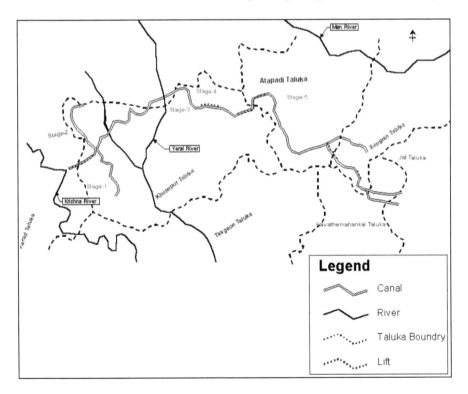

Map 1. *Tembu lift irrigation scheme.*

project area to demand their share of water. Atpadi, one of the six beneficiary *taluks* of the project, was at the forefront of the agitations. The government of Maharashtra (GoM) finally agreed to equitable water distribution in Atpadi as part of the TLIS on a pilot basis.

Atpadi is a drought-prone eastern *taluk* of Sangli district in south Maharashtra. It has eighty-four villages with 22,000 households and an area of about 90,000 ha. It is situated on the eastern side of the Krishna valley, which receives scanty and irregular rainfall.

The water crisis in the *taluk* has worsened over the years. The State did not take any concrete steps to mitigate the problem, except providing employment under the Employment Guarantee Scheme (EGS) during years of drought. In 1992, some prominent people from Atpadi came together to form a broader coalition to address the issue. Leading activists like Dr. Bharat Patankar and Sampatrao Pawar from the Mukti Sangarsh Movement (MSM)—a mass-based organisation involved in mobilising drought-affected people in the adjoining Khanapur *taluk* since the early 1980s—were invited to share their experiences. The SKSS came into being in 1993 in the aftermath of the demolition of the Babri Masjid when leading Left leaders and activists from south Maharashtra, like Nagnath Naikwadi and Dr. Bharat Patankar, organised a massive conference at Kini in Kolhapur to fight communalism. But very soon, the SKSS also started taking up other issues like drought, access to water, and rehabilitation of project-affected people, and became the fountainhead of people's struggles in the area. The SKSS also decided to take up the issue of restructuring the TLIS on equitable lines and the drought-affected people from Atpadi rallied behind the SKSS. A few years later, MSM merged with the SKSS.

Fig. 1. *A parishad organised by the SKSS.*

Carrying Water Beyond the Krishna Banks

The key demand that the SKSS took up was equitable distribution of Krishna water for all households, including the landless, in the thirteen drought-prone *taluks* in Sangli, Satara and Solapur districts. The movement was strongest in Atpadi and the organisation decided to make it a test case.

As per the government's plan, only 16,000 ha in sixty-three villages of the *taluk* stood to benefit from the TLIS. The organisation was concerned that this would keep nearly 75 per cent of families in the region in a permanent state of drought.

The 1975 inter-State water dispute tribunal's award, popularly known as the Bachawat Award, specified the quota of water from the Krishna river to be stored and used before May 2000 by Maharashtra, Andhra Pradesh and Karnataka. If the respective allocations were not utilised by then, the unused water would also be taken into consideration during the next round of negotiations. Maharashtra was awarded 16,833 Mm³.

The SKSS is of the view that only through the integrated use of local water (rainwater harvesting through micro-watershed development) and exogenous water (in this case, Krishna water) can drought be mitigated in a sustainable manner. The available water also needs to be distributed equitably so that each household gets the minimum quantity required to meet their needs.

Fig. 2. *A canal built under the TLIS in Atpadi taluk.*

Studies have shown that a family of five can meet its food, fodder and fuel needs, recyclable biomass for the farmland, and a certain amount of surplus to meet cash requirements if it can produce and/or get access to about 18 tons (dry weight) of biomass. About 6,000 m³ of water is required to produce 18 tons of dry biomass with a productivity norm of 30 kg per ha-mm of water. One has to also add about 400 m³ of water to meet domestic water requirements and the water required by livestock. Thus, the total water required by a family would be around 6,400 m³ (see, for details, Paranjape and Joy 1995, Datye 1997). The drought-prone area of the Krishna river basin gets an average rainfall of about 500 mm, and studies show that this translates into about 4,000 m³ availability for each family which, according to the SKSS, can be harvested through in-situ use and local watershed development programmes. But the people would still require another 2,400 m³ of water from outside sources at the point of use (keeping in mind conveyance losses, evaporation losses, etc. this works out to around 3,000 to 3,500 m³ per family). According to the SKSS, the quota allocated to Maharashtra by the Bachawat Tribunal is more than enough to satisfy the needs of the entire population in the Krishna river basin.

Chronology of Events

1992	People from Atpadi *taluk* come together to deal with the drought issue.
1993	SKSS takes up the key issue of equitable distribution of water to thirteen drought-prone *taluks*.
	Signature campaign, submission of resolutions by *gram panchayats* and other institutions demanding equitable access to water for all families including the landless in Atpadi; demonstrations, *jatha* programmes, etc.
1994	Government agrees to give water to Atpadi from Urmodi dam to be constructed in Satara district.
	SKSS declares non-payment of land revenue.
	Movement spreads to thirteen drought-prone *taluks* in Sangli, Satara and Solapur.
1995	Government announces Tembu Lift Irrigation Scheme, but the project design is not in keeping with people's demands.
	SKSS intensifies agitation.
1999	Peak of Mass mobilisation at its peak: organisation of a three-day-long *dharna* simultaneously in thirteen *taluks* with over 1 lakh people participating from dam-affected and drought-affected areas supporting each other's demands.
	Unprecedented *dharna* in south Maharashtra in terms of the nature of demands and huge mobilisation.
2000	Maharashtra Krishna Valley Development Corporation (MKVDC) agrees, in principle, to equitable distribution of water to Atpadi and makes it part of TLIS on pilot basis.
2002	SKSS submits proposal to MKVDC for water to all families in Atpadi.
	MKVDC and SKSS draw up a joint proposal to be submitted to the government.
2003	Agitation in Tasgaon *taluk* and the government is forced to make this *taluk* too a part of the Scheme.
2004	*Thiyya andolan* by 7,000 people (indefinite 'sit-in') in front of the MKVDC office in Pune. The authorities assure SKSS that they will meet some of its demands.
2005	*Thiyya andolan* in Mumbai. Government assures the agitators that they will all get water and asks SKSS to take the initiative to form WUAs.

Ever since 1993, on July 26 every year, the drought-prone people of these thirteen *taluks* gather a Atpadi in large numbers for an event known as the Atpadi Parishad (or Atpadi Conference) to pursue their demand for water.

Procrastinations

When the SKSS began the movement, the Congress party was in power at the state level but the government did not take any substantive steps to meet the demands of these people. The party lost the assembly elections in 1995 and the BJP–Shiv Sena alliance came into power. The Maharashtra Krishna Valley Development

Corporation (MKVDC) was established during the Bharatiya Janata Party (BJP)–Shiv Sena rule, with the primary aim of raising public funds and speeding up the construction of incomplete irrigation schemes. The Maharashtra government had lagged behind in fully utilising its share of Krishna water, and to rectify this situation the MKVDC decided to speed up the implementation of the massive lift irrigation schemes and build as many storages as possible, with the idea of using as much of the State's share as possible by May 2000. However, progress was slow. In 1999, the Democratic Front (DF) came to power and one of the first issues mentioned in its fifty-one point Common Minimum Programme (CMP) was equitable distribution of water on a per capita basis. Unfortunately the government failed to take any steps to initiate action in this regard.

Ray of Hope

The sustained efforts of the SKSS forced the MKVDC to agree in principle in 2000 to restructure the Atpadi section of the TLIS. In 2002, the SKSS submitted an alternative proposal suggesting that all the households in Atpadi should benefit from the scheme. The MKVDC scrutinised the document before submitting it to the government for approval.

Features of the Proposal

a) It is viable to allocate 5,000 m^3 of water per household in Atpadi *taluk*.
b) The restructuring of TLIS will require additional lift of up to 40 or 50 m, at an extra estimated cost of Rs. 49.5 crore, to serve some villages not covered by the original scheme.
c) Farmers will pay for the operation and maintenance cost in advance, provided the water and electricity are charged on par with other irrigation schemes.

The restructuring of the scheme has raised several issues regarding financial viability and energy burden. The SKSS leadership is aware of these problems and the people are willing to participate in co-management of water and energy, as also get involved in renewable energy generation options (see Joy and Paranjape 2002).

Difference of Opinion

The Government and the MKVDC

The government and the MKVDC are opposed to the idea of equitable distribution of water on the basis of households, and prefer gravity command area based irrigation. The authorities feel that the former scheme is not feasible in terms of cost-benefit ratio and resisted the demands till 2000. At present they have agreed to equitable distribution of water in Atpadi on a pilot basis.

SKSS

Access to water is a basic human right and everybody should get the minimum quantity required to meet their daily needs. Equitable water distribution is the only way to counter drought and will eventually lead to sustainable prosperity in the region. They are also willing to take on the responsibility of forming water users' associations, federating them and taking over the TLIS operation.

Independent experts

Some independent experts have been critical of TLIS and other similar high lift irrigation schemes. Madhav Godbole, a staunch critic of this project, has labelled it 'another Enron'.

SOPPECOM

This NGO supports the stand taken by the SKSS, but cautions that unless the people agree to a rational pricing policy and participate in co-management, the scheme will run into problems.

Implementation is the Key

The focus should now be on strengthening the process of restructuring the scheme as per the new agreement. The movement has to keep up the pressure on the MKVDC so that the corporation makes the financial commitment necessary to implement the programme. A lot of work needs to be accomplished at the ground level, for example forming WUAs in villages and federating them at the *taluk* level. Studies need to be conducted to understand livelihood patterns and water requirements in different households, explore the cheapest options for lifting water, take decisions on user fees and water allocation. Therefore, people's participation in the planning process is of vital importance.

SOPPECOM's study on 'Energy-Water Co-management Opportunities and Challenges in the TLIS' brought together the views of the various stakeholders. It highlighted that high lifts can be made viable provided:

- Energy use is prioritised and a part of it is set aside to provide basic water services to all.
- Water pricing should be on the basis of average costs worked out for the entire basin, and not on the basis of individual schemes, so that people who are disadvantaged by location are not further penalised.
- Renewable and dispersed sources of energy need to be tapped and enough area needs to be set aside for energy plantations.
- There is a need for water and energy co-management institutions.

Note

*Map 1 and figures by SOPPECOM.

References

Datye, K.R. 1997. *Banking on Biomass: A New Strategy for Sustainable Prosperity Based on Renewable Energy and Dispersed Industrialisation*. Ahmedabad: Centre for Environment Education.

Irrigation Department, Government of Maharashtra. 1994. *Revised Tembu Lift Irrigation Scheme (Proposed irrigation benefits to Khanapur, Tasgaon and Atpadi)*. Pune: Irrigation Projects and Water Resources Analysis Board.

Joy, K.J. and Suhas Paranjape. 2002. *Energy-Water Co-management Opportunities and Challenges in the Tembu Lift Irrigation Scheme, Atpadi Taluka, Maharashtra*. IWMI–SOPPECOM Partnership Study.

Maharashtra Krishna Valley Development Corporation. 2001. *Tembu Lift Irrigation Scheme: An Alternative Plan to Provide Irrigation to Outside the Command Area*. Sangli: Irrigation Department.

Paranjape, Suhas and K.J. Joy. 1995. *Sustainable Technology: Making Sardar Sarovar Viable*. Ahmedabad: Centre for Environment Education.

Patankar, Bharat. 1997. *Krishna Khorayache Pani- Rabnarya Janatecha Paryay*. Kasegaon–Sangli: Shramik Mukti Dal.

Phadke, Anant. 1994. 'Dam-Oustees' Movement in South Maharashtra', *Economic and Political Weekly,* November 18.

———, 'Anti-Drought Movement in Sangli District', *Economic and Political Weekly,* November 26.

Tail-End Discrimination in an Irrigation Project in Maharashtra: Quota Reductions for the Palkhed Left Bank Canal

S.N. Lele and R.K. Patil

The Upper Godavari Irrigation Project

THE UPPER GODAVARI IRRIGATION PROJECT IN NASHIK DISTRICT, MAHARASHTRA, IS A multi-storage, multi-canal system. There are two storages—Karanjwan and Palkhed on the Kadwa river, one each at Waghad on the Kolvan river and at Ozarkhed on the Unanda river. All these are tributaries of the Godavari. Each of these reservoirs has an independent canal system for irrigation and sundry utilisation. The balance water is pooled at Palkhed to feed the left bank canal (LBC), the longest and largest canal running through Niphad and Yeola *taluks*.

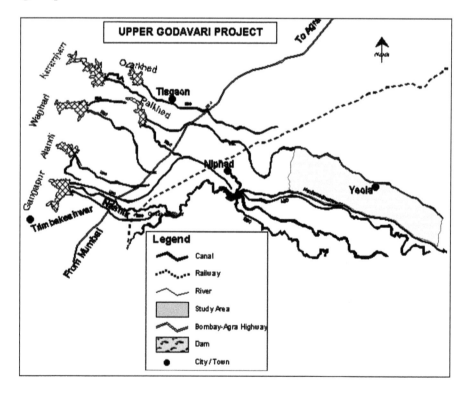

Map 1. *Upper Godavari project.*

Table 1: Sub-Projects and their Allocations in the Upper Godavari Irrigation Project

Sub-Project	Live storage (Mm³)	Canal length (km)	Culturable area (ha)
Palkhed	21.30	LBC—130	59,400*
		RBC—20	3,700
Waghad	72.23	LBC—15, RBC—45	2,357, 7,285
Ozarkhed	60.32	LBC—49	14,856
Karanjwan	166.22	LBC—16	2,284

*of which 22,000 ha are in Yeola *taluk* and the remaining in Niphad *taluk.*

The right bank canal (RBC), an ex-Palkhed dam, was in existence before 1947. Other sub-projects were completed between 1975 and 1990, and water was released for irrigation in 1990–91. The Palkhed left bank canal (PLBC) was the last to be completed.

The Upper Reaches Enjoy Abundance, The Tail Suffers

Since the dams were completed ahead of the canals, the areas commissioned in their upper reaches got abundant water pending the completion of the canal network. Because there was surplus water in the initial years, it was possible to lift some from the upstream sub-project canals, and such permissions were granted.

Fig. 1. *Typical installation of Dongala pipe in PLBC.*

Due to the increase in population and over-exploitation of groundwater for irrigation purposes, there has been a significant drop in the availability of groundwater for domestic use. Many villages and towns in the vicinity are demanding drinking water from the storages and canals. But since no storage for drinking water was ever planned, frequent water releases for this purpose have led to much greater seepage and loss through evaporation, reducing the water available for irrigation by larger amounts than is apparent.

For these reasons, the farmers on the PLBC command stopped getting water as per the plan. The problem became severe when the farmers in the upper reaches of the canal started siphoning off water illegally by installing pipes in the canal banks at bed level or even below bed level, and drawing water 100 to 200 m away from the canal to fill their wells directly in order to water their fields. These unauthorised pipes, locally called *dongala* pipes, are difficult to detect when the canals are flowing. Reportedly, there are more than 4,000 such pipes in the Palkhed system, because of which tail-enders do not get adequate water in most of the gravity canals; this is what happened in the case of the Yeola farmers of the PLBC *taluk*.

In the early 1990s, Participatory Irrigation Management (PIM) was taking root in the state. Under this system, water users' associations (WUAs) were formed at the minor level with operational areas ranging from 300 to 500 ha. (A 'minor' refers to a channel originating from the main/branch canal or a distributory that has a discharge capacity of less than one cubic metre per second.) Water was delivered to these WUAs on a volumetric basis, and water management was their responsibility.

In Maharashtra, the Water Resources Department (WRD) and WUA sign an agreement specifying their respective rights, responsibilities and functions. The WRD is supposed to allocate water and supply it to the WUA, keeping in mind the ratio of, operation area of the WUA to the total culturable command area (CCA) of the project, as fixed according to seasonal quotas (and water availability in) a normal year. This is indicated in the agreement. The WUAs, in turn, are expected to allocate and supply water to the farmers, maintain the system and recover the water fees from the farmers. The association has to pay water bills as per the volumetric rates fixed by the Maharashtra government for different seasons. The WUA has the freedom to grow any crop within the sanctioned quota.

PIM Forces the Issue

This process provided some solutions for reliable, equitable and timely supply of available water to all the farmers in the command area. Under the participatory irrigation management, the WUAs have to sign a Memorandum of Understanding (MoU) with the WRD, spelling out the terms, conditions, rights and responsibilities of both parties.

Having realised the benefits of the system, farmers in Yeola initiated about thirty-two WUAs by 2000–1, signed MoUs and started getting water reliably. The water quota allotted to these WUAs was 3,000 m^3 per ha at minor head or 4,478 m^3 at canal head. This quota was determined by the WRD on an ad-hoc basis, pending exact information about seepage/transit losses, drinking/domestic water demands, and commitments or sanctions for lifts in upper reaches.

The initial MoUs were for three years with a provision for extending/renewing the contract for a longer duration after mutual consultation. But when the WUAs requested a renewal they found that there was no response from the WRD. Meanwhile, there was steady shortfall in the supply. In 2000–1, even though the reservoirs were 90 per cent full, the water supplied was less than the previous years, when the reservoirs had much less water.

Fig. 2. *Area irrigated on lifts in the upper reach of PLBC in Niphad taluk.*

The Associations Federate

The WUAs, finding that the Water Resource Department had reneged on the agreement, came together under a single umbrella organisation called Balirajya Pani Vapar Sahakari Sangh (BPVSS) to represent their cause. The BPVSS served a legal notice on April 16, 2001 to the officials of the WRD and the Collector, Nashik. The allegation was that the water supplied was less than the sanctioned quota whereas the upper reaches of the PLBC in Niphad *taluk* and other sub-projects/canals received more. The Sangh requested the WRD to immediately restore supply to Yeola *taluk* but the Department replied that the losses in the canal were high (approximately 45 per cent), that there was no conscious injustice done to the WUAs (after all, the WRD has the rights to reduce water quota), and the Collector had set aside a sizeable quota to meet drinking water needs.

The BPVSS was not satisfied with this rejoinder and continued to negotiate with senior officers. Meanwhile, the WRDs introduced new conditions into the contract. Among these were: reduced quota; a clause that measurement of water would be done at the distributary head and not at the entry points of operational areas, as had been done earlier; a condition permitting freedom of crops grown within the quota, carrying over the balance from *rabi* to summer; and withdrawing the practice of storing water that was not used up in the *kharif* season. These changes were not acceptable to the BPVSS.

Just Short of a Solution

A meeting was held at the office of the Chief Engineer (CE) to resolve these issues. Both parties came to a consensus on the issues of (*a*) storing water in *kharif*, (*b*) carrying over of the quota not used in *rabi*, to summer, and (*c*) measuring the water at minor head. However, they could not agree on the demand for the restoration of quota. The BPVSS also spoke to the Secretary, WRD. On March 16, 2004, representatives of the Department and the Sangh met at the office of the CE, Nashik, to discuss the new water quota. The

Executive Engineer had worked out the net water available at the head of the PLBC after accounting for all upstream commitments. After taking into account the loss in the reservoirs and the loss due to seepage, the WRD could provide only, 1,537 m³ per ha at the canal head, which works out to about 877 m³ per ha at the minor head.

Table 2. WRD-Sanctioned Quota for Different Areas

Canal	Water allocated at the head of canal (Mm³)	Culturable area (ha)	Water quota per ha (m³/ha)
Karanjwan	8.21	2,284	3,594
Palkhed RBC	22.37	3,700	6,046
Palkhed LBC	91.35	59,400	1,537
Waghad	43.311	9,642	4,492

Note: The cropping pattern is evolved by the WUAs, keeping in mind, the sanctioned quota of water.

The reduced quota was rejected by the BPVSS. Their arguments were: (a) all losses such as evaporation from reservoirs, silting, etc. were assumed and not actually measured (b) quantities of water for lifts were considered according to the maximum area that could be sanctioned and not as per actual sanctions so far (c) nearly 4,000 unauthorised *dongala* pipes in the canal had been siphoning off water, thereby reducing the flow to downstream areas; this was camouflaged as losses (d) the upstream canal systems like Waghad, Karanjawan, Ozarkhed and PRBC were supplied their full quota as originally planned or demanded, and only the leftover water was allocated to the PLBC. This was a case of discrimination/bias. And most importantly (e) all seepage, silting, evaporation and water likely to be reserved for drinking water from the entire Upper Godavari project was deducted from the PLBC's water quota. A more equitable method of deduction should have been in place by which all sub-projects and sub-systems would have borne the burden of the losses. The water now offered by the Department was not adequate to meet the requirements for even one *rabi* rotation every alternate year.

The WRD refused to concede to the WUA demands and the associations were reluctantly forced to accept the terms for the much reduced quota. While water supply to these areas has resumed for the 2004–05 *rabi* season for one rotation, the BPVSS has not given up its struggle for a fair deal.

An Eight-fold Solution

The problem of shortfall in water supplied for irrigation and the conflict arising out of the situation are going to be a constant feature in the future. The government and the users will have to work out a long-term solution to effect sustainable service. There are various options they can explore:

(a) a dialogue between stakeholders; (b) maintaining transparency; (c) bringing precision and accountability into the whole business of measuring water; (d) determining loss through evaporation, seepage, etc. through periodical surveys, testing of canals, and regenerated flows from upstream irrigation; (e) adopting conjunctive use of surface and groundwater to increase the quantity of water supplied for irrigation, and also the frequency of irrigation; (*f*) recycling water used by industries/municipalities/*gram panchayats* to make it at fit least for irrigation; (*g*) supplying water for irrigation only (municipalities and *gram panchayats* should be told to make alternative arrangements for drinking water. Separate tanks should be constructed to hold such water for summer, and filled once to reduce transit losses due to continuous supply), (*h*) reducing or eliminating pilferage of water through *dongala* pipes.

Highs and Lows

The conflict started in 1994–95 when Yeola farmers stopped getting their share of water. It was at the lowest level between 1996 and 1999 when they began to get a little water after the formation of WUAs. The problem resurfaced in 2000–1 and was at its peak in 2003–04, when the WRD refused to accede to the terms of the Sangh and would not distribute water through the WUAs until they signed the new MoU.

The inequitable supply of water to the tail end of PLBC in Yeola *taluk* has affected about 17,000 farmer families, and their annual loss (*jowar*, sunflowers, vegetables) is approximately 31,000 tons, valued at Rs. 30.8 crores.

Note

*Map 1 from the Chief Engineer's office, North Maharashtra Region, Nashik; tables 1 and 2 prepared on the basis of data from Chief Engineer's office, North Maharashtra Region, Nashik; figures from SOPPECOM, Pune.

References

Legal notice issued by Balirajya Pani Vapar Sahakari Sangh Yeola (BPVSSY) to the Irrigation Department regarding short supply of water to Yeola WUAs, April 16, 2001.
Letter on the issue written by SOPPECOM to the Secretary, Irrigation Department and Command Area Development Authority (CADA) Mumbai, March 16, 2004.
Letter from one of the WUAs in Yeola (Gajanan Pani Vapar Sanstha Puranya Yeola) to the Executive Engineer, Palkhed Canal Division regarding renewal of MoU, May 25, 2004.
Letter from Chief Engineer, North Maharashtra Region, Nashik, July 15, 2004.
Paper submitted by BPVSSY to the Pani Parishad (Seminar on the Water Issue), May 14, 2005.
Rejoinder to the notice from Executive Engineer, Palkhed Irrigation Division, Nashik, May 15, 2001.

Mahad to Mangaon:
Eighty Years of Caste Discrimination
What Caste is Water?

Suhas Paranjape, Raju Adagale and Ravi Pomane

Time and Place

THIS CONFLICT ERUPTED BETWEEN THE DALITS AND THE SAVARNAS IN SEVERAL VILLAGES IN Mangaon *taluk* in Raigad district, Maharashtra, in 2003. All the villages are situated within 20 to 30 km of Mahad, the place where Dr. Babasaheb Ambedkar led the famous Chavdar Tank (lit. 'Tasty Water' tank) seventy-six years ago to establish the right of Dalits to public sources of water. What follows shows that not much has changed since that famous event.

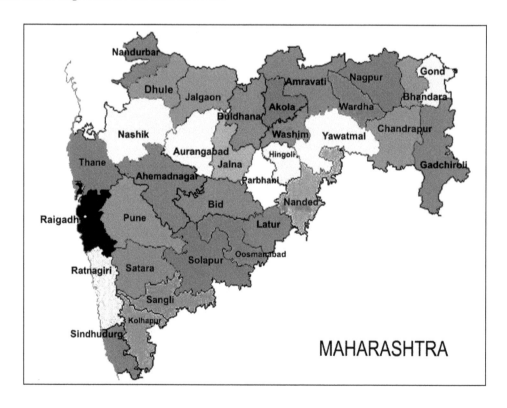

Map 1. *Location of Raigadh in Maharashtra.*

Cast(e) Out

The first in a series of incidents took place at Kuravade village, 6 km from Mangaon and 20 km from Mahad. The village has about fifty to sixty Savarna households and seven Dalit households. It has two public wells, one used by the Savarnas and one by the Dalits. Sometime towards the beginning of March that year, words were exchanged between the Dalit women and an old, ex-police *patil* (village functionary authorised to act on behalf of the police) whose field abuts the Dalit well. The old man became angry when he found excrement in the watercourse adjacent to the field and thought, who else but a Dalit would do this? He retaliated by asking them how they would feel if their well was similarly polluted. A couple of days later, the Dalits found excrement smeared on the parapet wall of the well. They were disturbed, but cleaned the mess and did not make an issue of it. The next day they found that someone had contaminated the well water.

A criminal case was registered and the old man was taken away and lodged in jail. The *bouddha panchayat* and a social activist intervened and appealed for the issue to be resolved in an amicable manner, and the old man was set free. The police *patil* who replaced him got the well cleaned, but the Dalits were not convinced that it was clean enough. The police suggested that they use the other public well for their water until the matter was resolved. The Savarna people agreed on the condition that the Dalits do not draw water directly from the well; a Savarna would do this and pour it out from a distance into the vessels of the Dalits.

The Sarvahara Jan Andolan (SJA) got wind of this and were incensed at this brazen return to untouchability. Their leaders and some of the Dalit women went to the other well and drew water from it. As far as the Savarnas were concerned, their

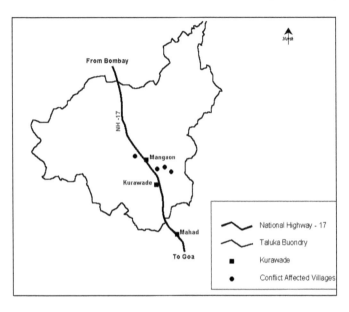

Map 2. *Location of Mangaon and conflict affected villages. Source: Prepared by authors.*

Fig. 1. *A public well used by dalits.*

well was now polluted. They could use the water for their cattle but not for themselves. For a while, many Savarna households resorted to fetching water from nearby canals and springs. Finally, the police *patil* had the Dalit well thoroughly cleaned and the latter went back to using it.

Hatkeli, Kavilvahal, Mugavali, Javali . . . the Conflict Spreads

While the immediate uproar subsided, the issue continued to fester. A few months later, the problem surfaced in other villages. Hatkeli too had two public wells, one in the Dalit *basti* and the other in the Savarna *basti*. The problem was that the well in the Dalit *basti* went dry in the summer, as a result of which the Dalits had to use the other public well for water. This had been happening for a few years and the Dalits were allowed water from the well, provided it was drawn by a Savarna and poured from afar into their vessels. But this year the atmosphere was charged, leading to an instant clash between the two groups towards the end of May. The Dalits drew water directly from the well. In Hatkeli, it is believed that the Savarnas actually pumped out all the water from the well and let it refill; there were also plans to dig two new wells so that they could be kept away from Dalit hands. It was also reported that they obstructed a water tanker from filling the well in the Dalit *basti*. By the end of May tension began to build as similar incidents took place in a number of neighbouring villages; the customary wells from where the Dalits drew water went dry, and there was inevitably a clash when they went to the other wells to fill their water. In some villages the wells were ceremonially purified.

The Social Justice Conference

All this prompted the SJA to organise a Samajik Nyaya Parishad or Social Justice Conference on August 9 (the same day that the Quit India Movement was announced by the Indian National Congress in 1942); it was attended by Dalit, progressive and left-leaning organisations and individuals. The Parishad was set up in July, but by then almost all political parties had taken note of the events and protested against the injustice meted out to the Dalits. Throughout May and June, leaders across party lines helped them draw water from wells in Savarna *bastis*, but as soon as the token ceremony was over, the conflict resurfaced and continued to simmer. Most leaders appealed to the two sides to resolve the issue peacefully. Finally, a timely monsoon intervened, filling all the wells and the Dalits went back to drawing water from their own wells. The August Conference was a huge success.

Strained Truce

In Kuravade, where it all started, drinking and domestic water are now supplied through pipes, and the Dalit *basti* has received a stand-post from which they draw water. The monsoon was good in 2005, and the situation is not as grave as it was in the summer of 2003. The Savarnas too have been making sure that the Dalits get adequate supply of water through their stand-post and tankers. The conflict is dormant though relations between the two communities have been strained. Economically this has been a setback for landless Dalit labourers who normally found work in Savarna fields. They have lost almost 40 per cent of their employment opportunities and the social impact has also been rather severe: Dalits are no longer invited to village functions, further straining the relationship between the two communities.

Of Different Minds

There were three distinct viewpoints in this entire conflict.

The Traditionalists

The majority of the Savarna households at the village level are orthodox. They are sympathetic to the plight of the Dalits and are quite ready to provide them with water, but they insist that they will draw the water themselves and pour it into the vessels while maintaining a distance. The disturbing thing is that they do not see this as untouchability. It was only when the police pointed out the fact to them and threatened them with imprisonment and punishment that they stopped interfering. In fact, they made all attempts to 'purify' their wells and dig other wells in order to 'cleanse' their water source.

The Accommodators

Most of the party leaders fall under this category. They regretted these incidents as a lamentable return to 'untouchability', but advocated *samanjasya* or accommodation. Effectively, this accommodation meant finding ways and means of getting water for the Dalits without causing 'undue hurt' to the Savarnas, that is, without making them break taboos that resulted from their practice of 'untouchability'. The idea was not to solve the problem but go around it, leaving the problem essentially the same.

The Militants

The militants, like the SJA, recognised that the problem was not that of water, but of untouchability and deeply ingrained prejudices. They felt that the only way to solve that problem was to use methods that would break rigid mindsets; their idea was to get water and show their opponents that they too were humans and entitled to equal rights. They demanded that they be allowed, like any Savarna person, to draw water directly from the wells.

Bone-deep Biases

It is sad that even eighty years after the Mahad *satyagraha* such incidents continue to occur. It raises serious and uncomfortable issues about persistent caste oppression. Have our efforts to eliminate caste only resulted in reinforcing caste identities? There has been a lot of progressive legislation aimed at eliminating caste discrimination in public life, just as there have been provisions for positive discrimination. But then, do these measures simply push caste from the public to the private sphere? Or, is caste so deeply rooted in our social and economic structure that there is no piecemeal solution; perhaps only a radical restructuring will eliminate it from our midst.

These are vexing questions. But they are the ones that matter in the long run. Kuravade settled for a temporary, technically and socially convenient resolution: a common piped delivery system with a separate stand-post for the *bastis*. It will bring everyone on level ground but will do little to eliminate prejudiced in the minds. Like ghettos, it separates warring factions but seals the boundaries, mental as well as physical. Biases remain, indeed rule, without seeming to do so. What will happen if the stand-post in the Dalit *basti* breaks down for some reason? It will be back to square one. We need a dialogue about deeper issues, about why water should have a caste and turn 'polluted' as soon as a Dalit touches it. Without that dialogue we, will only whitewash the fractures in our society, not heal them.

Eighty Years Ago...the Historic March

At the instance of veteran social reformer S.K. Bhole, the Bombay Council passed a resolution on August 4, 1923 to allow the Dalits, then 'untouchables', access to public places, to water from public tanks and to facilities in schools and *dharamshalas*. The municipal boards disregarded the resolution of the Bombay Council.

Four years later, Dr. Babasaheb Ambedkar decided to launch a water *satyagraha* in Mahad and a *dharma satyagraha* in Nashik to establish the rights of the deprived sections. A huge public meeting was organised at Mahad under his leadership on March 19, 1927, and on the following day about 5,000 men and women led by him marched to the Chavdar Tank and washed their hands in its waters. The incident sparked new hope among the deprived sections, and at the same time it impressed upon them that nothing can be secured without struggle.

There was a backlash. Upper caste Hindus 'purified' the waters of the tank by ceremonially pouring 108 pitchers of water into it. The Mahad municipality withdrew the government order allowing untouchables to collect water from there. Accepting the challenge, Babasaheb set up the Mahad Satyagraha Samiti and called for a new *satyagraha* on December 25, 1927. He set out for Mahad with 200 people on December 24, 1927, and was joined en route by thousands of Dalits. On reaching Mahad he found that a case had been filed in a court to the effect that the Chavdar Tank was private property. As a consequence of this he postponed the *satyagraha* so as not to break the law.

Babasaheb then filed a writ in the court and personally fought the case seeking for the Dalits the right to draw water from the tank. A decade later, on March 17, 1936, the Bombay High Court gave a verdict in favour of the Dalits. It was a historic victory for them, both on moral and legal grounds.

Fig. 2. *The Chavdar tank.*

Note

*Map 1 from SOPPECOM, Pune; map 2 and figures by authors.

References

Amachi Mumbai. May 29, 2003.
Krishival. March 17, 26, 28, 29, and May 30, 2003.
Lokmat, Raigad. May 28, 29, and 30, 2003; March 2004; January 6 and 14, 2005.
Mahanagar. May 26, 2003.
Maharashtra Times. May 28 and 31, 2003.
Ratnagiri Times. March 16, 2003; August 21, 2003.
Sakal. March 17, 2003; May 28, 2003.
Sagar. May 26 and 30, 2003.

A Dialogue Along the Jholapuri River, Coastal Gujarat: Addressing Water and Gender Conflicts through Multi-Stakeholder Partnerships

Sara Ahmed and UTTHAN

Once Upon a River

THE JHOLAPURI RIVER FLOWS FOR ABOUT 40 KM FROM ITS SOURCE IN THE HILLY REGION OF Mahuva *taluk* in the coastal district of Bhavnagar, Gujarat, across neighbouring Rajula *taluk* in Amreli district and into the Gulf of Cambay (Arabian Sea) near Pipavav port. The basin covers some 174 sq km with a population of about 26,135 (Census 2001) spread over twenty-five villages. The average rainfall in the region is 500 mm per year, though distribution varies—if the rainfall is intense and short in duration, flooding is not uncommon, as was seen in 2001. But for most of the year the Jholapuri remains dry, and it is hard to imagine that just thirty years ago the river was considered a lifeline by local communities—the basin flourished with rich fertile lands, abundant fruit orchards, cool gardens and flowing water.

Excessive deforestation in the upper catchment, over-exploitation of groundwater by large farmers for water-intensive crops in the middle reaches, and sand mining has led to salinity ingress in the flat, marshy plains comprising the river's last stretch. In addition, the acquisition of land for the development of the country's first private port at Pipavav has had an impact on the livelihood of small and marginal farmers. While competition over scarce water resources has affected the availability and quality of water for both domestic and other use, it has not yet led to a 'visible' conflict between different stakeholders. However, the impact of increasing water deprivation on poverty and

Map 1. *Location of Gujarat in India.*

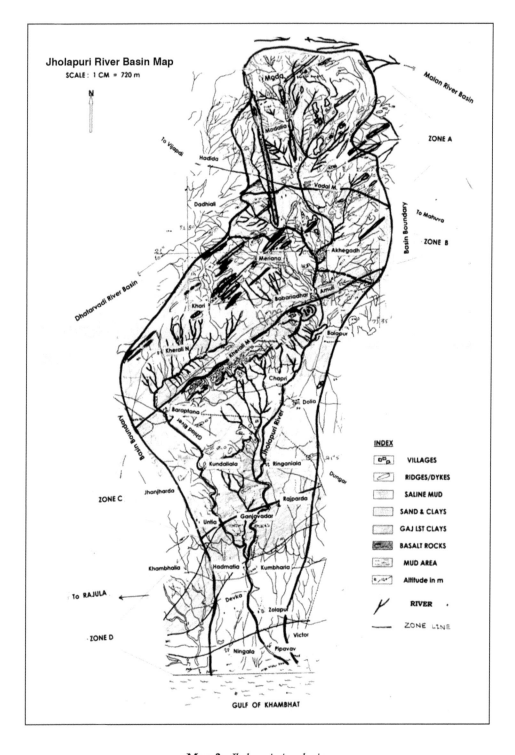

Map 2. *Jholapuri river basin.*

livelihood strategies and its implications, particularly for women and socially excluded groups (small and marginal farmers, landless, Dalits), are significant. It is precisely to avoid such latent conflict from becoming overt that Utthan—a voluntary, non-profit organisation with more than twenty years of experience in mobilising rural communities on questions of resource rights, gender equity and participatory governance—has been facilitating a multi-stakeholder dialogue in the basin.

A Dry River

A hydrogeological survey of the basin undertaken by Utthan in 2004 outlined four zones with distinct geological features that determine the physical boundaries of water availability, harvesting and recharging potential. This, in turn, has a bearing on access to water and competing claims at different institutional, sectoral and spatial levels.

Zone One constitutes three villages in the upper reaches of the Jholapuri that are part of Mahuva *taluk*. The terrain is hilly, surface runoff high, and recharging potential limited because of the underlying hard basalt rock. Cultivable area is about 60 per cent of the total land area and the dominant landholding castes are Patels and Darbars (upper castes). Apart from agriculture they also engage in raising livestock, mostly cattle, while semi-settled pastoral communities, Bharvads and Rabaris (pastoral communities), keep smaller ruminants like sheep and goat. Wells, both open wells and more significantly deep borewells, have had an impact on water withdrawal. In recent years, the construction of almost seventy small checkdams, supported by the Saurashtra Jal Dhara Trust under the government's '60:40 scheme'[1] has helped in well-recharging and irrigation.

However, these benefits are contested not only within the villages in this zone but also in other zones—for example, most of the wells which have been recharged are those near the river and they inevitably belong to the better-off farmers. Similarly, while farmers near the river/check dams have to some extent gained from increased irrigation potential, these gains have not 'trickled down' to those living away from the river, nor have they helped in recharging domestic water sources (handpumps). More importantly, many of the farmers in this zone feel that the benefits of the check dams are actually going to the farmers downstream, and would like their support to maintain the structures so that the benefits can be equally distributed and sustained. They have called for different water conservation structures, not just along the Jholapuri, but also along the ridgeline in the surrounding periphery so that the villages outside the basin can also benefit.

Another source of conflict is caste related—since this zone comprises a number of upper and middle caste communities, separate focus group discussions facilitated by Utthan revealed that there is reluctance to share water, either for domestic or productive purposes, with lower caste communities. In addition, women from 'marginal' (both in terms of caste/class and geographical location) households continue to go through the drudgery of water collection, with the exception of women from the Darbar community. Because of the practice of *purdah* (seclusion), it is the Darbar men who typically collect water, particularly during times of scarcity, by bullock carts rather than head-load.

Zone Two is characterised by hilly slopes and several small streams which contribute to the flow of the Jholapuri as does the Ghiyad, a small tributary which joins the main river in this zone. The area has hard black rock (basalt), which leads to aquifers with poor groundwater recharge. Open wells and borewells dominate the landscape, but according to data from a recent well survey, their irrigation potential is largely determined by the rainfall in the area. For example, 2003 was a good monsoon year after a series of drought years, and the open wells surveyed had reasonable levels of water even during February–March 2004. The caste composition in the seven villages of this zone is more diverse and includes traditional pastoral groups like the Bharvads who keep small animals such as goats instead of cattle because of the lack of fodder crops.

Conflict over access to water is similar to that in Zone One; for example, farmers who drill borewells along the banks of the river directly pump water while there are others who find it difficult to even access drinking water. Another source of conflict between the Bharvads and landowning farmers is land. In the past, the State had allocated a piece of land to the Bharvads for cattle grazing. Over the years, however, because of encroachment and the acquisition of land by vested interests through the *panchayat* (the local government) on the one hand, and landed farmers engaging in open grazing on the other, the Bharvads find that their rights to the commons are being gradually eroded.

Zone Three is, again, characterised by hard basalt rock with poor recharge capacity. River flow does contribute to some groundwater recharge by seepage, but the only wells that benefit are those near the river or in the river-bed. Old check-dams such as the one at Balapar village are not properly maintained—they need de-silting which some villagers see as a solution to their water problems. Salinity ingress and sand-mining have had a considerable impact on the availability and quality of drinking water. Difficulty in accessing water during the summer months means that most farmers cultivate only one crop, except for those near the river who can still afford to drill deep bores and use diesel pumps. Many young men have migrated to Surat and taken up jobs in the diamond-cutting industry or to nearby Alang where they work at the shipbreaking yard.

Women tend to fetch water from farm wells owned by large landholders during the summer months—often they are asked for small 'payments' such as diesel, or money for diesel to operate the pumps. Apart from competition over scarce water, conflicts also arise between the landed and the landless. Typically, landless communities depend on common resources to meet their biomass requirements, like fuelwood and fodder. During the summer they are compelled to buy fodder as there is little available from the *gaucher* (village grazing lands).

Zone Four, where the Jholapuri river flows out into the Arabian Sea, is a topographically gentle, undulating area comprising marshy coastal plains characterised by salinity ingress. There is hardly any agricultural potential here, and not surprisingly, many farmers and labourers migrate seasonally or on a long-term basis, or seek work at the Pipavav port or in the industries that have grown around the port, such as the Larsen and Toubro cement plant, or infrastructure development (roads) and salt-panning. Indiscriminate sand mining for construction purposes has had an impact on water retention in the aquifer, affecting the availability of water and the depth of the water table.

Migration has also had an impact on children's education and women's workload. Low levels of enrolment and high dropout rates, particularly for girls, are both common in this zone. During the summer months, women dig *virdhas* (shallow wells) in the riverbed for sweet, potable water. These are 'guarded' by male family members at night so as to prevent other villagers from 'poaching' their water. Occasionally, some of the water that collects overnight is shared with other women.

In sum, with the introduction of mechanised irrigation about thirty years ago, there has been a distinctive shift in agricultural patterns from food crops (*bajra, jowar*) to cash crops (cotton, onion, groundnut) within the basin. While landed farmers in the upper and middle zones have generally benefited in the short term, having access to capital to deepen wells, purchase fertilisers and pesticides and employ labour, the indiscriminate withdrawal of water coupled with decreasing yields has now begun to have wider and more noticeable impacts across caste, class and gender. Small and marginal farmers (0.4 to 1.2 ha) account for almost 35 per cent of the landholdings in the basin (Census 2001 and Utthan survey), while the landless stand at 33 per cent of the households. It is these groups which have increasingly been forced to adapt to water scarcity by diversifying their livelihood strategies, rather than confronting large farmers who account for one-third of the landholdings in the basin but command most of the available water in the absence of norms to govern groundwater extraction.

Dialogue in Progress

Although Utthan has been facilitating community-managed natural resource institutions (watershed committees, Pani Samitis) and women's participation through self-help groups (SHGs) in the area since 1997, the use of dialogue as a process for mobilising multi-stakeholder conflict resolution platforms was initiated in 2003. The dialogue process has been supported by the International Water Management Institute's Dialogue on Water, Food and the Environment (*www.iwmi.org/dialogue*) and the Indo–Dutch Programme on Alternative Development (IDPAD, ICSSR, New Delhi).

Significant milestones in this participatory process have included:

1. A situational analysis of the historical, socio-economic and physical dimensions of resource management in the basin through village surveys and a range of participatory rural appraisals as well as secondary data analysis.
2. A stakeholder analysis of primary and secondary stakeholders in the basin—the former are those who are directly impacted by water deprivation and the latter include those who may have a stake in supporting community resource management strategies.
3. A hydrogeological survey of the basin to demarcate the ridgeline (basin boundary) and map land-use patterns, water availability and recharging capacity. An inventory of wells and community-supported monitoring of water quantity and quality is currently in progress.
4. Institution-building through the formation of village river basin management committees with mandatory 50 per cent women's representation as well as the inclusion of vulnerable and marginalised groups. Currently, committees are evolving norms for water allocation, identifying all visible and latent conflicts, their root causes and possible strategies for resolution or negotiation. The size of village committees varies from ten to fifteen members and includes at least one representative from each caste group in the village. Two members from each village committee, a woman and a man, are invited to join the cluster or zonal committee, and discussions are on for the formation of an apex basin-level federation.
5. Strengthening women's participation through the parallel formation of separate women's SHGs. In patriarchal Saurashtra such separate spaces enable women to give voice to their priorities and concerns, develop self-confidence and engage in discussions that go beyond natural resource management to look at the larger question of gender rights (e.g. health and reproductive rights, gender violence). Women's participation, right from the inception of the dialogue process, has also ensured the representation of traditionally excluded groups, for example, the landless, Dalits and pastoral communities through the informal social networks that sustain rural women's everyday lives.
6. Communication strategies have involved the use of alternative media, songs and folk theatre on water and livelihood issues. In May 2004, the Utthan team undertook a *padyatra* (a long march on foot) along the entire length of the Jholapuri, facilitating information-sharing between upstream and downstream

Fig. 1. *UTTHAN team facilitating a participatory rural appraisal with the women of one of the basin villages.*

communities through daily *gram sabhas*. The *padyatra* culminated with a basin meeting where block and district level officials were invited to listen to people's strategies for the revival of the river and suggest how they could offer proactive support.

7. Capacity-building on conflicts and conflict resolution mechanisms has been an ongoing process for both the Utthan team and community leaders. Apart from perspective-building, the team and village committees have visited other river basin projects in the State and beyond, where lateral learning opportunities (people-to-people) have been significant in sharing different approaches to ensuring water security and livelihood sustainability. Workshops on water-saving agricultural practices, for example drip irrigation, have been held and discussions on building self-reliant community drinking water systems and sanitation services under the Swajaldhara provisions are a part of this effort.

There are a number of critical issues that have emerged from the Jholapuri dialogue process so far. The first concerns the diversity of voices: there are primarily landed farmers who have been advocating the repair of old structures (e.g. de-silting reservoirs to augment water flow) or building new checkdams, and those who have been stressing the need to forge a unified community—*prem setu* or a bridge of love, so to speak—between the villagers in the river basin as a means of resolving conflict.

At another level, the dialogue process has illustrated that perceptions of interest vis-a-vis water management are not only coloured by gender, but are also not necessarily 'visible'. For example, in many public forums men have blamed women for wasting water because, according to them, in their haste to get on with their chores, they do not shut the water taps or handpumps properly. But perhaps the devices need repair and inevitably, it is a male, private contractor who is in charge of maintenance.

Over the last year, the Utthan team has realised that while the potential for dialogue in understanding the multi-dimensional nature of conflicts and mobilising stakeholders has been important, it now needs to be supported by tangible social action that leads to conflict transformation and livelihood security. This does not mean only building structures or forming groups, but also enabling communities to access government programmes and negotiate with secondary stakeholders (the state, private sector) to ensure their basic livelihood security and livelihood rights. Such a process will need to be supported by a legislative or regulatory framework that not only establishes differential and equitable resource rights of all stakeholders, but also accords legitimacy to the emerging basin-level organisations.

Note

*Map 1 from SOPPECOM, Pune; map 2 and figure 1 from UTTHAN.

1. Sardar Jal Sanchay Yojana is popularly known as '60:40 scheme'; the government extends 60 per cent of the proposed expenditure towards water harvesting structures, like check-dams, *nala* plug, deepening of ponds, etc., and people contribute the balance 40 per cent.

Groundwater Irrigation in Northern Gujarat: Digging Deep for Answers

Jennifer McKay and H. Diwakara

Receding Groundwater

In PENINSULAR INDIA, GROWING GROUNDWATER SCARCITY DUE TO THE DEPLETION OF aquifers has led to negative externalities—frequent well failure (cumulative well interference), falling water table, saline water intrusion, and so on. Internalising these externalities is an onerous task. Getting institutions (rules) suitable for local needs is a challenging task for water managers in India, and much emphasis has been placed on self-governance.

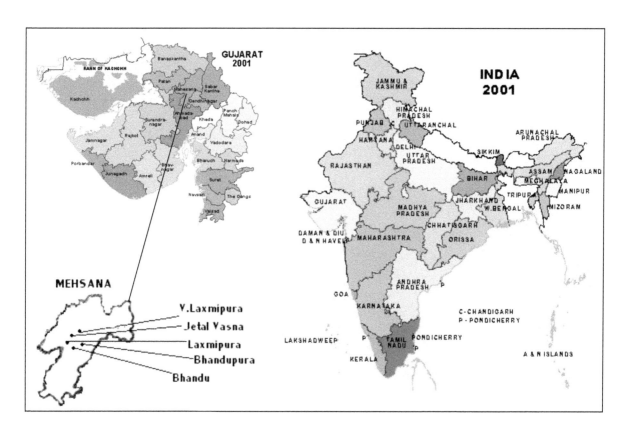

Map 1. *Study villages in Mehsana district, northern Gujarat.*

Fig. 1. Women bear the brunt when it comes to collecting water for the family, but are not involved in management decisions.

In the absence of surfacewater for irrigation, groundwater is under pressure as there is a race for exploitation of the resource by those who can afford to drill borewells and tap water from the deeper aquifer. This has resulted in inequitable access, and in some regions, conflicts between farmers.

Here we examine issues relating to conflicts between members and non-members of well organisations in Mehsana district, north Gujarat.

An Arid Land

There is a danger of over-exploitation because of the rate at which self-governed well organisations in the district have been extracting water. Gujarat depends heavily on groundwater for irrigation as rainfall is erratic and uneven. The State receives two spells of rainfall (south-west monsoon) between June–July and September–October, but precipitation has often been unreliable. The uneven rainfall has resulted in water scarcity; the southern districts receive high and assured rainfall, the central districts medium and less assured rainfall, while Saurashtra, Kutch and the northern districts receive scanty and irregular rainfall (Mathur and Kashyap, 2000).

In northern Gujarat, the agrarian economy has been under considerable stress due to the 'business as usual' approach to water management (IWMI-Tata 2001). The per capita water availability is 300 m^3 per annum, which puts it in the absolute scarcity category (According to the Falkenmark's indicator, absolute scarcity is a condition in which it is impossible to support human life). In northern Gujarat, farmers struggle to sustain agrarian and dairy economies; and in the absence of or very little access to surface water, groundwater has been the mainstay for agriculture, industry and urban and rural domestic consumption. Agricultural demand for groundwater has experienced a meteoric rise between 1950 and 2000. In 1997, of the 1.4 mha irrigated in north Gujarat, less than 100,000 were served by surface water and over 1.3 mha were irrigated by over 150,000 deep tubewells powered by electric pump sets of 15–75 hp. These tubewells accounted for over 3000 Mm3 of annual gross groundwater draft that sustains irrigated

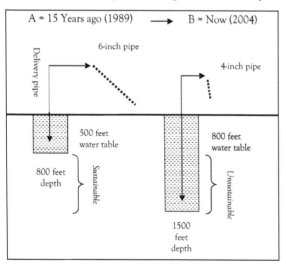

Fig. 2. Picturisation of depth and pressure of groundwater flow from irrigation borewells as felt by farmers in Bhandu village, Mehsana district.

agriculture in the region (IWMI-Tata 2001). Mehsana district has the highest gross groundwater draft of 1.420 km^3 and receives a total precipitation of 0.862 km^3. It is interesting to note that about fifty years ago bullock-bailers were used to lift water for protective irrigation. Now deep tubewells use roughly 0.35 to 0.5 kWh of electricity to lift one m^3 of water. Groundwater is being extracted from as deep as 370 m below the ground. It is no wonder that north Gujarat figures prominently in global literature on groundwater over-exploitation (Dubash 2000 & 2002a; Postel 1999; Mudrakartha 1999; Kumar 1999; TERI 2001; Shah et al. 2003).

The villages under study in Mehsana include Bhandu, Bhandupura, Jetal Vasna, Laxmipura, and V. Laxmipura. A total of 150 farmers who were interviewed owned tubewells. In addition, thirty group leaders (water managers) and fifty women were interviewed. While non-members were not formally surveyed, their concerns were noted.

When Water Costs the Earth

The first conflict deals with the members of the well organisation and concerns the timelines of water allocation. The second conflict is between the members of well organisations and non-members, and relates to the pricing of groundwater; the latter is the more serious issue. It was reported that there is inequity in groundwater pricing—non-members are charged double the price levied on members. Even within the non-members, there is a difference regarding their mode of payment of the water charge—crop-share (one-third of crop produce) or cash. Also, women are not involved at all in management decisions (all fifty women interviewed indicated this).

According to members of group-owned wells, the conflict between members and non-members is quiescent. Be that as it may, non-members who depend on groundwater for irrigation feel that the problem is a serious one because they suffer economically, forced as they are to sell their crops at a price that is lower than what they pay for the groundwater. There does not seem to be a way to resolve this issue. There have been informal meetings to try and come to a compromise, but these efforts have not been well-documented.

Non-members indicated that they wish to access groundwater for irrigation without having to go via the well partnerships that charge indiscriminately. At present, there are no mechanisms to resolve the conflicts between members and non-members.

Not Enough to go Around

Groundwater has been extracted and used in unlimited quantities. The field study indicates that the race for groundwater exploitation is connected with the profit associated with water sales by group-owned wells. This has already resulted in unsustainable groundwater utilisation. The fallout has been negative—in Mehsana the tubewells are deeper because the water table has fallen, wells have begun to fail, and the cost of extracting water has increased because huge sums are spent on deepening existing wells, drilling new ones and so on.

Looking for Solutions

The legal and political issues pertaining to groundwater allocation and management mostly uphold the tradition of treating access to groundwater as an unregulated private right under an apparently open access system. The private nature of this right has been a hurdle in any attempt by planners and policy makers to bring in regulations. An endeavour to regulate pumping and/or pricing would provoke significant political and legal opposition. The onus of managing groundwater resource falls on sustainable institutions that can be built around the rules, norms and procedures that determine access to the resource.

Potential Options to Resolve Water Conflicts

(i)   Defining property rights to groundwater in terms of volume of water extracted per year: this can be achieved through developing water allocation plans through community consultation. Once accepted, the plan would facilitate equitable and sustainable allocation of the resource. Since communities would be a part of such an initiative, a policy response would be relatively sustainable.

(ii) Two-part charging for groundwater extraction and use—marginal cost pricing of resource through pro-rata charges for electricity used to lift groundwater, and charges for groundwater. These will provide the necessary disincentive against extraction.

(iii) Members must maintain a system of rotation for extracting water instead of drawing from their wells as and when they please.

(iv) Depending on the need and urgency, a cost-sharing scheme can be considered to settle the dispute between water users. Such a scheme would compensate for any economic damage (for example, crop loss) due to delayed water allocation to members and non-members of well organisations. Funding arrangements can be made through mutual agreement to suit local conditions and the nature of problem.

(v)   A dialogue between farming communities and the state government will help modify the rules and norms created by the farmers so that use of the resource shall be relatively sustainable while the equity aspects are preserved.

(vi) Self-organised groups have well-developed social capital and are accustomed to community action. This would help State government authorities to devise some complementary rules with a negotiated approach with the community. Since development and enforcement of policy involves transaction costs, it would be pragmatic to assess them through research involving farmers and other key players including government officials, academics and researchers, as well as NGOs. Both short-term and long-term costs need to be worked out (for details, see Ostrom 2001) to adjust to the new rules and norms to reduce groundwater extraction. For example, the field study in Mehsana suggested that farmers are willing to incur some costs if the policies benefit them. They are willing to participate in meetings to devise the rules and norms governing allocation and pricing of groundwater; discuss growing crops that need less water; monitor water harvesting structures, and so on.

In conclusion, emphasis must be placed on collective action through user groups and cooperatively managed irrigation systems to complement state regulation, especially in the wake of economic and environmental externalities created by the unabated over-utilisation of the resource in many parts of the country.

Note

*Map 1 and figures by authors.

References

Carl, Boronkay and Warren J. Abbott. 1997. 'Water Conflicts in the Western United States', *Studies in Conflict and Terrorism*. 20(2): 137–66.

Colby Bonnie, G. and D'Estree, Tamra Pearson. 2000. 'Economic Evaluation of Mechanisms to Resolve Water Conflicts', *International Journal of Water Resources Development*. 16(2): 185–251.

Diwakara, H. and Jennifer McKay. 2003. 'Groundwater Conflicts: How WAMPS can Help', *Water*, Journal of the Australian Water Association. 30(8): 23–26.

Dubash, Navroz. 2000. 'Ecologically and Socially Embedded Exchange: Gujarat Model of Water Markets', *Economic and Political Weekly*, April 15, pp. 1376–85.

———. 2002. *Tube Well Capitalism: Groundwater Development and Agrarian Change in Gujarat*. New Delhi: Oxford University Press.

IWMI–Tata. 2001. *The North Gujarat Groundwater Initiative: IWMI–Tata Water Policy Program*; Institute for Rural Management, Anand; Gujarat Ecology Commission. Research Project Document.

Kumar, Dinesh. 1999. *Managing Common Pool Groundwater Resources: Identifying the Management Regimes*, VIKSAT. Ahmedabad: Nehru Foundation for Development.

Mathur, Niti and S.P. Kashyap. 2000. 'Agriculture in Gujarat: Problems and Prospects', *Economic and Political Weekly*, August 26–September 2, pp. 3137–46.

Mbonile, Milline, J. 2005. 'Migration and Intensification of Water Conflicts in the Pangani Basin, Tanzania', *Habitat International*. 29(1): 41–68.

Mudrakartha, Srinivas. 1999. *The Status and Policy Framework of Groundwater in India*, VIKSAT. Ahmedabad: Nehru Foundation for Development.

Naeser, Robert Benjamin and Mark Griffin Smith. 1995. 'Playing with Borrowed Water: Conflict Over Instream Flows on the Upper Kansas River', *Natural Resources Journal*. 35(1): 93–110.

Ostrom, Elinor. 2001. 'Reformulating the Commons' in J. Burger, E. Ostrom, R.B. Norgaard, D. Policansky, and B.D. Goldstein (eds), *Protecting the Commons: A Framework for Resource Management in the Americas*, pp. 17–41. Washington, DC: Island Press.

Postel, Sandra. 1999. *Pillar of Sand: Can the Irrigation Miracle Last*. New York: W.W. Norton. (She suggests that some 10 per cent of the world's food production depends on a yearly overdraft of groundwater of 200 km^3; of which 100 km^3 occurs in western India [including Gujarat].)

Shah, Tushaar, Aditi Deb Roy, Asad S. Qureshi, and Jinxia Wang. 2003. 'Sustaining Asia's Groundwater Boom: An Overview of Issues and Evidence', *Natural Resources Forum*. 27:130–40.

Sneddon, Chris. 2002. 'Water Conflicts and River Basins: The Contradictions of Co-management and Scale in Northeast Thailand', *Society & Natural Resource*. 15(8): 725–41.

Tata Energy Research Institute (TERI). 2001. Regulatory Framework for Water Services in the State of Gujarat, TERI Project Report No. 2000ER61.

Verghese, B.G. 1997. 'Water Conflicts in South Asia', *Studies in Conflict and Terrorism*. 20(2): 185–94.

http://business.unisa.edu.au/commerce/waterpolicylaw

http://secatchment.com.au

Problems at the Indira Gandhi Canal in Rajasthan: Conflict over Reduced Water Allocation

Binayak Das

Dust Flies in the Desert

THE AREA OF CONFLICT LIES WITHIN THE THAR DESERT, IN SRIGANGANAGAR DISTRICT IN north-western Rajasthan. On the north is the agriculturally prosperous state of Punjab, on the west, Pakistan, and on the south are Bikaner and Jaisalmer districts. The immediate area of conflict includes the peri-urban towns of Gharsana, Rawla, and Khajola and adjacent villages.

Despite being located in the Thar Desert, this region is no longer barren and sandy. With the construction of the world's largest irrigation canal project, the Indira Gandhi canal (IGC) in the 1970s, the desert landscape underwent a massive change and the region turned into one of the most productive areas in India. The IGC transformed Gharsana and Rawla into commercial hubs. The population of Rawla increased from 253 in 1971 to 12,325 in 2001. The population of Gharsana *tehsil* is about 15,000 (*Down to Earth*, November 2004). Farming became the new occupation and with enough water available, farmers began to concentrate on cash crops like cotton and mustard. Soon this *tehsil* and Rawla became trade centres and large *mandis* (markets) mushroomed. In terms of the produce it handled, Gharsana *mandi* was equal to

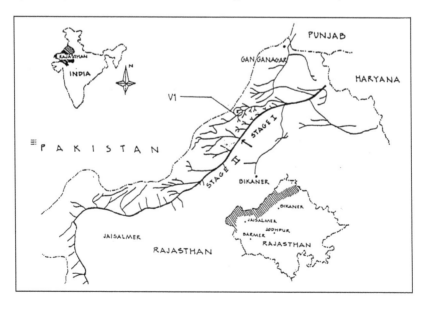

Map 1. *Location of the Rajasthan canal project*

Sriganganagar. The trading community prospered and agricultural products were sold to factories all over the country; meanwhile, the region around Gharsana–Rawla also turned into a small-scale industrial hub, with numerous cotton and oilseed processing units springing up in the vicinity. There was an influx of migrant labour from other parts of the country. Crops flourished, trade boomed and industry thrived. Standards of living improved dramatically because this had become an abundantly prosperous region.

Farmers Turn Militant

The conflict began about five years ago with reduction in the water allocated to this region—all of a sudden, there was only enough water to cater to drinking water needs. People here are entirely dependent on the canal for the water they need for both irrigation and domestic use because the groundwater is saline. In the original plan of the IGC, 400 cumec of water was allocated to this part of Rajasthan in two stages from the Pong dam in Punjab. Stage I included the districts of Sriganganagar, Hanumangarh and Bikaner with a total allocation of 0.37 cumec per 1,000 ha of land. Stage II covered Jaisalmer, Barmer, Jodhpur and Nagaur with 0.21 cumec per 1,000 ha of land. Stage I begins at the Hanumangarh head and terminates at 620 km reduced distance (RD) in Bikaner; Stage II follows from there. Gharsana *tehsil* falls in Stage I. The allocation was distributed in the ratio of 58:42 for a total command area of 11.2 lakh ha: 5.2 lakh ha in Stage I and 6 lakh ha in Stage II. More water was allocated to Stage I because it had more fertile land, its geographical location was more conducive to farming, and hence it was considered more suitable for irrigation. Most of the allocation for Stage II was meant to fulfil the drinking water needs of Jaisalmer and Jodhpur towns.

The water allocation to Stage I slowly started getting reduced, and five years ago the government passed an order stating that water from the canal could only be used for meeting drinking water needs. And, anyone found lifting water for irrigation would be liable for prosecution. The original allocation of 0.37 cumec was cut down to about 0.25 cumec. Earlier, during every rotation the canal ran for fifteen days at a stretch and there was a week's interval between rotations. Now, the canal runs for three or four days in a rotation, and rotations are spaced as much as a month or more apart.

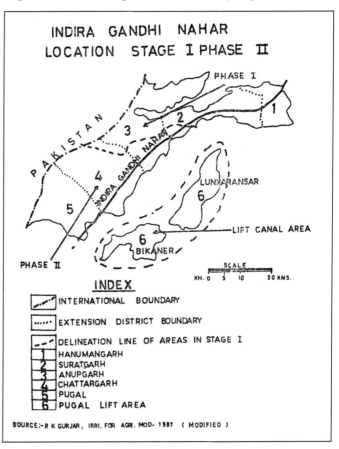

Map 2. *Location of the Indira Gandhi canal, stage 1, phase 2.*

Productivity decreased and the economy suffered a drastic loss. Cotton production plummeted from 100,000 to about 8,000 bundles per year. Mustard production fell to 20 per cent of its earlier value. (Table 1 details the produce handled for the last ten years in Gharsana *mandi*). As agricultural output dwindled, the flourishing *mandis* declined. The Gharsana *mandi* traded about 1,000 tons of cotton everyday, but current figures hardly reach 10 tons per day. The Rawla *mandi* almost shut down in 1998 when its turnover touched a dismal 80 crore tons per day, and fell even more drastically in the subsequent year. In the Gharsana–Rawla region, 35 per cent commission agents shut shop by 2004. Oil and cotton processing units were the next to down shutters—nine cotton and oil pressing units out of twelve were shut down or sold off. About 5,000 workers lost jobs, and almost 40 per cent of the population migrated. The state government lost approximately three crore in sales tax every year. Many of the farmers who had taken loans lost more than 500 crore and the merchants did not fare any better. Land was mortgaged and cost dipped.

Table 1: Annual Variation in Produce Handled in Gharsana *Mandi*

Year	Income from *mandi* (in Rs. lakh)	Agricultural produce handled (in '0,000 T)
1994–95	115.95	5.01
1995–96	157.00	9.13
1996–97	163.36	8.92
1997–98	182.74	8.64
1998–99	206.65	9.73
1999–2000	213.01	10.53
2000–01	196.04	10.10
2001–02	72.59	2.97
2002–03	84.38	4.59
2003–04*	119.62	4.59

*The figures for 2003–04 are higher than the previous year's because the Rawla *mandi* closed down, and all produce was diverted to Gharsana.

Then in September 2004, for the first time farmers, traders and labourers of the region came together under the umbrella of the Kisan Vyapari Mazdoor Sangharsh Samiti (KVMSS) and launched an agitation for their share of water. September–October is a crucial month for *rabi* harvest. The KVMSS was supported by politicians across all parties and the leadership comprised communists and congressmen. Thousands of villagers from various parts of Anupgarh gathered everyday in front of the *tehsildar's* office demanding the release of their rightful share of water. These protests continued for about a month; agitators in the Sub-Divisional Magistrate's (SDM) office held government staff hostage for a few hours. The agitators allege that the government ignored their demands. On October 26, there were some skirmishes and the police *lathi-*charged a group leading to many injuries. The protesters also turned violent and the next day they attacked the SDM's residence, the *tehsildar's* office in Gharsana and the police beat office in Rawla. The SDM's residence and a few police beats were gutted. Most of the records pertaining to the *tehsil* were destroyed

Fig. 1. *The Indira Gandhi canal.*

in this incident, and the police and para-military forces retaliated by firing on the mobs in Rawla and Gharsana, killing six and injuring sevral. Some of the dead had nothing to do with the agitation and were innocent bystanders or commuters. All the leaders of the agitation were arrested along with thousands of participants. Curfew was imposed in Gharsana, Rawla, Anupgarh and Suratgarh, and the army was called in. The agitators continued their protests demanding the release of their leaders. They went on a hunger strike and courted arrest. As news of these incidents spread, farmers from other places threatened to join the agitation if the arrested were not freed and their share of water was not released. The government gave a verbal assurance that the protestors' demands would be met, but the leaders wanted it in writing. Finally, an agreement was signed with the leadership of the farmers' association in Ajmer jail and water allocated to the two stages was released. This status quo will be maintained until a government appointed committee investigates the issue and submits its report. The government also declared an ex-gratia payment of Rs. 5 lakh each to the next of kin of the deceased and Rs. 1 lakh to those injured. It also promised a government job for one member of each family of those killed.

Various reasons have been cited for this conflict. First, the IGC had been receiving less than its share of water from Punjab. The capacity of the canal is 510 cumec, but according to the Rajasthan government, it receives only 147 cumec. The population of the region has also increased and domestic water demand has gone up to 59 cumec, leaving only 89 cumec for irrigation in both stages (the original capacity was 232 cumec in Stage I and 167 cumec in Stage II). Second, the immediate cause for the conflict was the attempt by successive governments to increase the command area from the original 11.2 lakh ha to 18 lakh ha, though many people maintain that this increase remained only on paper. Third, farmers allege that water was being diverted to Stage II for irrigating the constituencies of politicians from the area. They had promised free water to voters as part of their election campaign and this they did by diverting water from Stage I. Fourth, the IGC main canal gauge is also said to have been faulty, and this could be another reason why enough water was not being released. Fifth, there have been poor rains in the region and for the first time ever, three years ago, the Bikaner–Sriganganagar region was declared drought-hit. Finally, local people also feel that the receding Himalayan glaciers have resulted in less water in the dams (*Down to Earth*). This might have forced the Punjab government to extract more water from the canal. Punjab has recently passed the Punjab Termination Act that annulled all bilateral water agreements between Punjab and other states.

'All Quiet on the Western Front'

With the signing of the agreement between the farmers' union and the government, and the release of water in December 2004, the farmers have reaped a bumper mustard crop. For the moment, the government is able to allocate the required amount of water to Stage I and II. However, the state government plans to constitute a committee to look into the issue of future water allocation for this region, and a new command area commissioner has already been appointed in Bikaner to look into the matter. Crisis has been averted for the time being but a long-term solution is nowhere on the horizon.

Justifying their Stand

The protesters want the state government to ensure adequate and timely release of their water allocation. They reason that for the past several years, though the IGC has been getting less and less water, they have never protested. What bothers them is the fact that water is now being diverted to newly developed command areas, leaving them high and dry. The protestors are willing to settle for a fixed percentage of the water supplied to the IGC. If there is less water in the IGC, they too are willing to settle for less, provided the ratio is maintained. The government says that Punjab has been releasing less water and that this is what led to the crisis. They are firm that there has been no diversion of water to new command areas.

The impact of the conflict has resulted in mistrust between the local people and the administration. Politically, the opposition parties benefited so much while leading the agitation that for the first time ever in the history of Rajasthan, a communist who led the agitation was elected to the *panchayat* in February 2005. Needless to say, the economic loss has been massive.

Coming Together for a Common Cause

For the moment everyone is satisfied. While the government says that it is trying to work out a long-term solution, it has made no attempt to initiate a dialogue with the affected communities. They cannot hope to resolve the issue unless they take the stakeholders from both stages into account. Municipal corporations that have been allocated drinking water from the canal also need to be involved. The protests were limited to Stage I and, having tasted success, the local people are confident that they can achieve the same results by taking to the streets once again if the situation takes a turn for the worse. The government seems to be playing safe for the time being and is not raking up the issue, though one can see that bold measures are called for. There is no hope that the canal will get more water and studies have to be initiated to see how the region is going to cope with increasing demand. People have to be made aware of these realities and face new developments. On their part, the government should stop pandering to politicians and not allocate water illegally; the development of new command areas too must cease. And lastly, Rajasthan must negotiate withy Punjab and work out readjustments in its allocation. The situation calls for compromise and good sense from all sides. Let us hope it prevails.

Note

*Maps from New Mandi, Gharsana Tehsildar's office, Ganganagar; figure 1 by author.

References

Parashar, Vikas. 'Farmers Uprising in Rajasthan', *Down To Earth*, New Delhi: Centre for Science and Environment. November 2004.

Dutta, Saikat. 'Dust Bowl', 2004. *Outlook*. New Delhi. November 2004.

PART 3

Water Quality Conflicts: A Review

Paul Appasamy

Introduction

CONFLICTS OVER ALLOCATION OF WATER RESOURCES—AMONG UPSTREAM AND DOWNSTREAM users, among States, and even among countries—are well-known, but in recent years conflicts have arisen among water users over issues of water quality. When one user, generally an industry, takes water for processing and then discharges the used water in the form of effluents, the receiving water body (river, canal, tank or lake) or groundwater may become polluted. The next user of the water is affected according to what it is used for: agriculture, drinking water for human beings or livestock, or fisheries. It may also affect the aquatic ecosystem and biodiversity. In theory, if the effluents are treated to the required level, the respective water bodies should be able to assimilate the wastes. However, as many of the case studies in this section demonstrate, the mere existence of treatment plants (individual or common) does not guarantee that the effluents will not cause degradation of water quality. In most of the cases, the users first protested and later took up the issue in court through petitions or public interest litigation. In a few cases the court has ordered closure of the polluting units, while in some cases they have been asked to pay compensation to the victims. In other cases there have been attempts to create a multi-stakeholders' forum and initiate a dialogue. In India, environmental laws treat pollution as a criminal offence. The way the present legal system operates often precludes the possibility of mediation since it pits the polluter against the victim as adversaries. In what follows, we will discuss the common features of the twelve case studies (Table 1), the legal/institutional provisions, and possible ways to resolve the conflicts.

Common Features

Rapid industrialisation has taken place in India in the last two decades. The industries that use large quantities of water, such as chemical units, textiles and tanneries, discharge most of the used water as effluents. Unlike agriculture, which is largely consumptive[1], industries consume only 15 to 20 per cent of the water they use and discharge the rest as effluents. If the effluents could be treated and reused, the stress on the available fresh water resources (which are often scarce in the arid and semi-arid areas) is reduced. This is often described as a 'win-win' situation. However, if the effluents pollute the receiving water body or the groundwater, even existing fresh water resources become unusable, i.e. pollution puts greater stress on the limited water resources of a basin or region. This basically means that the other sectors, such as agriculture, drinking water or fisheries, are affected. Pollution, thus, is not merely non-compliance with environmental laws or standards; it also creates a situation where a user is adversely affected, i.e. a case of environmental damage.

Table 1. Case Studies of Water Quality Conflicts

No	Basin	State	Source of pollution	Affected sector (s)	Action taken
1.	Musi River (Hyderabad)	Andhra Pradesh	Industries (CETP) Sewage	Farmers Residents	Protests; Court Case Pending
2.	Noyyal River (Tiruppur)	Tamil Nadu	Textile Industries (CETP)	Farmers; Water Supply	Court case pending; Loss of Ecology Award of Compensation
3.	Palar Basin	Tamil Nadu	Tannery Industry (CETP)	Farmers; Water Supply	Supreme Court Decision; Loss of Ecology; Award of Compensation; MSD
4.	Kolleru Lake	Andhra Pradesh	Aquaculture Agricultural Runoff; Industrial Effluents	Fishermen Ecosystem	Wildlife Sanctuary
5.	Pandasozhanallur Village	Pondicherry	Berger Paints (ETP)	Groundwater Pollution (Irrigation, Drinking Water)	Complaint to PWD
6.	Kolhapur (Chipri Village)	Maharashtra	Ghodavat Oxalic Acuid Plant (ETP)	Health; Drinking Water	Protest; Factory closed and Reopened; Factory now closed
7.	Kanpur (Ganga Basin)	Uttar Pradesh	Tannery Effluent Sewage (CETP & STP) Sludge	Irrigation Drinking Water Health; Ecology	PIL; Ganga Action Plan
8.	Hootgalli Village (Mysore)	Karnataka	Chemical Units		Alternative Supply; Protests by villagers
9.	Arkavali Subbasin	Karnataka	Textile Units Chemicals	Irrigation Fisheries	PCB orders closure;

Table 1 contd...

Table 1 contd...

					Appellate Authority dismisses appeal by industry
10.	Chaliyar Basin	Kerala	Grasim Industries (Rayon/pulp)	Health; Drinking Water	Factory closed
11.	Eloor Island (Periyar Basin)	Kerala	Insecticides & other checmical units (247) (Hazardous wastes)	Health; Fisheries Ecology	Representations; Supreme Court Monitoring Committee Local Area Environmental Committee
12	Khari River (Sabarmati Basin)	Gujarat	Three Industrial Estates (CETPs) Discharge to river, canal and aquifer	Health; Water Supply Agriculture Livestock	Representations, Petitions, PIL Fund for villagers Diversion of Narmada Water; High Power Committee; Stakeholder Forum

In western countries, pollution is viewed mainly as an ecological problem as the ecology of rivers, lakes, etc. is affected. In India, it is also an economic problem because pollution affects the livelihood of farmers, households and fishermen since different water users live in close proximity and a substantial proportion (80 to 90 per cent) of our water resources is used for irrigation. Consequently, the effect of water pollution is most often immediately felt in agriculture. Moreover, when effluents are discharged on land (rare in western countries), groundwater gets polluted and the effects are felt in agriculture and drinking water. In rural areas drinking water comes mainly from groundwater sources, which need to be protected from pollution. Consumption of contaminated water may damage health. However, the impact of pollution on health is difficult to prove due to other complicating factors.

Most water quality conflicts are inter-sectoral conflicts; conflict rarely occurs within the same sector. Intra-sectoral conflicts, however, are common in water quantity conflicts; for example, head and tail-end farmers in an irrigation system)[2]. The inter-sectoral dimension makes conflict more difficult to resolve. Industries are run by managers and engineers who usually have little in common with farmers or fishermen. Often, they may not be from the same area, not even from the same state[3]. It is therefore very difficult to create a common forum for conflict resolution—although some attempts have been made in the Palar and Sabarmati basins.

It should also be noted that most of the conflicts are in the semi-arid areas of the country (peninsular and western India) where water is scarce (the Kanpur and Kerala case studies are exceptions). There are two reasons—one is that industrialisation occurs rapidly in areas where the potential for irrigated agriculture is limited. Second, the effect of pollution is more apparent where the water resources are scarce. In areas where there are abundant water resources, it is possible to dilute pollution (if it is of the non-toxic kind), or it may be possible to provide an alternative water source as in the Kanpur (Ganga) and Mysore cases (Kaveri). When there was a diversion of Narmada water to the Khari river in Ahmedabad, the problem of pollution was solved to some extent.

The ecological dimension of the conflict as such has been raised only in a few cases—the Kolleru lake (Andhra Pradesh), designated a wildlife sanctuary is the only case study, which could be truly classified as an ecological conflict[4]. The Supreme Court decided on the Bhavani river case (not included in this compendium), mainly on ecological grounds. In both the Palar and Noyyal cases, the Loss of Ecology Authority is engaged not only in the compensation of the victims of pollution, but also in attempting to reverse the ecological damage through remedial and clean-up action.

One of the striking features of most of these case studies is that they describe the impact of pollution on other water users, and the protests or legal measures which the victims have taken to redress the problem. The role played by the respective State Pollution Control Boards is sometimes mentioned. But, the viewpoint of the industries is almost never ascertained. Some of the case studies admit that the concerned industries may be major employers and that the local economy may be adversely affected if they close. Why have industries failed to ensure that their effluents are properly treated? The implication seems to be that industries are interested only in their bottomline profits, and lack social or environmental consciousness.

It is interesting to note that in virtually every case, the industrial effluent is being treated in an individual or a common effluent treatment plant before it is discharged. The complaint is that the treatment is not satisfactory, whereas the industry and the respective Pollution Control Board claim that the effluent standards are met. This divergence in views could be due to the following reasons:

• The treatment plants are constructed, but not operated in order to save operation and maintenance costs.
• The treatment plants are not designed to treat certain pollutants—total dissolved solids, in the case of the tannery and textile effluents in the Palar and Noyyal basins in Tamil Nadu.
• The effluents may meet the permissible standards, and yet cause environmental damage. This reason is peculiar to the Indian context. Discharge standards in respect of surface water bodies is based on the assumption of a certain amount of flow in the river. In semi-arid parts of India, there may be no flow for several months in a year. Consequently, there is no opportunity for the pollutants to be assimilated, and often the so-called flow observed in a river or canal is largely treated effluent. Obviously a downstream user cannot use this as fresh water. Or, the flow may be too small to assimilate the effluent. This is a difficult situation because both parties are technically correct. The industry might be meeting the standard, yet the downstream user may still be adversely affected.

Discharge on land is even more complex because the pollutants accumulate in an aquifer, unlike surface water flows which carry away the pollutants. The transport of pollutants in aquifers is complex and needs to be modelled to understand the impact on the environment. It is unfortunate that land disposal of effluents is permitted in India, ostensibly for irrigation; it often irreversibly pollutes the groundwater. It is very difficult and expensive to decontaminate aquifers once they are polluted. Another problem mentioned in some of the case studies is the disposal of sludge. The treatment plants may clean the effluent streams, but generate sludge which often has to be disposed of in a landfill, otherwise it will again get washed into the river/canal and cause downstream damage, negating the whole purpose of treatment.

The most serious environmental problem is that of hazardous substances, as in the Eloor Island case study in Kerala. Some of the units produce pesticides that come under the category of persistent organic pollutants (POPs), which are hazardous substances regulated under the Stockholm Convention to which India is a signatory. The Supreme Court Monitoring Committee has taken the industries and the Kerala State Pollution Control Board to task for not controlling pollution. A Local Area Environment Committee has been set up to monitor compliance.

In two of the case studies, the industries—Grasim Industries in Kerala and Ghodavat Industries in Kolhapur—have been forced to close down due to local pressure. This was also the case with South India Viscose in the Bhavani basin. The Coca-Cola plant in Plachimeda in Kerala is also temporarily closed. In all these cases, the question to be asked is whether closure is the solution. Can industries not co-exist with agriculture and other water users? What is the long-term solution?

Legal and Institutional Aspects

Water pollution is regulated under the Water Pollution (Prevention and Control) Act of 1974 and the Environment Protection Act (E. P. Act) of 1986. Hazardous wastes are regulated under the rules promulgated under the E. P. Act. The E. P. Act has very stringent penal provisions and managers can be sent to jail for non-compliance, water and power supply to the unit can be cut off, etc. Under the Water Cess Act, industries have to pay a cess on the basis of their water consumption and the purpose for which the water is used. The Water Pollution Act of 1974 provides for the creation of State Pollution Control Boards (PCB) which can issue a consent order, which has to be periodically renewed, to both establish and to operate a plant. The consent to operate can be withdrawn if the unit does not comply with standards and a show-cause notice can be issued to the plant. Generally, based on a court order, the PCB has the power to close the plant if it is not in compliance. Under the E. P. Act, citizens too can file a case against the PCB after giving it due notice if they feel that the Board has not taken appropriate action. NGOs and other environmental organisations can also file a public interest litigation on behalf of a large number of affected persons.

The courts, particularly the Supreme Court, have taken water pollution issues very seriously and have rendered many landmark judgments. The Vellore Citizens Forum Case (Palar basin) was one such judgment which led to the creation of the Loss of Ecology Authority. In this case, the court upheld both the precautionary principle (of abating pollution) as well as the 'polluter pays' principle (of compensating victims of pollution). In the Palar and Noyyal cases, the Loss of Ecology Authority has forced the industry to create a fund to compensate victims, as well as to reverse the ecological damage. In the Khari case, the Gujarat High Court also interpreted the 'polluter pays' principle in a similar way and ordered the units to contribute to a fund, which would be used for the socio-economic uplift of the villages. However, unlike the Loss of Ecology Authority, the fund will not be used to compensate individual victims of pollution.

The setting up of a High Power Committee under the Chief Secretary as well as a Stakeholders' Forum are the institutional measures taken to deal with the problem in Gujarat. The multi-stakeholder dialogue promoted by the Madras Institute of Development Studies was an important step forward in the Palar case. However, the sustainability of such an initiative is possible only if government agencies, including the Palar River Basin Board, take an active interest in it. Similarly, the Local Area Environment Committee in the

Eloor case is an attempt to bring in the local stakeholders. The Madras High Court has appointed an Expert Committee to look into the Noyyal problem, particularly the cleaning up of the Orathapalayam reservoir. Thus, there have been civil society initiatives as well as appointment of expert/local area committees to help resolve the conflicts. It is now increasingly recognised that civil society can play a useful role both in monitoring compliance and resolving conflict. Ironically, it was the implementation of the Ganga Action Plan of the Ministry of Environment and Forests to clean up the river which itself led to the conflict with the Jajmau villages near Kanpur[5]. This has led to a very complex institutional tangle regarding who is responsible for the problem. The agencies concerned and the courts have to sort it out.

One of the major legal issues in water quality conflicts stems from the fact that water pollution is a criminal offence. In other words, if an industry even marginally violates a single effluent standard out of the numerous water quality parameters, then it is technically not in compliance and could lose the consent to operate. Moreover, the courts can impose penal provisions if they choose to do so. The draft National Environmental Policy Act of 2004 recommends that the pollution laws be changed to a mix of both civil and criminal provisions. The problem with the criminal provision is that the industry and the Pollution Board become less transparent, and it provides opportunities for rent-seeking behaviour. NGOs, on the other hand, view the recommendation as a dilution of the existing provisions aiding the polluter.

When an industry has a genuine problem of meeting effluent standards, it should be in a position to seek the available technical expertise in the country to help it rectify the situation. Research institutions in the country will have to come forward with the necessary technological options, and financing institutions with the finance needed for this purpose. The government provided subsidies for common effluent treatment plants (CETP) to enable small industries to take advantage of the economies of scale of effluent treatment. It would also be convenient for the pollution boards to monitor a few CETPs rather than a large number of individual effluent treatment plants. Unfortunately, for a variety of technical and institutional reasons, the CETPs have rarely been effective. In most of the cases discussed here, CETPs have been constructed but have not solved the problem. In some cases, the CETPs may actually have worsened the problem by concentrating the pollution at one location, as it did in the Kanpur case.

One of the major lacunae in our environmental legislation is the emphasis on effluent standards. Pollution boards rarely study the impact of the industry on the ambient environment. For example, if there is no flow in the river, the effluent standard does not ensure that the ambient environment is protected. The land-based discharge standard (for irrigation) has ended up in pollution of the groundwater. The goal of our environmental laws should be to protect the quality of the environment. Unfortunately, the 'instrument', namely effluent standards, has become the goal, even when the standards are totally ineffective in protecting environmental quality. Most of the case studies testify to the gross failure of our institutions to protect water quality and the livelihood of individual water users who rely on the common water resources in the respective basins. Alternative ways have to be explored to resolve these problems.

Policies for the Future

a. Legal Framework

The concept of conflict resolution is virtually absent in the case of water quality conflicts. The polluter is categorised as the guilty party subject to criminal prosecution if (s)he is not in compliance with effluent or emission standards. The legal system in India is unequivocal when it puts the burden on the source of pollution (usually industry, but could also be a municipality discharging sewage or an aquaculture farm). The benefit(s) provided by the economic activity are not considered relevant in any assessment of pollution. The cost of meeting effluent standards is also not considered, since most courts go either by the precautionary principle or the 'polluter pays' principle. For example, the Madras High Court issued an order that all the bleaching and dyeing units must install 'reverse osmosis' technology regardless of affordability in order to meet the total dissolved solids standard. Units which failed to do so would be closed down by the Pollution Control Board.

The recommendation in the draft National Environmental Policy (2004), that a mix of civil and criminal penalties be used, seems to be a step in the right direction. Currently, industries are not forthcoming about their problems since any admission of non-compliance means inviting criminal penalties. It also leads to rent-seeking behaviour—so much so that the general public has lost faith in the whole regulatory process, except for the judicial remedies. The courts are then overloaded with petitions and public interest litigations. It is clear that the judicial route is an expensive, time-consuming process which cannot be employed in every case involving pollution. Other institutional mechanisms have to be put in place, and judicial procedures used only as a last resort. In the present situation, affected parties immediately approach the court to seek redressal. Civil liabilities, such as fines, could be set high enough to act as a deterrent. Criminal penalties could be imposed when there are serious health implications.

b. Environmental Mediation

The United States was among the first countries to institute environmental mediation. The US Environmental Protection Agency found that in some cases local communities did not want their local industries to close down because the cost of environmental compliance was too high. In the case of non-toxic pollutants, it may be possible to relax certain standards if the impact on the ambient environment is not serious. Such an approach becomes essential with non-point source pollution, such as agriculture, cattle yards, construction, urban runoff, etc. One has to examine the benefits and costs of enforcing certain standards.

Mediation has been suggested in other spheres of life, such as property disputes and other civil litigation. If environmental disputes could be considered under civil law, it may be possible to use the services of a mediator (individual or institution) to resolve such conflicts. Even if there were a large number of people affected by pollution, they could be represented collectively in the mediation process.

c. Voluntary Compliance

European countries are stressing the need for industries to comply voluntarily without pressure from regulators. Large industries in India already realise the benefits of having a 'green' or clean image, including the impact on their stock price. There might also be spin-off benefits, such as reuse of valuable waste products (chrome recovery in tanneries), savings on water and chemicals, better quality of output, etc. In some cases these benefits/savings may offset the cost of compliance. Industries in the export business (leather, textiles) may lose their markets if the importing countries find that they are not in compliance with domestic environmental regulations. Eco-labelling, ISO 14000 and other initiatives are increasingly being adopted by industries.

Water quality conflicts are bound to increase in the future with higher industrial growth rates. The quantity of effluents generated will undoubtedly increase in the coming years. While the policy framework may need to be changed to permit mediation, industries must also comply voluntarily to ensure that they do not cause damage to the communities and the environment in which they are located. In the long term, closure is not a desirable option for anyone.

Notes

1. Almost 70 to 80 per cent of the water delivered to its root zone is taken up by a crop for evapo-transpiration, and only the balance is discharged as drainage or return flow.
2. There are some exceptions, such as the Noyyal case, where the pollution of the local groundwater by the textile industry has made the water unsuitable for use by the industry itself.
3. Sometimes the conflicts have taken a communal colour, when industries like tanneries are run by the minority community (Palar and Kanpur cases).
4. Kolleru is a complex case of a lake affected by multiple sources of pollution— aquaculture, agriculture, sewage and industry.
5. These villages, which were previously using sewage mixed with fresh water for irrigation, now find that industrial wastes from the Kanpur tanneries are combined with the sewage that they receive from the effluent treatment plant.

A Toxic Hotspot on the River Periyar in Kerala: Corporate Crimes in God's Own Country

M. Suchitra

Troubled Island

ELOOR, A TINY 11.21 SQ KM ISLAND-VILLAGE ON THE RIVER PERIYAR, IS SITUATED 17 KM north of the port city of Cochin, Ernakulam district, central Kerala. The village, with a population of 40,000, is home to the largest industrial cluster in the State—the Udyogamandal Industrial Estate. There are 247 industrial units including the only DDT (Dichloro Dipheny Trichloroethane) producing factory in India (Hindustan Insecticides Limited). Of these, 106 are chemical units that manufacture fertilisers, pesticides, petro-chemicals, rare earth elements, rubber processing chemicals, zinc/chrome products and leather products. Most of the units have been operating here for the last fifty years and use outdated and highly polluting technologies.

A Disaster Waiting to Happen

The resulting pollution has caused extensive damage to soil, water and air in and around Eloor, threatening the health and the very existence of its people and ecology. The factories use large amounts of freshwater from the Periyar and discharge untreated toxic effluents into the river. Gas emissions range from acid mists, sulphur dioxide, hydrogen sulphide, to ammonia and chlorine. In 1999, Greenpeace declared the Eloor industrial area one of the top toxic hotspots in the world, all its water resources and wetlands having become highly contaminated.

The Periyar, the largest river in the state and the drinking water source for approximately 40 lakh people in Cochin and nearby Aluva town, is alarmingly poisoned. The river often dons different shades—red, orange, green and yellow. Studies reveal that downstream stretches of the river are almost dead.

For several decades now, the residents of Eloor and neighbouring villages have been agitating against the factories under the banner of the Periyar Malineekarana Virudha Samithy (PMVS). Their demands include environmentally

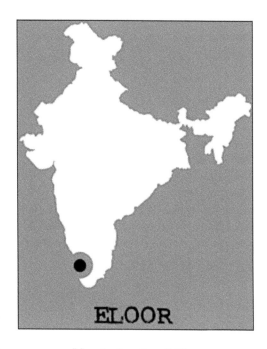

Map 1. *Location of Eloor.*

Fig. 1. *The global toxic hotspot.*

safe production at the industrial estate, zero discharge into the river and compensation/medical rehabilitation for all the people affected by industrial pollution. They also want the environmental impact to be assessed immediately, and have been asking for a comprehensive chemical disaster management plan for Ernakulam district.

Ills of Pollution

A study conducted by Greenpeace on the occasion of the fifteenth anniversary of the Bhopal gas tragedy revealed that incidence of death and disease has increased among the residents here. Titled 'Status of Human Health at Eloor Industrial Estate, Kerala', the document points out that the chances of Eloor residents contracting cancer are 2.85 times higher compared with the less polluted Pindimana village in the same district, upstream Periyar; children are at a higher risk (2.63 times) of malformation due to congenital and chromosomal aberrations; chances that children may die due to birth defects are 3.8 times higher; chances of death due to bronchitis are 3.4 times, and of death due to asthma, 2.2 times higher. Approximately 80 per cent of the people are suffering from respiratory diseases. Also, incidents of miscarriage and congenital defects have increased.

Watery Grave

Pollution has devastated the farms and the aquatic life in the river. The fisherfolk complain that their catch is drastically reduced. Many women fish-hawkers have become housemaids as there is not enough fish to sell. Often massive fish kills occur in the river due to the toxicity of the water. An analysis of three species of edible fish has revealed significant radioactive contamination due to thorium. Sixteen species of fish are believed to have disappeared from the Periyar within a span of fifty years.

A team of scientists from Cochin University of Science and Technology (CUSAT) studying the region say that Eloor's ecology and environment have almost reached moribund stage. Not only has the health of the population undergone a drastic decline, but fish species in the river have come down in the last twenty-five years from thirty-five to twelve. The average catch from this part of the river has come down from 2.7 to 0.33 kg/hr.

Fig. 2. *Polluted water in the Eloor.*

Deadly Dangerous Toxin

Hindustan Insecticides Limited (HIL), a Government of India enterprise, has been manufacturing pesticides including DDT and endosulfan at its Eloor plant since 1956. The plant, located adjacent to a wetland, apparently discharges effluents into an open creek. The water samples from this creek contained more than a hundred organic compounds, thirty-nine of which were organochlorines, including DDT and its derivatives, endosulfan and several isomers of hexachlorocyclo hexane (HCH).

Persistent organic pollutants like DDT and related compounds are of particular environmental concern because not only are they toxic, but are also highly resistant to degradation and also liable to bioaccumulate. DDT is the most notorious of the twelve chlorinated chemicals identified for total elimination by the Stockholm Convention on Persistent Organic Pollutants, 2001. HIL is one of the few remaining DDT manufacturers in the world. Traces of DDT and its derivatives have also been detected in the wetlands surrounding the industrial estate. Sediments collected from the HIL site contained high levels of heavy metals like cadmium, chromium, copper, mercury and zinc.

A recent study conducted as part of a global campaign by public interest groups to pressurise respective governments to implement the decisions of the Stockholm Convention has revealed that concentration of dioxins and DDT in chicken eggs collected from the HIL neighbourhood is four and three times higher respectively, than European Union (EU) norms. The level of hexachlorobenzene (HCB) in the sample is seven times higher than EU norms.

No Escape

The location of the factories is such that they trap the entire island. The only bridge to the mainland is far from the residential area. In the event of a chemical disaster similar to Bhopal, it would be impossible for the residents to escape. In July 2004, when a massive fire gutted the endosulfan plant of Hindustan Insecticides Limited, emitting toxic gases into the air, hundreds of people rushed to the ferry point to cross the river but there was no late-night boat service. The factory did not have any off-site emergency plan. Many people were hospitalised with complaints of giddiness and nausea. In short, there are no arrangements to evacuate people in the event of a disaster.

Fig. 3. *Trying to cope with the toxic waste.*

Government Plays the Fiddle

Representations submitted by activists and NGOs to various government departments have gathered dust while the industrial units have persistently denied allegations of polluting the river. The factories have been insisting that all necessary steps have been taken to check pollution. The Kerala State Pollution Control Board (KSPCB) too has consistently maintained that a majority of the companies have adhered to environmental standards on waste water discharge into the river. The Board claimed it had succeeded in getting most of the factories to install effluent treatment plants and that they have been regularly monitoring the quality of water near the industrial estate. In 2001, people protested against the industries by blocking a stream that carried waste into the river. The conflict will only be resolved if the factories move towards non-toxic raw materials, safer products and processes. But there is no denying the total absence of industrial planning strategy. How was an industrial estate allowed to mushroom in a thickly populated area, that too on the banks of the largest river in the State? Why doesn't the government or pollution control agencies take into consideration the people's right to live, to have safe food, water and air?

Enough is Enough

In August 2004, the Supreme Court Monitoring Committee (SCMC) on Hazardous Wastes assessed the situation in Kerala and observed, 'The ground realities in Kerala are terrible. Kerala is one of the States that have miserably failed to act on hazardous wastes. The PCB has wilfully and callously disregarded the Supreme Court's orders on hazardous wastes'. The SCMC found that several industrial units had been operating without the authorisation required by the Hazardous Waste (Management and Handling) Rules 1989. It also found that provisions of the Air (Prevention and Control of Pollution) Act and Water (Prevention and Control of Pollution) Act are being openly flouted. The SCMC threatened the KSPCB officials with contempt of court unless effective action was taken.

In October 2004, the SCMC completed a second round of review on the facilities available within the industrial units for the treatment, storage and disposal of hazardous wastes, and observed that both the state governments and the PCB have totally disregarded the directives of the committee. In a bid to inject transparency and make the process of pollution control participatory, the SCMC set up a Local Area Environmental Committee (LAEC) with representatives from the PCB, industry association/industrial units and local environmental groups. The monitoring committee further allowed the State government six more months to set up a landfill. The factories too have been asked to set up proper treatment, storage and disposal facilities.

A Small Beginning

Since November 2, 2004, the LAEC carried out an environmental audit of all the 247 factories in Eloor. It completed checking over seventy factories for raw materials, products, production process, waste generation, compliance with environmental laws, unauthorised disposal of wastes, etc. The Committee submitted its Environment Impact Assessment (EIA) Report (2004–2006) to the SCMC in March 2006. The EIA report gives a very deplorable picture of the Eloor-Edayar Industrial Estate and demands immediate action to arrest further deterioration. There are 25 general recommendations including a Comprehensive Chemical Disaster Management Plan involving the district administration, the factories, the KSPCB, the local bodies and the NGOs. The report recommends that the KSPCB should endeavour to achieve zero discharge by the industries within three years. It also recommends to revise the parameters issued to each chemical industries in this region. The report points out that the parameters mentioned in the consent letters given by the PCB to the industries fall short of many important factors. The LAEC has recommended immediate action including closure against some industries which keep on discharging pollutants indiscriminately even after being warned many times. The LAEC has asked the KSPCB to ensure water supplies to the affected families in five wards of the Eloor Gram Panchayat by the companies identified as the culprits for groundwater contamination. The report recommends to provide the affected people with free medical facilities. The LAEC's term was over by March 2006. Since there is no monitoring committee now, the industries continue to pollute the river.

However, there are signs of positive change. With the SCMC ultimatum hanging over their heads, some of the factories have introduced new technologies, though the LAEC is aware that many of the companies have taken up a new night-time activity—they have begun to discharge concentrated effluents directly into the river through illegal outlets. The committee also found out that there was no arrangement for stack monitoring in most units. The Kochi office of the KSPCB does not have adequate equipment or trained staff to do this; further, the PCB has been depending on private agency reports for data on air quality. The committee has pointed out that the Board is not willing to allot funds for the high-tech equipment needed for stack monitoring. On July 8, 2005, the Eloor region was filled with poisonous smoke, forcing the LAEC to recommend closure of FACT, an acid plant, since the emission was much beyond the approved parameters. The LAEC was concerned about recurring incidents of fugitive emission.

More Power to People

Eloor is a glaring example of corporate crimes against neighbourhood communities. It also highlights the apathy of the authorities towards the protection of ecology and people. There is an obvious imbalance of power between people and industry. While people have been protesting for decades, the industry, the government, and the pollution control agencies have all ignored their demand for justice until threatened by a high-power judicial committee.

There are stringent laws that regulate the handling and transport of hazardous material; provide for the relocation of industrial plants that are based in populated areas; bestow on workers the right to demand information about health and safety at work; oblige employers to disclose to the public all information pertaining to the danger of chemical operations, health hazards, and measures to overcome such dangers; and have in place an up-to-date and adequate on-site and off-site emergency plan in case of a chemical disaster. In the case of Eloor, all these regulations were flouted.

Even though the SCMC's suggestion to set up a landfill to dump waste from all the industries is certainly a well-intentioned intervention, it seems to be a simplistic solution, especially taking into account the fact that landfills typically develop leaks.

A continuous dialogue has to be initiated between labour organisations and community groups. Trade unions, in general, tend to view the people's struggle as being against the workers' interests; they are yet to realise that the people's struggle for a clean environment and their own fight for a safe workplace are one and the same. It is time that the unions began to look beyond old agendas like fair wages and working hours, and joined hands with the community to fight for a better tomorrow.

Note

*Map 1 and figures from the Periyar Malineekarana Virudha Samithi.

References

Action Plan for the Reduction of Reliance on DDT in Disease Vector Control. World Health Organization, 2001.

Corporate Crimes- The Need for an International Instrument on Corporate Accountability and Liability Bhopal Principles–The Indian Cases. UK: Greenpeace, August 2002.

Joseph, M. L. *The Declining Trend of Biodiversity and Fish Production in Consequence of Pollution in the Lower Reaches of Periyar River*. Kochi: St Albert's College, Department of Zoology.

Kurup, B. Madhusoodana and T. G. Manoj Kumar. *Index of Biotic Integrity of Eloor and Edathala Industrial Zone of the River Periyar and Management Plans for the Restoration of Fishery Health*. Cochin: School of Industrial Fisheries, Cochin University of Science & Technology.

Labunska, I., A. Stephenson, K. Brigden, R. Stringer, D. Santillo, P.A. Johnston, and J.M. Ashton. 1999. *Toxic Hotspots: A Greenpeace Investigation of Hindustan Insecticides Ltd, Udyogamandal industrial estate, Kerala, Persistent organic pollutants and other contaminants in samples taken in the vicinity of the Hindustan Insecticide Ltd plant*. University of Exeter: Greenpeace International.

Status of Periyar's Health at the Eloor Industrial Estate, Kerala, India. A Compilation of New Evidence from the Periyar Riverkeeper and Greenpeace Research Laboratories, September 2003.

Stockholm Convention on Persistent Organic Pollutants. United Nations Environment Programme, 2001. Accessed from *http://www.pops.int/*.

Stringer, R., I. Labunska, and K. Brigden. 2003. *Pollution from Hindustan Insecticides Ltd. and Other Factories in Kerala, India: A Follow-up Study*. University of Exeter: Greenpeace Research Laboratories.

www.greenpeace.org

www.keralapcb.com

Bidding Farewell to Grasim:
Lessons that Remain Unlearnt

Abey George and Jyothi Krishnan

Grasim is Invited to Kerala

THE RICH FOREST WEALTH AND RIVERS OF KERALA THAT HAVE ATTRACTED PEOPLE FROM all over the world have long sustained its ecology and economy. In the past, foreign traders, colonial administrators, as well as local kings and chieftains have exploited this natural wealth to consolidate their power. After independence, the already degraded natural resource base was targeted for various developmental

Map 1. *Location of Kerala in India.*

activities. In tune with the national objective, the Government of Kerala headed by the late E.M.S. Namboothiripad, decided in 1957 to utilise the natural resource wealth of Kerala for industrial development. The motive was political and meant to rebuff notions that this first communist ministry in power in the country would repel prospective industrial investors and cause industrial stagnation in the state. The government thus invited the Birlas, one of the largest industrial houses in the country, to set up a pulp and rayon manufacturing factory in Kerala.

Grasim and the River Chaliyar

The Gwalior Silk Manufacturing (Weaving) Co. Ltd., also known as Grasim, was set up in 1962 at Mavoor, 20 km east of Kozhikkode city, and continued to function there until 2001. It provided direct employment to about 3,000 and indirect employment to about 10,000 people. The primary raw materials consisted of forest produce such as reed, bamboo, eucalyptus and water from the river Chaliyar. The government had agreed to provide it with forest resources and electricity at highly subsidised rates,[1] along with 'free' water from the Chaliyar. The factory produced rayon-grade pulp using up more than 50,000 m³/day of water from the river. The factory effluents released into the river resulted in severe water pollution, which affected the livelihood of a large section of people while the gaseous effluents became a source of air pollution.[2]

Water and air pollution were so severe that there were local protests in 1963 itself, the year the factory commenced operations; agitation continued until the factory closed down in 2001.[3] Local people and environmental activists alleged that the air and water pollution caused by the factory was responsible for the high incidence of cancer and respiratory diseases in nearby Vazhakkad and Mavoor *panchayats*.[4] Independent studies conducted on this issue lent further credence to the widespread suspicion about the factory's role in the disturbing morbidity pattern.[5]

The Conflict over Water and Forests

The primary conflict precipitated by the Grasim factory was over the use of common pool resources, namely water, air and forest. The intense State-wide industrial extraction of forest resources affected the catchment areas of many streams and rivers, but it was the issue of water pollution that took centrestage in the long-drawn-out battle against the factory, perhaps because it was the most visible and directly felt trauma.

The Story of Forest Extraction

The signing of the initial agreement with the factory on May 3, 1958, committed the government to supply it a minimum of 1.6 lakh tons of bamboo per year from the Nilambur forest division. Subsequent negotiations led to a gradual increase until a considerable area was dedicated to bamboo extraction. Within the first ten years, the entire bamboo stock in the assigned areas was depleted. Eucalyptus replaced bamboo. The high elevation *shola* grasslands of the Western Ghats were thought to be 'wastelands' and converted into eucalyptus plantations.[6] By the 1970s, almost the entire bamboo and eucalyptus plantations in the state were opened up for extraction by Grasim industries. Extensive bamboo extraction and simultaneous deforestation for eucalyptus monoculture had a serious impact on the already fragile and fragmented forests of the Western Ghats in Kerala. The government's commitment to supply raw material to Grasim ended up removing the vegetal cover in the catchment area of many rivers, reduced river flow and resulted in water scarcity in the state.

The Slow Death of the Chaliyar

The Chaliyar river was the source of drinking and domestic water supply for the people living along its banks. It was known for its aquatic wealth and was the lifeline for hundreds of fishermen. The river was the primary mode of transport during the 1960s and 70s, and a large number of people earned their livelihood by plying country boats that carried goods, people, and even mobile tea and grocery shops. The release of toxic, viscous effluents changed the colour of the river and also gave rise to a foul smell. On many occasions, a large number of dead fish were found in the river.

The reduced flow during the summer months further increased the concentration of pollutants, and saline water intrusion in the Chaliyar created a cesspool of pollution, leading from Mavoor right up to the river mouth. Finally it began to threaten the water intake point for the factory itself. Even the drinking water pumping station of the Kozhikode corporation, located further upstream at Koolimadu, began to face the threat of pollution. The company sought to resolve this problem by constructing a temporary bund across the river between the intake and outlet point, at a place called Elamaram. This was done with the sole intention of protecting the company's water source. However, the construction of this bund aggravated the already severe problem of water pollution, as the wall blocked the meagre flow of water. It also affected the people living downstream of Mavoor, who relied on the river for drinking water, bathing and irrigation.

People Take Action

In 1987, when the problem assumed serious dimensions, hundreds of local people and activists of the Jala Vayu Samrakshana Samithi (JVSS) gathered at Elamaram and demolished the bund in the presence of the police and government officials. This was one of the most notable events in the anti-pollution struggle. As the polluted waters reached the intake point of the factory, the company was forced to suspend operations. However, even this violent protest proved ineffective when it came to persuading the factory management or the government to take strong measures to address the problem of air and water pollution.

The Ramanilayam Agreement was signed between the Grasim management, the government, trade unions and representatives of the anti-pollution groups on December 16, 1974, and can be considered the most significant initiative taken by the government. However, even this did little to reduce pollution; it only shifted the effluent outlet point from Mavoor to Chungapally, which is 7 km downstream, by a pipeline laid out specifically for this purpose.[7] Even such mutually agreed upon, minimal mitigative measures were not taken seriously by the company management and the pipeline was finally laid only after six years of prolonged protests.

In 1974, the passing of the Water Act led to the setting up of the Kerala State Pollution Control Board (KSPCB). However the Board, due to various institutional problems, was ineffective in assessing and controlling pollution. Timely action was not forthcoming despite numerous research studies conducted by institutions such as Calicut University, Calicut Medical College and other independent bodies which confirmed the pollution of the Chaliyar water.

The Struggle to Protect Air and Water

The most important features of the anti-pollution struggle were that it was initiated by local people under the leadership of local leaders such as the late K. A. Rehman, and that there was only a thin line separating the victims from the beneficiaries within the affected areas (most importantly, Mavoor and Vazhakkad *panchayats*). A vast majority of the company's casual workforce belonged to the affected *panchayats* and the company employed many close relatives of the victims, as a result of which the local agitators found it difficult to take a tough stand or to demand closure of the factory.

K.A. Rehman's role was a unique feature of this conflict. One of the initiators of the struggle, he consistently fought the factory till he succumbed to cancer in 1999. When many local leaders were dissuaded from taking the protest forward, Rehman continued to put pressure on the factory. Later, as President of the Vazhakkad *panchayat*, he initiated a health survey that revealed the high incidence of cancer and respiratory diseases in the surrounding areas. It was an eye-opener on the implications of the air and water pollution precipitated by Grasim.[8]

Despite all this, the JVSS, led by Rehman, never demanded the closure of the factory. Rather, they asked for more accountability and transparency in the process of pollution control. Only after repeated failure by the management to enforce agreed upon pollution control measures, and the inability of the government and the KSPCB to get it to do so, did the JVSS call for closure of the factory.

The struggle centred on the issue of air and water pollution and did not address the larger issue of raw material extraction which had equally important social and ecological consequences across a much wider geographical area. An alliance between local protestors and affected people from areas where raw material was being extracted was not adequately explored. Such an alliance would have strengthened the movement and also exposed the social and ecological implications of industrial resource extraction. In such a scenario the closure of the factory would not have been viewed as the end of the conflict; on the contrary it would have helped review similar ventures such as the Hindustan Paper Corporation at Velloor in Kottayam district, which also exploits forest resources and pollutes the Moovatupuzha river; the proposed revival of the rayon factory at Perumbavoor in Ernakulam would also have come under the scanner.

The government, the company management and the trade unions prefer to forget that the factory precipitated severe air and water pollution and contributed to the incidence of cancer and respiratory diseases in Vazhakkad and adjoining areas. Hence, it is not surprising that in the final rounds of negotiations that preceded the closure, none of the parties involved discussed critical issues related to pollution, its impact on health or the compensation owed to the victims of pollution. Given such indifference, it seemed unlikely that the company or the government would think about restoring the damaged ecosystems.

Conflicting Positions

The company, the government, the trade unions, the JVSS and environmental groups like the SPEC and KSSP, were the parties involved in the issue. The company's position was that it had been invited to manufacture pulp and rayon, to provide employment and to make profit, and that some amount of pollution was inevitable in the process. The trade unions argued that technology upgradation would reduce pollution. However, the price of improved technology would have decreased the economic viability of the venture and therefore the management was not interested in such alternatives. With the threat of unemployment looming large in the event of closure of the factory, the trade unions were willing to settle for any compromise, especially on the issue of pollution. Due to the influence exerted by their respective trade unions, both the CPM-led Left government and the Congress-led Right government turned a blind eye to the issue of pollution. The victims of pollution were not organised enough to put sufficient pressure on the government, company management or the trade unions. The Forest Department and the State Pollution Control Board were always soft on the company. Not only were raw materials made available at an incredibly cheap rate, no procurement procedures were put in place to monitor their extraction. Ironically, the same forest department was strict about the illegal extraction of bamboo by tribals and others, but it was only in 1976, eighteen years after signing the agreement with Grasim, that the government formulated specific rules and regulations regarding the allotment, felling, collection, pass system, auditing and verification with respect to the bamboo supplied to it. Until then Grasim had uncontrolled and unregulated access to the bamboo resources in the State.

Three lakh traditional artisans including tribals from various parts of the Western Ghats who depended on bamboo and other forest produce for their livelihood were completely forgotten. In spite of the fact that raw material extraction had a widespread ecological impact, there was no attempt to organise the affected people in this sector, or to represent their plight before the government. The same government and political parties who had expressed concern about the loss of employment of 3,000 odd factory employees in the event of closure of the factory were silent over the impact of deforestation on the livelihood of traditional artisans.

What Did We Learn from the Grasim Story?

The thirty-eight-year history of Grasim has thrown up a number of important issues: the impact of industrial use of natural resources, the polluting nature of pulp and rayon manufacturing, and the impact of pollution on health. According to one assessment, the total subsidy the company enjoyed since its inception in 1962 was a whopping Rs. 28,936 crore. But even this atrocious sum does not cover the cost of the irreparable damage caused to the riverine and forest ecosystems, and to wells and paddy fields in the area.

Cancer took away more than 200 lives in the pollution-affected area. The factory management may not like to see a correlation between their activities and such diseases, but can society turn a blind eye to this issue? What about the crucial role of the Pollution Control Board, medical colleges, the Regional Cancer Centre and other research institutions in understanding this relationship? Mavoor is not an isolated example. There are many Mavoors in the making in Kerala—industrial pollution in Eloor in Ernakulam district, contamination and depletion of groundwater reserves by the Coca-Cola factory in Plachimada in Palakkad district, the iron smelting units in Kanjikode in Palakkad, the pollution caused by Hindustan Newsprint Limited in Velloor in Kottayam district—all these examples of the problematic relationship between industrial pollution, ecosystem health and human life.

For how long can we, as a society, remain insensitive and pretend ignorance in this regard? Why are our governments not taking a proactive stance? Is it because the objective of industrial progress is too sacred to be subjected to scrutiny, since pollution is viewed as its inevitable component? That the government of Kerala never imposed pollution control as a precondition while renewing contracts for the subsidised supply of raw materials to the company, even when the anti-pollution struggle was at its peak, indicates this unfortunate reality.

The Grasim story demands a change in the way in which we view our natural resources. We should not consider forests as mere stockyards of bamboo and eucalyptus, or rivers as mere sources of water. The aim of setting up industries cannot be confined to employment generation and profit-making. The right to life implies the right to a dignified livelihood for all sections of society. More importantly, this right to life extends to all elements of the natural ecosystem as well. The Grasim story tells us that when the profit motive holds sway over all others, both the human and the natural world will be under stress.

Notes

*Map 1 from SOPPECOM, Pune.

1. The Kerala government had agreed to provide bamboo to the factory at the rate of Re 1 per ton, one of the major reasons why the company was inclined to set up the factory.

2. The pulp and fibre unit and associated units such as the carbon disulphide and sulphur dioxide plants of the Grasim Industries all contributed to the pollution.

3. Various groups like the Society for Protection of the Environment Calicut (SPEC), the Kerala Shastra Sahitya Parishad (KSSP), and the Prakruti Samrakshana Samity were involved in the struggle in different capacities. However, the affected people's organisation Jala Vayu Samrakshana Samithi (JVSS) was the main group involved in this struggle.

4. A fact that was revealed in local surveys.

5. One of the first studies carried out by the State Health Department in 1976 revealed that respiratory diseases, anaemia and skin diseases were on the increase in this area. Out of 2,343 people examined by the medical team, 856 (37 per cent) had contracted respiratory diseases and 515 continued to suffer from the same. Another study in 1981 by Dr T K Gopinathan of Calicut Medical College had reported that cardiopulmonary diseases affected 23 per cent males and 20 per cent females in the area.

6. As per the Administrative Reports of the Kerala Forest Department, the area of eucalyptus plantations increased from 266 hectares in 1966 to about 30,000 hectares in 1975.

7. According to the KSSP, this was "one of the most unscientific and illogical solutions one can think of to solve the problem of river water pollution".

8. Cancer death as per the death register of the Vazhakkad Panchayat

1990	1991	1992	1993	1994	1995	1996	1997	1998
10	12	14	16	21	21	23	18	32

Noyyal River Basin:
Water, Water Everywhere, Not a Drop to Drink

N. Jayakumar and A. Rajagopal

Multi-Coloured Water

NOYYAL RIVER IS A TRIBUTARY OF THE KAVERI. IT ORIGINATES IN THE VELLIANGIRI HILLS in Coimbatore district and meets the Kaveri at Kodumudi in Tamil Nadu. It is a seasonal river that flows during the northeast monsoon and the return flow from the Lower Bhavani Project (LBP) canal also drains into it. The river passes through Coimbatore, Erode and Karur districts of Tamil Nadu. It has twenty-three anicuts and twenty-eight system tanks irrigating 7,920 ha of land. Apart from canals and tanks, dug wells, and borewells are also used for irrigation in the area.

The Noyyal–Orathapalayam project is an important surface water project in the Noyyal river basin. It consists of a dam across the river near Orathapalayam village. The Orathapalayam reservoir was completed in 1991, and it irrigates 8,200 ha of land.

However, in recent years the irrigation sources in the basin have been affected by effluents discharged by dyeing, bleaching, and other units in Coimbatore and Tiruppur. The effluents affect not only the surface water but also the soil and groundwater. The water from the dam has not been used for irrigation since 1997 due to this contamination. Two major studies have been conducted on the status of surface and groundwater quality in the basin (Palanivel and Rajaguru, 1999; Environmental Status Report, Environmental Cell, Coimbatore, 2000). This report is mainly based on these studies and field visits to the area.

Pollution Cocktail

The main reason for water conflict in the region is the enormous growth of bleaching and dyeing units after independence. In 1941 there were only two such units but by 1997 the number had increased to 866. They consume a huge quantity (about 90 Mld [million litres per day]) of water and release around 87 Mld of effluents into the river. Over a period of time, there was a build up of 11.3 m of contaminated water in the Orathapalayam dam and the water could no longer be used for anything. Farmers in the nearby area were against the release of water since it affected their lands. The storage in the dam also affected groundwater sources in a 15 km radius, and the local people stopped using well water for either irrigation or domestic purposes. Since 1995, despite there being enough water, no one has been able to use it because of contamination. This has had a harsh impact on livelihoods in the region.

Available studies, particularly the one conducted by the Madras School of Economics in 2004 on 'Damage Assessment' in the basin, clearly show that the accumulated effect of pollution in this area is high. All groundwater studies indicate that open and borewells in and around Tiruppur exhibit high levels of total dissolved solids (TDS) and chloride, and that the available groundwater is not suitable for domestic, industrial or irrigation purposes (Janakarajan, 1999). The studies on surface water in the basin show that water in the Noyyal river, the dam as well as the irrigation tanks downstream, have been affected by industrial pollution, making it unfit for use in any capacity (Palanivel and Rajaguru, 1999).

The pollution load between 1980 to 2000, according to the Pollution Control Board (PCB) data, suggests that the accumulated TDS load during this period was 2.35 MT (million tonne), of which chloride was 1.31 MT and sulphate 0.12 MT (Madras School of Economics, 2002). Both in the textile and tanning industries, the salts or total dissolved solids are a serious problem. Salts like chlorides may not be toxic, but can cause salinity problems. On the other hand, removal of TDS is an extremely expensive proposition.

Who Pays for Pollution?

Agriculture, drinking water and fisheries in Tiruppur area and downstream villages on the Noyyal river have been affected. The water is classified as injurious (electrical conductivity—EC is greater than 3 mmhos/cm) to agriculture in a 146 sq km area and critical (EC between 1.1 and 3 mmhos/cm) in a 218 sq km area. As a result, crop productivity has declined substantially, which has ultimately affected the welfare of the farmers. The estimated damage to agriculture is approximately 210 crore rupees. Drinking water for Tiruppur town as well as the villages downstream has had to be sourced from outside. While the municipality has been bringing 32 Mld of water from the neighbouring Bhavani basin to meet the demands of the town, the State Water Board has introduced special supply schemes in the villages. Many villagers have to fetch potable water from distant places. All this effort costs the state 10 crore rupees annually. Pisciculture in the Noyyal river, tanks and reservoir has been halted since 1997. The loss on this count is estimated at 63 lakh rupees (Madras School of Economics, 2002). Another study undertaken in 2004 (also by the Madras School of Economics) estimated that the overall damage is in the range of 350 crore rupees. This is in addition to the unquantified impact on the ecosystem and the loss of biodiversity.

Off to Court

Farmers in the downstream area (Authupalayam) filed a case in 1995 against the opening of the dam for irrigation. The High Court ordered its closure in 1997. The farmers near the dam (including those in upstream areas) filed another case against the government in 1999, pertaining to the destruction of their land due to seepage. In 2002, the court appointed a Loss of Ecology Authority (LEA) to assess the damage and compensation.

The LEA appointed by the Madras High Court's green-bench has estimated the compensation amount as 104 crore rupees and ordered the government to collect it from the polluting industries. The cash compensation to the Orathupalayam farmers has been decided on the basis of a damage assessment study by a team from the Centre for Environmental Studies, Anna University, which was engaged by LEA, and it covers the loss incurred by the farmers in the Orathapalayam dam *ayacut* (irrigated area) for a four-year period between 1999 and 2002. However, the number of people affected by this issue is far greater, and has not been covered by the LEA.

Chronology of Events

1991: The Orathapalayam dam is completed
1992: Water released for irrigation purposes

1995: Downstream farmers from Karur object to the release of polluted water
1996: Karur farmers file case against the Tiruppur dyeing and bleaching industries
1997: Madras High Court orders that dam water not be released for irrigation
1997: Fisheries department closes its fishery unit at the reservoir
1999: Orathapalayam (upstream) *ayacut* farmers file case for compensation as their lands are also affected
2002: Madras High Court appoints Loss of Ecology Authority to assess the damage
2004: LEA submits its report

Tricky Situation

Due to pressure from NGOs, farmers' organisations and the order of the High Court, the State government has ordered the Pollution Control Board to make it mandatory that all polluting units are either connected to a common effluent treatment plant (CETP) or they install their own individual effluent treatment plants (IETPs). Since then, 424 units have constructed IETPs and 278 units are connected to eight CETPs in Tiruppur. Another 164 units that were not connected either to CETPs or IETPs have been shut down by an order of the Madras High Court. Despite these steps, however, it has been observed that the treated effluents do not entirely meet the standards prescribed by the Pollution Control Board, especially when it comes to parameters like TDS and chlorides.

According to PCB rules, the red and orange category units should be situated 1,000 metres away from a river/stream or any other water source. However, in Tiruppur, according to a study by the Madras School of Economics (MSE, 2003), around 239 units are located at a distance of less than 300 metres from the Noyyal river. The possibility of polluting the river remains high, especially since approximately 83 per cent of the IETPs discharge their effluents directly/indirectly into the water bodies.

The Flip Side of Development

Initially, when there were a small number of bleaching and dyeing units, the effluents were easily assimilated by the environment. It was the spectacular growth of the hosiery industry, particularly in the 1990s, that led to the current problem. It was the cumulative effect of the discharge from 700–800 units located in a cluster that ended up polluting the surface and groundwater; the Pollution Control Board recommended that the units join common effluent treatment plants, but only 278 of them did so. Moreover, the CETPs were not designed to remove salts like chlorides and sulphates. These salts can be removed/reduced in one of two ways: use of technologies like reverse osmosis, and/or clean production technologies.

Unfortunately, many of the smaller units were not in a position to implement either of these options because of the lack of technical/financial capacity. It should also be stressed that the bleaching and dyeing units are an integral part of textile production. Closure of these units will affect the functioning of the entire hosiery industry that has more than 9,000 units in Tiruppur, and provides employment to more than two lakh people. Recently, the dyers' association went on a day's token strike to prove the point. Hence, it should be the collective responsibility of the entire hosiery industry, and not just its bleaching and dyeing segments, to ensure that pollution abates.

The rapid growth of the hosiery industry was largely due to trade liberalisation and opening up of the export market. However, the importing countries are now increasingly concerned about the environmental compliance of these units. Ultimately, pressure from consumers and buyers alone might force the Tiruppur industry to control pollution.

Trying to Strike a Balance

For the time being, the problem continues. All those affected have not been compensated as per the Loss of Ecology Authority. The textile industry provides employment to a large number of people in the area and is an important foreign exchange earner for the country; therefore, the solution should not be detrimental to any of the parties concerned. There is a need for an integrated approach whereby a mechanism for coordinated development of water resources in the basin can be worked out with the participation of all the stakeholders—especially the government departments, farmers, industrialists and NGOs. This could be undertaken by an external agency by organising dialogues and negotiations through the multi-stakeholders' dialogue approach or the formation of multi-stakeholders' platforms aimed at sustainable development of water resources in the basin.

References

Appasamy, Paul and Padmanabhan. 1997. 'Methodology of Damage Assessment: A Case Study of the Palar Basin'. Unpublished report. Chennai: Madras Institute of Development Studies.

Central Groundwater Board Report (CGWB). 1993. *Groundwater Resources and Development Prospects in Coimbatore District, Tamil Nadu.* Hyderabad: CGWB Southern Region.

Central Groundwater Board. 1980. *Inventory Data of Permanent Observation Wells in Noyyal River Basin.* Chennai.

Economic Assessment of Environmental Damage in Noyyal River Basin. 2002. Research report. Mumbai: Indira Gandhi Institute of Development Research.

Environmental Impact of Industrial Effluents in Noyyal River. 2004. Research report. Chennai: Institute for Water Studies.

Environmental Status Report. 2000. Coimbatore: Environmental Cell.

Environmental Status Report on Noyyal River Basin. 2002. Coimbatore: Public Works Department, Environmental Cell (WRO).

Janakarajan, S. 1999. *Conflicts over the Invisible Resources in Tamil Nadu: Rethinking the Mosaic.* USA: Nepal Water Conservation Foundation and Institute for Social and Environmental Transition.

Madras Institute of Development Studies. 1995. *Water Resources Management in Tiruppur* in SIDA Report.

———. 1995. Proceedings of the Seminar on Urban Environment of Tiruppur, July 22.

Palanisamy, K. 1995. 'Rural and Urban Interaction: Water Market in Tiruppur'. Paper presented at the Second Water Conference. Chennai: Anna University.

Palanivel, M. and P. Raja Guru. 1999. '*The Present Status of the River Noyyal'.* Proceedings of Workshop on Environmental Status of Rivers in Tamil Nadu. Coimbatore: Bharathiyar University.

Raja, Guru P. and V. Subramani. 2000. *Groundwater Quality in Tiruppur: Environmental Awareness on Quality Management of Irrigation Water.* Research report. Coimbatore: Department of Environmental Sciences, Bharathiyar University.

Swaminathan, Padmini and J. Jeyaranjan. 1995. 'The Knitwear Cluster in Tiruppur: An Indian Industrial District in the Making'. Working Paper 126. Chennai: Madras Institute of Development Studies.

Conflict over Water Pollution in the Palar Basin: The Need for New Institutions

S. Janakarajan

The Palar River Basin

Palar, one of the major rivers in Tamil Nadu, has a catchment of about 18,300 sq km. The average annual rainfall in the basin is 970 mm and the climate is tropical and highly humid. The major irrigated crops in this basin are paddy, sugarcane and groundnut; coarse cereals and groundnut in water-scarce areas are the major unirrigated crops. Tanks have been the most important source of surface irrigation in the basin. Besides a large number of non-system tanks and the 606 spring channels that provide water for thousands of hectares, there are about 700 system tanks connected to the river channels with a combined command area of 61,000 ha (IWS, 1992 and 2000). However, the emergence of groundwater as the major source of irrigation has led to the neglect of these traditional sources and they are

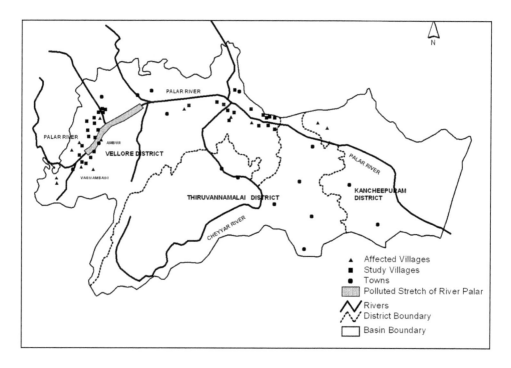

Map 1. *Pollution in the Palar basin.*

Fig. 1. *Palar anicut/weir with stains of tannery pollutants.*

today in a dissipated state (Janakarajan, 1993). Besides irrigation, groundwater is a major source for drinking and industrial water, and it accounts for as high as 92 per cent of all water use in the basin. The quality of groundwater, however, varies a great deal across the basin (TWAD Board, 1997, and Madras School of Economics, 1998).

The leather industry is the most important industrial activity in this basin. Its export earnings shot up from a mere Rs. 0.32 billion in 1965 to Rs. 100 billion in 2001. Today it accounts for 7 per cent of the country's export earnings and employs over 2 million people. Fifty-one per cent of leather exports originate from the southern states and 70 per cent of the tanning industries are concentrated in this region. Tamil Nadu contributes Rs. 50 billion to exports, which is about 90 per cent of the total from the southern States, and 75 per cent of its tanning industries are concentrated in the Palar basin. They contribute over 30 per cent towards the country's total leather exports.

However, on an average, the industry also discharges 35 to 45 litres per kg of raw skin/hide that it processes. The 847 tanneries in the basin use a minimum of 45 to 50 Mld (16.5 to 18.3 Mm^3/yr) of water and discharge 37.5 Mld. (13.7 Mm^3/yr) of effluent. The quantity of effluent discharged from these tanneries, which are supposed to be connected to one of the eight common effluent treatment plants (CETPs) installed in the Palar basin, works out to 37,458 m^3/day. When compared with the water used for agriculture, 50 Mld might appear a small amount. However, the toxic elements, salts and heavy metals in the waste discharged cause irreparable damage.

The industry processes 1.1 million kg of raw hides and skins per day; and for each 100 kg processed, it generates between 38.5 and 62 kg solid waste. A study of pollution loads in the Palar, sponsored by the

Fig. 2. *A spring channel being used by the tanneries for discharging effluent.*
In the background is the beautiful temple, which was renovated and to which 'generous' contributions
came from tannery owners. Village Gudimallur near Walaja Pet.

Asian Development Bank and carried out by Stanley Associates and the Tamil Nadu Pollution Control Board
(TNPCB) (ADB, 1994), shows that pollution loads in the Palar river are extremely high: total suspended
solids (TSS) 29,938 kg per day; total dissolved solids (TDS) 400,302 kg per day; chloride 101,434 kg per
day; sulphide 3818 kg per day; biological oxygen demand (BOD) 23,496 kg per day; chemical oxygen
demand (COD) 70,990 kg per day: total chromium 474 kg per day; and cyanide 22 kg per day. Groundwater
quality data collected by various government and private agencies also indicate a very high level of
contamination. More than 60 per cent of the wells in the affected villages are defunct due to water
contamination, and the investment in those wells is a complete loss.

The Conflict

There is a severe conflict of interest between tanners and farmers in this basin, but it did not surface until
the late 1970s mainly because the industry used supposedly ecofriendly vegetable tanning, and mostly
exported semi-finished leather. However, when the Government of India banned the export of semi-finished
leather in the late 1970s, tanners were compelled to shift to processes using chrome and other chemicals
in order to export finished leather. The real damage to the environment started then. It has now created an
explosive situation in the basin: groundwater contamination has adversely affected agricultural yields,
employment and income, while health hazards due to air pollution and consumption of contaminated water
have led to a high incidence of diseases such as asthma, gastroenteritis, dermatological problems, and kidney

Fig. 3. *This is a process that helps turn fresh water into effluent and dumping it back to the river. Water being pumped into trucks round the clock from the Palar river—catering to the needs of tanneries, domestic and other commercial users near Ranipet.*

malfunctioning among the general population. Thousands of families have left their homes and sought employment elsewhere.

Tannery owners have yielded somewhat to the pressure of civil society initiatives. They have recently installed a few reverse osmosis plants in order to treat high TDS levels. Also, eight CETPs were installed in the basin after the intervention of the Supreme Court, but they have hardly been a solution to the problem of pollution because there is no effective monitoring mechanism. Therefore, the conflict remains largely unresolved.

The Shape of the Conflict

About the conditions in the Palar basin, the Second International Water Tribunal says, 'As a result of the uncontrolled discharge of waste water from the tanning industry, both surface and groundwater have been seriously contaminated. The water is no longer suitable for drinking and has to be brought in from other areas at a price beyond the means of the poor. In addition to the above, these practices, resulting in contamination with salt and chemicals, have rendered useless large areas of once fertile land' (International Water Tribunal, p. 220, 1992).

In 1991, the Vellore Citizens' Forum, a civil society organisation, filed a public interest litigation against tanners in the Supreme Court. The final judgment was in their favour, and thousands of affected farmers demanded an immediate closure of tanneries. But the Supreme Court said that the tanneries could resume their operations if they set up treatment plants. However, these treatment plants were not properly designed to treat the pollutants generated by tanneries; in particular there were very high levels of total dissolved solids. Obviously, the responsibility rests with public-funded institutions the like Central Leather Research Institute (CLRI), National Environmental Engineering Research Institute (NEERI), TNPCB etc., all of which should have taken proactive measures to protect the environment. The fact that the problem worsened over years unambiguously demonstrates the fact that law enforcement and monitoring mechanisms have ensured that the pollution problem—rather than the environment—remains safe and sound (Biswas, 2001).

The progress of such a conflict may be described through a diagram comprising four blocks (Figure 4). (The shape should not be viewed as a parabola but as a general convex shape.) Block 1 indicates a low level of conflict with a low level of people's participation; the problem of pollution has not yet surfaced. Block 2 indicates increased level of conflict and increased people's participation; meaning, the conflict has now surfaced. Block 3 indicates a very high intensity of conflict—almost a crisis situation. When the curve falls down into Block 4, people's participation in the conflict is steeply reduced because the crisis has turned into catastrophe, and feeling helpless and disillusioned, people have withdrawn. The Palar basin conflict is still in Block 3, facing very high intensity of conflict, but fortunately has not yet fallen into Block 4.

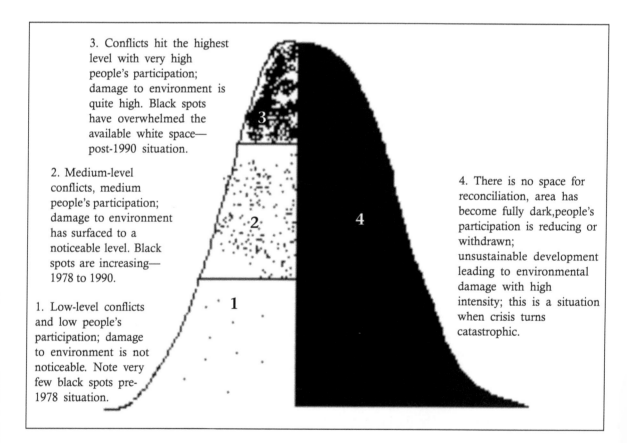

3. Conflicts hit the highest level with very high people's participation; damage to environment is quite high. Black spots have overwhelmed the available white space— post-1990 situation.

2. Medium-level conflicts, medium people's participation; damage to environment has surfaced to a noticeable level. Black spots are increasing— 1978 to 1990.

1. Low-level conflicts and low people's participation; damage to environment is not noticeable. Note very few black spots pre-1978 situation.

4. There is no space for reconciliation, area has become fully dark, people's participation is reducing or withdrawn; unsustainable development leading to environmental damage with high intensity; this is a situation when crisis turns catastrophic.

Fig. 4. *Degrees of conflicts and sustainable development.*

The Dialogue Initiative

Multi-stakeholder dialogue (MSD) is an approach that has immense utility precisely in a Block 3 kind of deadlock situation. The approach has emerged in response to the deficiency and failure of conventional socio-economic and institutional tools to tackle such a situation. It provides a useful framework and platform to explore ways to prevent further degradation of natural resources, and to work towards sustainable development with a common agenda within a framework acceptable to all stakeholders. In fact, the Palar basin has reached a threshold level of crisis where all institutional and legal mechanisms have failed to deliver (including the highest judicial authority), which provides an ideal situation for facilitating a dialogue process among stakeholders.

Multi-Stakeholders' Meeting: Conflicting Parties Sit Together

The first MSD meeting of the Palar river basin was held in Chennai on January 28 and 29, 2002. It was attended by over 120 participants including tannery owners, farmers, NGOs, bureaucrats, effluent treatment managers, media persons, lawyers, doctors and academics.

The basic objectives of this meeting were:

1. To take stock of use and abuse of water in the Palar basin in the overall context of urban and industrial expansion, and in the context of poverty, food security and hunger.
2. To assess, examine and pinpoint the defaulters of law, their positive and negative contributions to society and economy.
3. To bring together various stakeholders for a fruitful dialogue with a view to hear, debate, document and make public their voices.
4. To find ways to prevent further degradation of water resources in the basin and to work towards sustainable development with a common agenda within a framework acceptable to all stakeholders.
5. To find ways to turn situations of conflict and distrust into opportunities for mutual aid and cooperation.

On the first day, there were twelve presentations made by various stakeholders, covering a wide variety of issues pertaining to the use and abuse of water in the Palar basin. During the dialogue sessions, arguments were impassioned and lively. Initially the discussion was quite intense. One tanner's outburst was: 'We [tanners] are treated like Afghan refugees; what sin have we committed except involving ourselves in this dirty business?' By the afternoon of the first day, however, the variety of perspectives had begun to generate common points of understanding. Tanners began to acknowledge the huge impact the pollution generated by them was having on the livelihoods of people and on the environment; farmers and other stakeholders recognised that closure of tanneries is not the solution to the problem. Mutual concerns were expressed, as can be seen from this comment of the tanners: 'So far we [farmers and tanners] were meeting only in the courts; for the first time we are meeting on the same platform with a view to share the concern.'

Remedial measures to counter the problem of effluent discharge and environmental pollution were extensively debated and discussed. Towards the end of the meeting everyone drew a sigh of relief. All the participants widely acknowledged that MSD has to be a process and not a one-off meeting. They decided to form a committee so that the dialogue process could be carried further. The result was the birth of the multi-stakeholders' committee with twenty-four members drawn from different stakeholder groups. The two-day meeting was given wide publicity in the media.

The Committee's Gains

1. The committee unanimously agreed that closure of tanneries is not the solution; members committed themselves to finding solutions not only for pollution, but also for restoring the ecology of the basin; they acknowledged that prevention of further pollution is the first step towards ecological restoration.
2. Different stakeholders agreed to share information to aid useful and concrete decision-making; in particular, tanners who earlier denied access to information agreed to share all details pertaining to tanneries and CETPs, and were also willing to allow committee members to visit their units. This has been one of the most positive outcomes of the meeting.
3. The committee explored a variety of potential solutions: (*a*) the possibility of handing over effluent treatment to a private company; (*b*) the use of a mobile cold storage system to collect and transport raw hides and skins so that pickling, which is the main source of TDS accumulation in the effluent, is avoided.

The two-day meeting was recorded and transcribed along with an introduction on the conditions of river basins in Tamil Nadu. It has already been published in Tamil by the Madras Institute of Development Studies and Kalachuvadu as a 270-page book with photographs, and was released by the Governor of Tamil Nadu on December 26, 2003. The book has been received well by all the stakeholders, and a second edition is on its way.

Looking into the Future

Institutions such as the PCB, CLRI, NEERI, and Council for Leather Exports played a crucial role in promoting the tanning industry, but did not sufficiently heed the issue of pollution. The ongoing dialogue initiative is striving hard to develop a rapport with these agencies to (*a*) get access to official information, (*b*) influence policies, and (*c*) fulfil the objectives of the committee with government endorsement and financial support so that the initiative receives official recognition.

Unresolved Questions

Though the MSD initiative has succeeded in bringing the conflicting parties to the table, the sustainability of dialogue depends very much on the continued support of an impartial facilitator. Secondly, it is not clear as to what extent tanners will be willing to bear the cost of treatment of effluents, given that the international market for finished leather has become very competitive. These are issues that will need to be resolved if the dialogue process is to succeed.

Note

*Map 1 and figures by author.

References

Biswas. 2001. 'Environmental Legislation: Challenges of Enforcement', Eastern Window E-mail, Vision, p. 1.

'Economic Analysis of Environmental Problems in Tanneries and Suggestions for Policy Action', a study supported by UNDP. Madras School of Economics, 1998.

Hashim, M.M. 1996. 'Indian Leather Industry', *The Hindu*, July 26, 1996.

Janakarajan, S. 1993. 'In Search of Tanks: Some Hidden Facts', *Economic and Political Weekly*, 28(26).

Rajagopal, A. and A. Vaidyanathan. 1998. 'Conditions and Characteristics of Groundwater Irrigation in the Palar Basin: Some Preliminary Results'. Madras Institute of Development Studies. Unpublished.

Report on the Palar Pollution studies. Tamil Nadu Water Supply and Drainage Board, Government of Tamil Nadu, 1997

'State Framework Water Resources Plan: The Palar River Basin, Chennai'. Institute for Water Studies, Water Resources Organization, Government of Tamil Nadu, 2000

'Tamil Nadu Environmental Monitoring and Pollution Control'. Final Report, prepared by Stanley Associates for Tamil Nadu Pollution Control Board, Canada. Asian Development Bank, 1994.

'Water Pollution Due to Leather Industries in Tamil Nadu'. Paper presented at a conference held at Amsterdam by Muthu, Trust Help, 1992. International Water Tribunal, p. 220.

'Water Resources Assessment and Management: Strategies for Palar Basin'. Institute for Water Studies, Water Resources Organization, Government of Tamil Nadu, 1992.

A Factory in a Paddy Field in Pondicherry: Is Berger Paints Polluting Pandasozhanallur?

Benjamin Larroquette and Gaspard Appavou

Planting a Factory

Pandasozhanallur village is situated approximately 20 km south-west of Pondicherry town. The area comprises prime agricultural land where traditional crops such as paddy, groundnut and sugarcane are cultivated. It is one of the main suppliers of rice and other produce to Pondicherry.

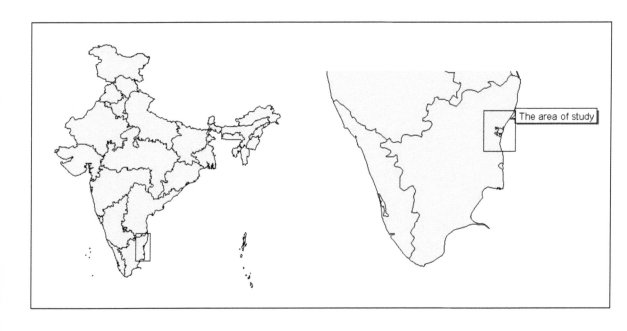

Map 1. *The area of study.*

In 1998, Berger Paints set up a production plant in the village. The land was bought legally from a local farmer and all authorisations were granted by the Pondicherry Pollution Control Board (PPCB). The plant employs around 200 local unskilled labourers and is situated close to the main irrigation tank. The command area is approximately 125 ha, and provides livelihood to around 1,500 people.

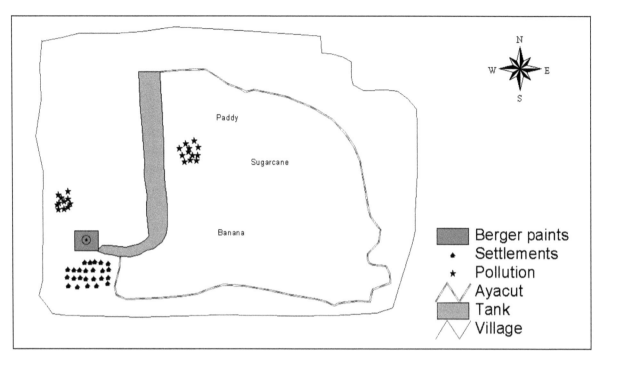

Fig. 1. *This is a sketch of Pandasozhanallur village with the location of the company, pollution levels claimed by farmers, and the cultivated areas. The sketch (not to scale) was derived from participatory discussions with the villagers.*

Water Colours

There have been complaints from agricultural landowners that the paint factory has been releasing untreated waste directly into the aquifer, polluting the water used for irrigation and drinking purposes. Initially the company assured the villagers that all waste water would be treated and that it would not harm the environment, but according to the farmers, sometime in 2002 the water started showing signs of pollution in the form of white deposition on the surface of well water.

Berger Paints, on the other hand, maintains that it has kept its promise and that it has complied with PPCB norms. Its officials say that the waste is treated in an effluent treatment plant (ETP) on the premises and is used to irrigate the gardens on the Berger Paints campus. Solid waste such as carbonates of calcium and magnesium are soaked with residue of alkyd resin and stored in concrete pits.

The conflict is dormant as the farmers never took the issue to court owing to lack of evidence. There has not been any organised attempt to deal with the problem either, and even the local government, which is very concerned about the quality of groundwater, has not taken any action so far. There really has not been a low point since no attempt has been made to take the matter any further. The only attempt to talk out the problem was initiated by the village youth group which tried to meet the manager of the plant. However, he refused to see them.

Cautious Truce

The farmers believe that Berger Paints is pumping waste directly into borewells without treating it. Their demand is that this should be stopped at once. They are not thinking of taking any action at this point of time and are quite clear that they do not want the company to close down because it has generated local employment. According to Berger Paints, the waste water is treated adequately in the ETP and they have documents from the PPCB to back its claims. The management also says that the treated water is used to water plants in their compound; the plants are healthy, meaning that the water is fit for irrigation. Berger Paints seems to feel that the farmers are attempting to use the pollution issue to get compensation or donations from the company. (The company does occasionally contribute to temple festivals in the area.)

If the situation worsens, the impact on ecology and economy will be severe. If indeed there is some truth in the farmers' claims, then there is clear danger of the water eventually becoming unfit for irrigation, leading to loss of livelihood for several thousands. Approximately 125 ha of paddy is being grown in the area adjacent to the paint company. If we assume that 50 bags of rice are collected from one hectare of cultivated land, we have a seasonal production of 6,250 bags, which amounts to approximately Rs. 25 lakh.

On the other hand, if the company is in the right, having to shut down will mean a loss of over 200 jobs for the local population, especially for the Dalit community. If Rs. 2,500 per month is the average salary for an unskilled labourer and approximately 200 are employed by Berger Paints, then the loss of income to the community is approximately Rs. 5 lakh per month. It is, however, difficult to properly evaluate the cost of treating the water.

Without Proof

No evidence of pollution has been seen in the various rounds of water testing that have taken place so far. Furthermore, when asked about the impact on yield or growth rate of crops, the farmers are unable to point out changes.

However, the company too is not transparent about the treatment of water. A request to inspect the ETP was turned down. Even though Berger Paints has the requisite permission from the PCB there is scepticism and cynicism about the corporate world.

The problem is that even if the company is indeed pumping waste directly into the aquifer, it may be difficult to track the pollution because we are unaware of the water flow in the deeper aquifers. It must be mentioned that water samples were always taken in the vicinity of the company; but because of the nature of underground water flow, the symptoms may surface at quite a distance from the discharge point.

To have located a factory in the middle of prime agricultural land is itself a mistake. But to accuse the company of harming the environment without evidence is not acceptable either. The company should abide by the norms and be transparent in its dealings with the community. This will go a long way in reassuring the farmers. Perhaps an independent body can undertake regular monitoring of the quality of water and look into the functioning of the ETP. This would not only allay the farmers' fears but would also be a good public relations move for Berger Paints.

Note

*Map 1 and figure by authors.

The Arkavati Sub-Basin in Karnataka: Industrial Pollution Versus Rural Livelihood Systems

D. Dominic

A River Runs Dry

THE ARKAVATI RIVER IS ONE OF THE PRINCIPAL TRIBUTARIES OF THE CAUVERY IN KARNATAKA with a catchment area of approximately 4,351 sq km. Its origin is traced to the southern series of the Nandi hills, 3,000 m above the sea, though the real source lies in a series of about twenty-six tanks that are connected to the Nagarakere reservoir in Doddaballapur town. The river flows through Bangalore's rural districts of for a distance of 190 km, and while its catchment includes a total of 150 big and 1,084 small tanks, the biggest is the Hesaraghatta tank with a catchment area of 490 sq km. The total surface water potential created in the *taluk* is about 2,330 ha, though it has almost always remained dry in recent times. The other dams are the T.G. Halli reservoir, the Manchnabele tank, the Biramangala tank, and the Arubhihalli reservoir.

The conflict zone under discussion is in Doddaballapur *taluk*, a region with an urban and rural population totalling 2,68,332 according to the 2001 Census. The *taluk*, 36 km from Bangalore city, is at an altitude of 892 m above sea level. There are 152 tanks spread over five *hoblis* (cluster of villages). Traditionally, this area has been known for silk production and horticulture; recent urbanisation has led to a scarcity of drinking water, especially in the town. The Nagara tank that once supplied water to the town now looks forlorn, and whatever little water it manages to collect is contaminated by industrial pollution.

Growing Pains

Rapid urbanisation and industrialisation in and around Bangalore has put immense pressure on

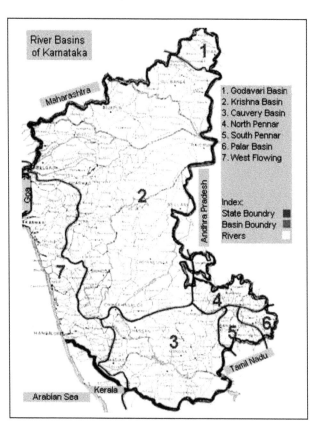

Map 1. *River basins of Karnataka.*

the land and water resources of rural Bangalore. During the 1990s, large chunks of agricultural land around Doddaballapur were sold to industrialists and resort owners. With government approval the area near the Doddaballapur railway station was earmarked for industry. The Bashettihally industrial area (popularly known as Doddaballapur industrial area), hosts garment export units like Go-Go and Hi-Images (also a dyeing unit), apart from a fruit-concentrate manufacturing concern, an animal-feed processing factory, and chemical producers such as Vetcare Brindavan Phosphate and Deepak Enterprises.

The town itself has more than 50,000 power looms and 80 small dyeing units that consume about 2,40,000 lpd of fresh water. Needless to say, the effluents flow into the Nagarakere tank.

In Troubled Waters

In May 2002, Bashettihally gradually began to feel the adverse impact of this industrial growth. Groundwater level dipped below 275 m. The Go-Go factory (set up in 1995) dug nine borewells to extract groundwater, of which six failed. The factory used the dry wells to disgorge toxic waste that slowly began to affect the quality of groundwater within a 6 sq km radius. Crops began to fail in neighbouring villages and livestock

Fig. 1. *Waste dumped on public grounds by the Phosphate industry.*

Fig. 2. *Industrial effluents dumped in private fields.*

died after consuming the contaminated water. All these settlements—Bashettihalli, Aradeshhalli, Veerapura, Bisuvanahalli, Marasandra, Kadathanamale, Majarahosalli and Chikkatumkur—and the surrounding fields form part of the catchment area of the Arkavati river.

This was just the beginning. To add to the misery of the local population, the Brindavan Phosphate Factory suddenly dumped about 250 truckloads of chemical waste on public grounds. The fruit-processing unit too began to dump its waste there.

Chronology of Events

January 2, 2002:	Janajagruti Samiti comes into existence. 'Janajagruti Samiti' literally means people's awareness committee. It began with a group of concerned villagers, part of the villages affected by the Go-Go effluents discharge. Later on, this group was officially formed to represent the pollution case at the court.
January 13, 2003:	The KPCB Chairperson constitutes a Watchdog Committee, involving the local representatives from various organisations. He is immediately transferred.
January 18, 2003:	KPCB issues Notice of Proposed Directions under Section 33 (A) of The Water (Prevention and Control of Pollution) Act 1974 to M/s. Go-Go Exports Pvt. Ltd. for violation of Section 25/26 of the Act.
January 24, 2003:	The Janajagruti Samiti catches Go-Go unit still discharging effluents and documents it.

April 21, 2003:	Various organisations protest against the KPCB's inactivity.
November 3, 2003:	The High Court asks for five consecutive days' test of effluents. The tests prove that Go-Go is guilty.
April 30, 2004:	The guilty factories violate the High Court's order of closure of the factories.
May 10, 2004:	Magasaysay awardee Rajendra Singh visits the area and praises the people's committee for the water harvesting structures constructed to reclaim groundwater.
December 4, 2004:	The Chairperson and two technical members of the Appellate Authority of the Court inspect the Go-Go unit and surrounding areas.
May 27, 2005:	The Appellate Authority dismisses the appeal by Go-Go. In the strongly worded order, the Authority squarely blames Go-Go of knowingly destroying the eco-system in the area, criticises KPCB of neglect, and orders the factory to be closed and sealed.
June 2, 2005:	People pressurise the KPCB to act on the Court's order. At 6.30 PM power to the factory is disconnected.
June 3, 2005:	Go-Go unit is closed and seized by the Assistant Commissioner.
June 9, 2005:	Writ petition filed by Go-Go in the High Court comes up for preliminary hearing. The KPCB fails to defend itself. The Court directs KPCB to conduct fresh tests within six weeks. Till then, the factory is allowed to run.

The Struggle

The conflict that had been sparked off by the digging of borewells at Go-Go resulted in the local people organising a protest against the factory. This caught the attention of various environmental organisations in the region as well as the *taluk* Raitha Sangha (farmers' union). Bolstered by the support of the Green Peace Foundation, the protestors managed to achieve temporary closure of the factory. However, this shutdown lasted only four days and the unit resumed operations after company officials promised that they would not repeat their mistake. But matters did not end here. On the contrary, problems continued to increase and in order to mobilise public opinion, the villagers, supported by environmental groups, staged a *dharna* in front of the factory on February 12, 2001. It came to their notice that the factory did not have an effluent treatment plant on the premises and was instead releasing waste into the defunct borewells. Once again company officials promised to resolve the problem, but a few days later when a people's delegation approached them to check if they had taken any steps to keep their word, they were denied entry into the premises.

On October 6, 2002, the Janajagruti Samiti held a meeting during the course of which villagers described the hardship they had been facing in recent times because of pollution levels. The Arakere village *panchayat*, Bangalore north *taluk*, lodged a complaint with the Pollution Control Board (PCB); their case was strengthened by the fact that several eminent persons including the late B.J. Sridhara, Bette Gowda, H.G. Dinakar, and a representative of the Department of Forest, Environment and Ecology, had already registered independent complaints with the PCB. Officials visited the site and confirmed that the toxicity of effluent emitted into the water exceeded permissible limits; they issued closure notices to Go-Go and two other factories. Despite this, the factories continued with their manufacturing and dumping activities over the next few months.

A report from the office of the Commissioner of Agriculture declared the water in the area unfit for irrigation. The fact that even fish could not survive in the water pushed the committee—formed by concerned local persons—to report the matter to the police, and a fresh complaint was made to the PCB.

A tribunal ordered that the factory be shut down but the owners managed a stay order. The battle has now moved to the High Court as a result of a writ petition filed in 2003. In December 2004, a judge from the Court visited the spot and submitted his findings. Meanwhile, the factory management continues to deny all allegations.

An interesting aspect of the conflict was the revelation that a farmer in Aradeshhalli was actually accepting Rs. 15,000 a month from the polluting units for allowing the waste to flow into his field. Since he could not grow anything on the land any more, he was content to receive his monthly 'fee'. Ironically, pollution seems to actually have supporters.

Facing Reality

The Janajagruti Samiti, in the meantime, took its job seriously and sought to educate the public on rainwater harvesting. The people pooled their resources and started building check dams. Their efforts did yield results. The ponds were briefly used for pisciculture, but in November 2004, several months of hard work came to naught when effluent from the Hi-Images factory made its way into one of the check dams, killing the fish instantly.

Keeping in mind that inordinate delay might well mean justice denied, the Samiti has now entered into negotiations with the factory management. While the discharge of waste has now been discontinued, the question of compensation remains. The manager has shown his willingness to clean up the pond water.

The Arkavati and Kumudwati Rejuvenation Coordination Committee is involved in finding solutions to rejuvenate the Arkavati water resources. Formed in February 2005, this committee is a federation of various organisations and groups that look into the health of the sub-basin system in the north of Bangalore. The fact that a workshop conducted by the coordinating committee to assess critical issues pertaining to the river managed to ensure the participation of the government secretary and the PCB, shows that the group has come a long way. An outcome of the event was that in 2002 the Karnataka government issued a notification forbidding industrial activity within a 2 km radius of the river. A vigilance committee would monitor the progress and seek measures to protect it.

Hope for the Dying River?

Doddaballapur industrial area is a microcosm of rural India. The women in the *taluk* have been working to produce staple foodgrains that do not need too much water. They have also been experimenting with organic farming. As they go back to the traditional roots of farming, they are slowly beginning to realise that all production has to be an effective partnership with nature.

The struggle of the people of the Arakavati river basin is not just about agitation against pollution, but a battle for day-to-day survival. This case study is an attempt to document a valiant battle to resuscitate a dying river, which might well be reduced to a trickle in the not too distant future.

Note

*Map 1 from *www.wateresources.kar.nic.in*; figures by Mr. Seetharam.

Pollution in Hootgalli Village, Mysore

S. Manasi and N. Deepa

Hootgalli is about 15 km from Mysore City. Its area is 1,600 hectares and there are 2,500 households with a population of about 10,000. The village is divided into five main blocks inclusive of a housing board colony that has 1,000 families. The village is afflicted by water contamination—a result of discharge of effluents from industries [1] manufacturing potassium nitrate. The dispute is confined to four blocks of the village as surface water is provided to the housing board colony block.

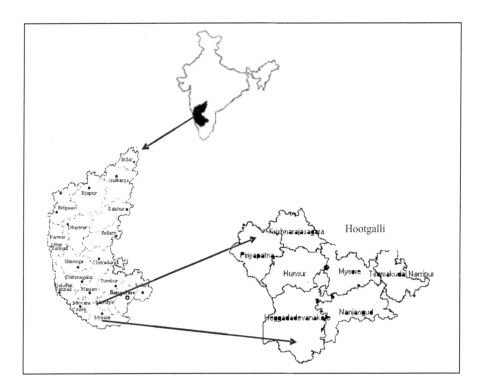

Map 1. *Location of the study area in Karnataka.*

According to the EPA, 'Only two substances for which standards have been set pose an immediate threat to health whenever they are exceeded: bacteria and Nitrate.' (US EPA, 1991). Studies have proved that Nitrate is a potential health threat especially for infants; it causes the 'blue baby syndrome' and fatal diseases like cancer. It is also known for its teratogenic effect (Bruning-Fann 1993). At Hootgalli, industries adopt various methods to discharge waste—they dump it into dried borewells, dig pits (13–17m deep) to put the

Fig. 1. *The destroyed water tank.*

waste in or dispose of it in the Bommannahalli tank half-a-km away. The contaminated water tastes bitter and has a pungent odour.

Although the locals realised that something was wrong, it took some time for them to link their problems with water. The village youth noticed that the fish they reared in a small tank started dying suddenly. Various health problems such as skin rash, stomach upset and hair loss began to show up; then, some livestock died. Gradually the reason became clear.

Chronology of Events

- July 2001: The *gram panchayat* (GP) members write to the *zila parishad* (ZP), drawing its attention to groundwater contamination.
- February 14, 2002: People protest and demand closure of industries.
- November 28, 2002: Officials from the Department of Mines and Geology (DMG) visit the village to study the situation.
- December 2, 2002: Two cows die as a result of the contamination. People write to the ZP again and demand the closure of the industries.
- December 4, 2002: The Deputy Commissioner and the ZP Chief Executive Officer visit the industries for spot investigation.

- December 15, 2002: People demand potable water and block roads during the Deputy Chief Minister's visit.
- January 13, 2003: Protests take place outside factories' premises demanding their closure.
- February 18, 2003: Senior chemist from Department of Mines and Geology DMG tests two water samples on the spot and reports high concentration of potassium and nitrate.
- November 28, 2004: Political leaders highlight the problem during the annual industrial meet.
- December 17, 2004: Karnataka Industries Area Development Board (KIADB) officials inspect the industrial units.

Dying of Thirst

The conflict has two dimensions. Firstly, the villagers have been demanding an alternate arrangement to supply them with safe and adequate drinking water. Secondly, they also want the industries to be shut down to avoid further contamination. At present, water is supplied in limited quantities—twice a week for 2–3 hours—from the Cauvery. People also fetch water on bikes and motorcycles from a kilometre away. The contaminated water continues to be used for other purposes. In the summer, even the polluted water is not available because of power shortage. To make things worse, the Cauvery water and contaminated groundwater are both supplied through a common pipeline. The situation often gets desperate enough for people to consume boiled contaminated water. When water is boiled, an oily layer floats over the surface and the vessels in which it is stored acquire a rough white layer. The ZP made red markings on the borewells to warn against their use due to high contamination. The government sanctioned alternate drinking water from the Rajiv Gandhi Drinking Water Scheme. But at this point, the conflict took a turn for the worse.

A suitable spot was identified for the construction of the overhead tank, but a resident of the village claimed that the land was his, although others argued that it was common property. The man then went ahead and filed a case against the GP at the Karnataka Appellant Tribunal Court, Mysore. The construction came to a halt, and the scenario turned violent with the alleged owner inflicting damage on the structure that had been built thus far. He then decided to construct shops at the site but the enraged villagers demolished the buildings. In December 2004, the court decided in favour of the resident but the GP members now want to petition the High Court. In the process, the plan to make alternate arrangements for water has been stalled, though separate pipelines have been laid throughout the village. Several attempts have been made by local leaders—both through protest and requests—to allow the construction to go on. Meanwhile, two of the polluting units that had been shut down temporarily have been reopened, while two others that claimed to have ceased operations continue to work seasonally but at irregular times.

'Good for Flowers and Firecrackers, What's your Problem?': Industrialists

The conflict specifically involves two groups: the people and the factory owners. The local politicians have been trying to fight the battle on behalf of the people and have made attempts to resolve the problem by trying to get them an alternate water supply scheme. The factory owners, of course, claim that the contamination is not due to waste discharge. Various groups have now got involved in the case and their views are summarised here:

Residents

The villagers are keen that Cauvery water be supplied in adequate quantity; the present supply is insufficient and irregular. They also demand a permanent closure of the industrial units. They would rather protect their environment and water source.

Politicians

Many ministers and the Deputy Commissioner, KSPCB, have promised to resolve the issue by providing the village an alternative supply of water at the earliest.

Local Leaders

They stand by the people's demands and want the industrial units guilty of contamination to shut down; their plea is for adequate drinking water supply at the earliest. There are practical constraints when it comes to tracking the activities of the industries because they work on a seasonal basis. They also believe that there is a nexus between important officials, industrialists and political parties, which adds complexity to the situation. Irrespective of this, they grant that some serious attempts have been made on the part of the local leaders to resolve the issue.

For example, the then party in power (Congress-I) highlighted the problem by bringing it to the notice of the ZP, and placing the results of water samples sent to two laboratories (SJC and NIE Engineering Colleges) that confirmed the contamination. The Janata Dal president mobilised the people, particularly women; the people blocked roads and did not allow a State minister to pass through the village when he was on his way to a neighbouring town. The minister in question had taps installed to allow people access to surface water from the pipeline. The follow-up action included the laying of a main pipeline drawn from the KHB to supply water to the village. This happened in July 2003; a 6-inch pipe linked to a network of six pipelines was connected to fulfil this aim. A BJP leader took up the people's cause by demanding the closure of chemical industries in the vicinity. He joined the people in protesting outside the factories' premises. Thus, political leaders, irrespective of party affiliations, have been consistently working to keep the issue in the limelight.

Industrialists

The factory owners claim that potassium nitrate is harmless, saying that it is a natural salt used to make toothpaste and firecrackers. According to them, it does not leave any residue, and local leaders are only persisting with the issue for political gain. They claim that water contamination occurred as a result of inadequate drainage facilities and prolonged drought. They say they have all the requisite permissions from the KSPCB and are working within stipulated norms.

Impact on Livestock and Crops

High nitrate levels in water and feed lead to reduced vitality and cause stillbirth, low-birth weight, and slow weight gain in livestock (Committee on Nitrate Accumulation). All these effects have been observed in the study area. Even livestock have been turning away from the water in the local lake; crops have been failing in the fields adjacent to the industrial area; and people have been suffering from various ailments. However, no studies have been undertaken to observe the long-term effects. What we know so far seems like the tip of an iceberg; much remains under the surface.

Impact on Society

People have been traumatised by the worries caused on account of having insufficient drinking water. Continuous rationing of water and the tiring trips to salvage a few pots of water have taken their toll on the physical and mental health of the people. Many of them are sick and plead for help—they offer water to every visitor to show them how hard it has become to survive on what they drink everyday. They are helpless and are confused about whether to blame the politicians, industrialists or officials from monitoring agencies.

Impact on the Economy

Investments made towards drilling borewells turn out to be a waste, as water is simply not potable. The destruction of the overhead tank meant for supplying water from an alternate source has caused further losses. The greatest loss is to health and peace of mind.

Issues to be Addressed

1. The process of resolving the conflict is on. However, there is need for coordination between the local leaders belonging to different political parties, because this is a serious problem that concerns the welfare of people. While one party has been calling for the closure of polluting industries, another has been demanding an alternate supply of water. While politicians have lent their voice to the struggle, some of them suspect their counterparts from other parties of being hand-in-glove with the factory owners.
2. It is unfortunate that a land dispute has stalled what could have been a solution to the problem. It is important that the court resolves the issue without further delay.
3. Detailed studies on the negative effects of potassium nitrate are necessary to drive home the point about its danger to human beings and the environment.
4. There has to be greater coordination between agencies like the KSPCB, KIADB, ZP and GP. They should be able to get the industrialists and the people on a common platform so that both sides can express their views and initiate a dialogue.

Note

*Map 1 from authors and Gram Panchayat office, Hootgalli; figure 1 by authors.
1. Canara Chemicals, Ugama Chemicals, Krishna Chemicals, Pavana Chemicals, Swasthik Chemicals, Mysore Chemicals.

References

Bruning-Fann, C.S. and J.B. Kaneene. 1993. 'The Effects of Nitrate, Nitrite, and N-nitroso Compounds on Human Health: A Review'. *Vet Human Toxicology*, 35: 521–38.

'Hazards of Nitrate, Nitrite, and Nitrosoamines to Man and Livestock' in *Accumulation of Nitrate*, pp. 46-75. Washington, D.C.: National Academy of Sciences, The Nitrate Elimination Co., Inc., Committee on Nitrate Accumulation, Agricultural Board, Division of Biology & Agriculture, National Research Council, 1972. Report accessed from an article from *www.nitrate.com/nitrate3.htm*

'Is Your Drinking Water Safe?' US EPA (Environment Protection Agency) Office of Water (WH-550), EPA 570 9–91–005, September 1991.

Kolleru Wildlife Sanctuary: Pollution Through Aqua Culture

J. Rama Rao, Jasveen Jairath and Umesh P.

A Sanctuary Under Threat

KOLLERU IS ONE OF ASIA'S LARGEST FRESHWATER LAKES. IT IS LOCATED IN ANDHRA Pradesh, and is a famous habitat for a number of resident and migratory birds, including the vulnerable grey pelican (*Pelecanus philippensis*). Situated between the Godavari and the Krishna river basins, it is an invaluable wetland ecosystem. The lake spans 90,100 ha, and the water shrinks or expands depending on the rains; many rivulets drain into the Kolleru and surplus waters runs off into the Bay of Bengal. Abutting the lake

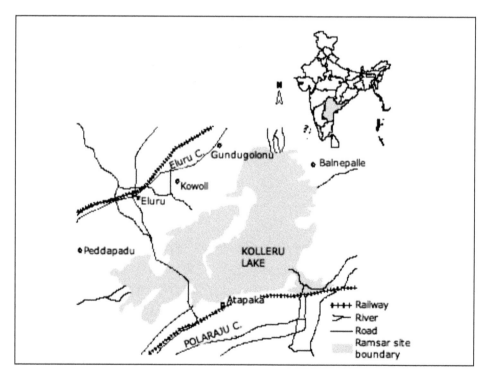

Map 1. *The location of Kolleru lake.*

Coordinates: 16°30'–16°45 'N latitude and 81°05–81°20' E longitude
Area: 90,100 ha
Elevation: 0–5 m above MSL

are about seventy-five villages spread over nine *mandals* in Krishna and West Godavari districts with a population of 3.5 lakh, as per the 1991 Census. Twenty-five years ago the water surface of the lake was 918 sq km. At the Ramsar convention[1] held in Spain in November 2002, Kolleru was designated one of the eleven new Indian wetlands of international importance.

Wilful Destruction

Rich in aquatic life, Kolleru has for a long time provided a habitat where there is a harmonious coexistence of birds, people and life-supporting water. The resources of the lake—many rivulets such as Tammileru, Ramileru and Budameru—that brought in the floodwaters necessary to sustain it were being used by the local communities for fishing, agriculture, bird catching, etc. Records of fishing licenses exist since 1956. People from Orissa and other nearby places used to migrate to the region to make a living. The government had assigned lands in the lake area to Scheduled and Backward Castes (SCs and BCs respectively). While the BCs—who are mostly fishermen—converted their lands to fish tanks, the SCs used theirs for agriculture. There are various small islands in the lake that were inhabited by fisherfolk from lower castes.

The fishermen would fish during the rains and take up seasonal cultivation of paddy on the land vacated by receding waters around the islands and the edge of the lake during the winter and the summer seasons. During the late 1970s, under the Chief Ministership of J. Vengal Rao, the fishermen were encouraged to form registered cooperative societies and loans were sanctioned to members for seasonal cultivation of one hectare of dry land per family. Because of repeated floods, the banks and the government encouraged them to convert agricultural land into fish ponds and tanks. The beneficiaries were to practice collective cultivation and their remuneration was in proportion to their share. At this point, the better-off sections of the community entered the scene and took the land/water area on lease from the society members for periods up to five years. This continues till date; it means that the land is in the name of the poor 'beneficiaries', but is, in fact, used by the well-off sections of society while the real fishermen work for a salary of Rs. 20 a day for women and Rs. 40 for men. Ironically, those legally entitled to the benefit have been reduced to wage earners on their own land/water; the rich have not only taken over all the cooperative societies but have also started illegal encroachments. Until 1990, this influential class, also comprising political leaders and policy-makers who successfully sidelined the real beneficiaries, were only involved in fishing—an activity that requires sweet water.

Aqua Boom, Induced Death

The fishermen—who were dependent on traditional fish capture until the lands were assigned—were encouraged by the government to go in for aquaculture. Some rich enterprising farmers, taking advantage of the government policy, reaped rich harvests through the scheme, and their success and prosperity in the 1990s encouraged outsiders to invest in fish tanks leading to an aquaculture boom and over-exploitation of the lake.

Between 1992 and 1993, aquaculture was practised in a big way. The problem was that it needed saline water to flourish, and borewells were sunk in the lake bed to pump out saline water for the aqua ponds. Because of this, the lake bed level sank, and the tides brought in more and more saline water into the lake since the banks also sank. Prawn seed is cultivated for one month in small ponds and then transferred to larger ponds with saline water. Both fish and prawn cultivation require chemical fertilizers, *gobar* manure, chicken waste, etc. Once the harvest is over, this water stagnates and pollutes the surrounding water. Unscientific and illegal aquaculture coupled with agricultural runoff from the area, that also contains chemical residue, untreated water from neighbouring industries and domestic sewage from areas like Vuyyuru, Hanuman Junction, Gudivada, Eluru, and Tadepalli Gudem, etc. flow into the lake and contribute to its pollution.

Hopeless Situation

At present, aquaculture is carried out on over 80,000 ha, producing more than 7 lakh tons of products with an annual turnover of Rs. 4,000 crore. However, reports say that the area under aquaculture is much higher than official estimates since there is a lot of illegal activity as well. Many politicians from the two districts have hundreds of hectares of land under their control.

The consequences of this pollution are there for all to see. There are frequent fish kills, there is scarcity of safe drinking water in dozens of island villages, and contamination of groundwater around the lake in Krishna and West Godavari districts. This once glorious and picturesque freshwater body is now doomed to become a dead lake. It has undergone several chemical and biological changes that have contributed to its depletion and pollution. The water has turned saline, fish are contaminated with pesticides, and polycyclic aromatic hydrocarbons and heavy metals have entered the lake, making fish and prawns unfit for human consumption. The Kolleru is now clinically dead.

Government Intervention

In 1982, the Andhra Pradesh government established the Kolleru Lake Development Committee (KLDC), which set up a Rs. 300 crore master plan for Kolleru. The plan suggests that the lake level be maintained at 5 m above mean sea level (MSL), and irrigation and drainage regulators be constructed across the Upputeru channel from the lake to the sea. It also calls for the creation of a Kolleru Lake Development Authority to check encroachments, regulate and monitor pollution, and clear the lake of weeds and use them as compost and raw material to produce biogas. Pisciculture, a bird sanctuary and tourism have been on the cards. The government, however, is yet to allot funds for the project.

The KLDC's scientific laboratory is equally tardy. Equipment worth Rs. 3 lakh gathers dust for want of staff. In a personal interview in 2003, the administrative officer, Satyanarayana Rao, said, 'I cannot review anything in the field as I have inadequate technical and executive staff. Even all the committee members have not been appointed. We have written to the government to wind up the office if matters don't improve.' As of now, the KLDC staff just pull out weeds from the lake.

In 1999, the State government (by a Government Order Miscellaneous [G.O.Ms.0] No. 120 dated October 4) declared Kolleru a wildlife sanctuary. The Kolleru Wildlife Sanctuary comprises all the area up to 5 m above MSL and is spread over 30,855.20 ha or 308.55 sq km. The notified sanctuary area at present spans seven *mandals* of West Godavari district and two *mandals* of Krishna district, and includes a total of seventy-four villages.

When Water is Poison

The degradation of the Kolleru has many implications for the weaker sections of the community. The poor blame the rich aqua farmers who were responsible for the shift from traditional freshwater fishing to saline prawn farming. It is ironic that the main problem is the dearth of drinking water in the neighbourhood of the largest freshwater lake in the country. It has been found that the government does not supply drinking water to the island villages, and no investments have been made to rectify this situation in spite of the fact that every year this region contributes nearly Rs. 4 crore to the election fund. The few taps that exist supply polluted water. Ironically, fifteen years ago the villagers used to drink only lake water, and now they walk 3–6 km to get water for domestic use. To combat this problem, many buy sachets of water—a thriving local industry—but there are no quality checks; there is no expiry date on the packets and no treatment of water before packaging. But the illiterate villagers have no idea about the risk they incur and suffer from various water-related diseases such as diarrhoea, typhoid and amoebiasis, etc. During the last two years prawn and fish have become prone to various diseases, and some farms have been abandoned. These lands are useless for agriculture as well since the soil and water have been contaminated. The young men of this area have started migrating in search of jobs while old people have taken to begging. This, when 'prawn dons' earn up to Rs. 22.5 lakh per hectare in four months.

The fish pond owner ('illegitimate') is different from the actual owner ('encroacher'), whose name is reflected in the revenue records. But the name of the real encroacher is not mentioned anywhere in any document and the court does not find recorded evidence to convict the actual encroacher who gets away with all manner of illegal activities.

In July 2001, the AP High Court passed an order directing the authorities to ensure removal of encroachments in the lake bed and see to it that all possible endeavours are made to bring Kolleru lake back to its pristine glory. This was a major victory for the backward classes, and it came after a prolonged battle by various community-based organisations (CBOs).

At the same time, the delay in implementation of these orders due to various reasons has been a major setback. An attempt was made to set right the land records by appointing an official who was specially assigned the task of preparing fresh records with complete details of all encroachment of lands, ponds, etc. The day before he was scheduled to submit his findings he was murdered while out on his evening walk, an indication of how high are the stakes involved in prawn/fish cultivation in the area.

Encroachments continue to take place unchecked even five years after the sanctuary was notified. It is now difficult to actually identify the boundaries of the sanctuary since it is a series of fish ponds. The gravity of the situation can be judged from the fact that more than 20 per cent of the area—supposedly part of the protected region—is full of illegal settlers.

Pleas and Threats

The native fishermen want a way out of this downward spiral by reverting to freshwater fish culture so that the lake slowly regains its past glory; but they are unable to do much due to high indebtedness, lack of unity, and and political strength of their opponents. The prawn cultivating rich farmers, of course, believe that they are doing the poor fishermen a favour by providing them with jobs and wages. The fishermen and local NGOs are trying to get the authorities to enforce the wildlife act and prevent further degradation while the prawn farmers do their best to legalise the encroachments in the lake and the protected area.

Cry for Help

There is a need for strong advocacy to highlight these issues and mobilise public opinion from across the globe to support and assist local efforts to stop further damage to the lake, restore its ecological health, and make it sustainable again. The struggle cannot continue with local resistance alone. The prawn/fish lobby is far too strong and is capable of violently eliminating any potential threat. The local groups are particularly weak and vulnerable, and the solution has to come from a platform that can guarantee a level playing field for all stakeholders. The situation calls for a public dialogue that can bring about practicable and workable solutions. Water networks can play a useful role in mediating and connecting local groups with similar outfits that are willing to participate in mutual exchange of ideas that can take the process forward.

Note

*Map 1 from *www.geocities.com/uravikumar/*

1. The Convention on Wetlands, signed in Ramsar, Iran, in 1971, is an inter-governmental treaty that provides the framework for national action and international cooperation on conservation and wise use of wetlands and their resources. There are presently 146 contracting parties to the Convention with 1458 wetland sites, totaling 125.4 million hectares.

References

Prasad, M.K. Durga and Y. Anjaneyulu (eds). 2003. *Lake Kolleru: Environmental Status (Past and Present)*. Hyderabad: BS Publications.

Rao, T. Shivaji. 2003. *Conflict between Development and Environment of Kolleru Lake* (A Case Study of Environmental Impact Analysis). Hyderabad: BS Publications.

Seshagiri, B.V. 2003. *The Impact of Fisheries on Lake Kolleru, A Wetland Ecosystem*. Hyderabad: BS Publications.

http://www.rainwaterharvesting.org/Crisis/Kolleru_etc.htm

http://www.vedamsbooks.com/no30532.htm

http://www.worldlakes.org/activities.asp?activityid=18

http://www.worldlakes.org/lakedetails.asp?lakeid=8621

http://www.wwfindia.org/programs/fresh-wet/koleru.jsp?prm=110

Pollution of the Musi in Andhra Pradesh: River Metamorphoses into Drain

Jasveen Jairath, Praveen Vempadapu and Batte Shankar

The Making of a Toxic Concoction

MUSI IS AN IMPORTANT TRIBUTARY OF THE KRISHNA RIVER. THE RIVER RUNS OVER A gently rolling area in a shallow valley dotted with low craggy ridges, hills and boulders. Its source is in the Anantagiri hills in Ranga Reddy district, west of Hyderabad. It emerges near Vikarabad town, about 90 km west of Hyderabad, and enters the city near Rajendranagar flowing west to east, thus dividing the city into old and new. Before joining the Krishna at Wadapally in Nalgonda district, it runs about 20 km within the city limits and traverses downstream for about 150 km. Total catchment area at its confluence is 11,082 sq km. Musi has three tributaries, namely, Easa, Aleru and Paleru. While Easa is upstream of Hyderabad, Aleru and Paleru join Musi downstream. Another river, the Alum, joins Musi via Bhongir *taluk*.

Sewage system I, Hyderabad, was started in 1921, when the city's population was 4.7 lakh. The total length of the sewage line laid since then is 1,630 km. Needless to say, the existing lines are inadequate for the city's population, resulting in overflows. Rapid urbanisation and industrialisation around Hyderabad have led to a huge increase in waste generation. The sewage is just dumped into the Musi without being treated, thereby causing a deterioration in the quality of water. The two main sources of pollution— domestic sewerage and industrial effluent—have reduced the river to a sewage drain.

Domestic sewerage refers to waste water that is discarded from households. Such water contains a wide variety of dissolved and suspended impurities, disease-causing microbes, residues of cleaning products such as synthetic detergents that come from the petrochemical industry. Most detergents and washing powders contain phosphates, which are used to soften water. Such chemicals have an adverse effect on all forms of life.

The drainage area (for domestic sewerage) includes not only the entire Municipal Corporation of Hyderabad (MCH) but also part of the surrounding municipalities, Secunderabad cantonment, Osmania University, etc. As far as the MCH is concerned, only about 62 per cent of it is covered by a sewerage network. Amberpet Sewage Treatment Plant (STP) is the only one of its kind on the Musi river. The technology used in the plant is outdated—it was installed during the Nizam's rule and was meant for a population of 3 lakh. It is, in effect, only a primary treatment facility. Of the 400–500 million litres per day (MLD) of sewage that the plant receives in a day, it only has the capacity to process 115 MLD. Analysts at the Andhra Pradesh Pollution Control Board (APPCB) discovered that waste water is merely stored between the four walls of the STP to allow the dirt to settle and then released without any treatment. There are no aerators in the treatment plant.

In addition, industries in the Musi catchment area too discharge waste into the river without putting it through a secondary treatment system. The sewage contains solids like sulphates and lead, and their

Fig. 1. *Sewage entering riverbed is used for growing paragrass.*

volumes are much higher than the limits prescribed for discharge into inland surface water. The coloured effluent does not allow sunlight to penetrate the water, thereby hampering the growth of aquatic biota needed for self-purification. The problem has become particularly acute since 1994. There are several industrial estates in Hyderabad, Medak and Ranga Reddy districts where there are about 300 factories. Not surprisingly, maximum groundwater contamination is reported from this area. (Kiran et al., 2002)

Along the banks of the Musi in urban and peri-urban areas, poor landless farmers grow Para-grass, which is sold to milk producers in the city. Their source of irrigation is the untreated waste water (UWW) from the Musi. Once planted, Para-grass can be harvested for more than twenty years without replanting, and fetches the growers an income of Rs. 5,000 per month. But the presence of heavy metals like lead and chromium in the UWW can be harmful for the soil as well as living organisms. They enter the plants via the roots and find their way into the food chain. Traces of these carcinogenic elements have been found in the Para-grass grown in the area.

Government Looks the Other Way

There is now a conflict between communities that are affected by the pollution, organisations that support them, and representatives from industries, APPCB, Hyderabad Metro Water Supply and Sewerage Board (HMWSSB) and the Government of Andhra Pradesh (GoAP). While the APPCB is supposed to be an independent regulatory authority that monitors pollution in the river, it has been virtually co-opted by government agencies to look after their interests and therefore rendered ineffective. The HMWSSB too

Fig. 2. *Polluted water spilling over a causeway.*

is technically an autonomous body, but the GoAP controls it as well because the Chief Minister is Chairman of the Board. The relationship between the industries and these major decision-making and executing agencies is that of allies. Concerned citizens and organisations have been demanding closure of specific industries till adequate measures are taken to control pollution. Protests and campaigns also target government agencies and there has been a concerted effort to create pressure so that judicial decisions are implemented.

There is also an intra-community conflict among those who are dependent on the Musi waters; they belong to the poorer classes. The farmers are opposed to any attempt to control inflows into the Musi, as the river sustains their livelihood despite being a serious health hazard. For, even the slushy, frothing fluid with the intolerable smell and dark colour is a blessing. They obviously feel threatened by any counter-initiative by NGOs/other community organisations that have been fighting for the cause because of ailments brought on by sustained exposure to pollutants. Local protests against industry are to some extent neutralised as a result of non-cooperation of the farmers.

Judicial intervention has met with limited success. Some industries in Pattancheru industrial area have been shut down and the locals are now protesting against the laying of an 18 km long pipeline that is meant to carry more sewage from Pattancheru to Musi. This proposal to lay a pipeline to facilitate discharge of treated effluents into the Musi is hazardous and illegal as per Rule 5 (8) (ii) of the Hazardous Wastes (Management and Handling) 1989. Besides, no proper Environmental Impact Assessment (EIA) has been conducted according to the guidelines specified by the Ministry of Environment and Forests (MoEF). The

APPCB is involved in 113 cases pending before the Andhra Pradesh High Court, and a few others in the Supreme Court *www.appcb.org* (legal cell link).

In accordance with the directions of Supreme Court [Writ Petition (Civil) No.1056/1990], the Central Pollution Control Board and AP Pollution Control Board submitted a joint action plan recommending measures to restore the affected areas to their original state (*http://www.cpcb.delhi.nic.in/legislation/ch15dec02a.htm*). The issue of the pipeline from Pattancheru Effluent Treatment Limited (PETL) to Amberpet Sewage Treatment Plant (STP) has been taken up by an NGO. A writ petition has been filed in the Supreme Court challenging the implementation of the joint action plan. PETL was established as a cooperative venture to treat effluent from a cluster of small and medium enterprises (SMEs). According to the joint action plan, the member industries are supposed to segregate waste into organic and inorganic streams. The plant proposes to carry only the organic, nutrient-rich effluent into the proposed 18 km pipeline that will be linked with the HMWSSB line. The treatment is to be done at the STP in Amberpet. There is a proposal to upgrade the facilities, but at the moment PETL discharges partially-treated effluent into the Nakkavagu stream. Moreover, laying a pipeline to discharge treated effluent into the Musi is not allowed under Rule 5 (8) (ii) of the Hazardous Wastes (Management and Handling) Rules (1989), and no proper EIA has been conducted as per the guidelines of the MoEF.

PETL is a highly subsidised venture and was meant mainly for SMEs. But as things stand now, seven large units out of a total of about ninety industrial units contribute three-quarters of the polluting waste. These seven units can afford to install their own treatment plants.

Court to the Rescue

The Supreme Court Monitoring Committee (SCMC) issued a restraint, on February 2005, against carrying effluent through the 18 km pipeline, a restraint that is binding on the State. The APPCB requested the committee in April 2005 to lift the restraint but was refused in June 2005. Meanwhile, the Industries Ministry in a letter dated May 11, 2005 allotted rupees 4 crore for extending the pipeline to Amberpet. These developments reflect their unwillingness to comply with court orders and also show the lack of coordination between government departments.

Opposing Views

The widespread feeling among civil society groups and communities is that the current situation has arisen out of corruption within the APPCB, weak implementation of environmental laws and inordinate delays in judicial processes, allowing industries to get away with their polluting activities in most cases. The pipeline plan is a cleverly camouflaged scheme that aims to transfer toxic waste from Pattancheru to Musi.

Consider, for example, that the APPCB failed to implement the 1998 orders of the Supreme Court that all members of the CETP should bring levels of pollutants down to specified limits. In July 2000, the Supreme Court observed that in spite of its efforts to monitor the situation there was no change. The greatest threat is from water-soluble non-volatile halogen compounds, some of which exhibit high toxicity and bioaccumulation. The APPCB has been misleading the state government on pollution levels. The Board has been suppressing the fact that 350 tons toxic non-degradable substances are being discharged into the river everyday. The APPCB has been asked to explain the high arsenic level (1.33 mg/l) in the villages on the banks of the Musi, as well as the large-scale fish kill in Enkiriyal Cheruvu that is fed by the river hardly

30 km from Hyderabad. The argument that the UWW contains all the nutrients required by crops is a dangerous and misleading assumption. There are cases where it can have an adverse impact on productivity.

Government agencies, on the other hand, quote paucity of funds as a major problem, but it is obvious they know on which side their bread is buttered. During a public protest, a member-secretary of the PCB remarked that industrial pollutants in the Musi do not comprise even 1 per cent of the total sewage in the river. He argued that the bulk of waste comes from untreated sewerage—a problem, he said, that would be rectified once the STP at Amberpet is upgraded.

The concerned industrial units have organised themselves into groups such as the Organisation of Pharmaceutical Manufacturers, the Federation of AP Small Industries Associations, the Bulk Drug Manufacturers Association (India), the AP Plastic Manufacturers Association, and so on. They 'talk' to the government and the APPCB through these bodies.

Some academics believe that it is all right to use UWW to irrigate the Para-grass; they believe that such small-scale low-cost farming is a livelihood option for the poor who have no other alternative. They also feel that the presence of nutrients in the water can enhance productivity of some crops.

The Bearer of Death and Disease

There have been reports that the toxicity of the Musi is the cause of cancer among those living in the area. Contamination of drinking water in Hyderabad, Ranga Reddy and Medak districts is also a major concern (Reddy, 2002). In some villages, the borewells supplying drinking water are now lying unused due to contamination. There are instances of skin ailments among those who work in the fields. Even animals have begun to show signs of disease. Those who cannot afford to buy water have to use the contaminated groundwater and suffer from ailments relating to skin, bone and nerves. Stunted growth and problems with the reproductive system have also been noted. Traditional livelihood has been severely affected in rural areas. Apart from drastic reduction in agricultural output, the quality has also decreased and it fetches a low market price. The fishing community too has been badly affected. The Yadavs, a community involved in cattle rearing, have been forced to change their occupation due to scarcity of drinking water and fodder for their animals. The washermen have migrated to the city and work as casual labourers. The weavers, agricultural labourers and toddy-tappers have also been affected.

Control Garbage Production

The focus of the state agencies is on coping with waste generated rather than attempting to regulate polluting industries and reducing waste.

The current struggle for a clean Musi is characterised by strategies that are costly and unpredictable. They rely heavily on conflict, litigation and lobbying. All the strategies are administration centred, with the NGOs and community organisations targeting government agencies and politicians in courts. The stalemate could be resolved through special courts that can facilitate speedy justice. The Right to Information Act must be made functional to enforce accountability and transparency. Public awareness campaigns can be strengthened through media participation and strengthening of existing community organisations. Bringing the industry to the negotiating table is important if viable solutions are to be found.

More attention needs to be focused on reducing waste rather than handling and managing waste. Source reduction offers many benefits including reduced waste management costs and lower liability risks, and because it often means upgrading or modifying production processes, it typically improves efficiency and productivity. The idea is not to create waste—one does not have to manage what one does not generate.

Note

*All figures by authors.

References

Kiran, D. Sai, M. Purnend and Kausalya Ramchandran. 2002. 'GIS for Non-Point Source Pollution Studies in Ranga Reddy and Medak Districts of Andhra Pradesh'. Report by GIS Lab. Hyderabad: Central Research Institute for Dryland Agriculture (CRIDA)–Indian Council of Agricultural Research (ICAR).
'Musi Jalam..Halahalam' (Telugu), Editorial, *Eenadu,* November 11, 2000.
Reddy, Muthyam. 2001. 'Predatory Industrialization and Environment Degradation: A Case Study of Musi River'. Hyderabad: Osmania University.
Rao, J. Rama. 2003. *Report of Pipeline Transfer of PETL Effluents to Amberpet STP.* Hyderabad: Forum for Better Hyderabad.
Supreme Court Monitoring Committee (SCMC) Directives Order No. 14.10.2003–WP 657, 1995.
Vempadapu, Praveen. 2004. 'Management of River Waters in Context of the Musi River, Hyderabad'. Report. Hyderabad: ICFAI Business School of Management and the South Asia Consortium for Interdisciplinary Water Resources Studies (SaciWATERs).
www.apspcb.org.
www.cpcb.delhi.nic.in/legislation/ch15dec02a.htm

Water Turns to Sludge in Kolhapur: Villagers Ransack Industrial Unit

Binayak Das and Ganesh Pangare

An Oft-Heard Story

THE VILLAGE OF CHIPRI IS NEAR THE TEXTILE TOWN OF ICHALKARANJI IN KOLHAPUR DISTRICT, Maharashtra, and has a population of 5,740 people. Groundwater is the only source of water for the village, both for drinking and irrigation. The conflict here is over the contamination of groundwater because of discharge of effluents from an oxalic acid manufacturing unit located nearby.

In 1997, a local industrial house, the Ghodawat group of industries, established an oxalic acid manufacturing unit in a slope upstream of Chipri. A *gutka* factory (chewable tobacco) belonging to the same owner was also located in the vicinity.

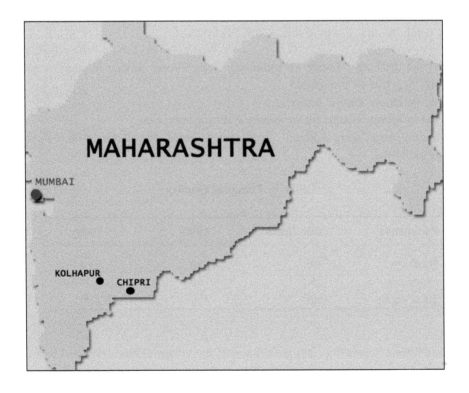

Map 1. *The location of Chipri in Maharashtra.*

The Ghodawat group has, over a period of time, set up many factories around Jaysingpur town that have generated employment in the area. Many villagers work in these units. Additional businesses, like eateries and hotels catering to the factories, have also come up. The people of Chipri had no complaints about these factories. However, after the establishment of Star Oxychem Private Limited, the oxalic acid factory, villagers began to notice that the quality of water had begun to deteriorate.

Oxalic acid is used in its diethyl oxalate form by the pharmaceutical, leather, dyeing and bleaching industries. Nitric acid and sugar are the raw materials used in the manufacture of oxalic acid and its derivatives. Despite its numerous uses, oxalic acid is highly toxic and can cause kidney damage, abdominal pain, low blood pressure, tremors and convulsions. The Star Oxychem factory produces about 25 tons of oxalic acid and 5 tons of diethyl oxalate annually. The plant generates about 15,000 litres of acidic effluents daily, plus 2–4 kg of hazardous waste. An effluent treatment plant (ETP) was installed in 2000 with a capacity of 10,000 litres a day and was upgraded in 2002.

The villagers were apprehensive about the chemical unit right from the beginning. When the factory was set up in 1998, the local people had the water tested and found the quality good. But a year later when they repeated the test it had deteriorated (see Table 1), and they came to the conclusion that the oxalic acid plant was responsible for the change. There was, by this time, a high level of air pollution too. Though the people started appealing to district authorities to run a check, there was little response from them. Then, in April 2002, protests intensified under the banner of Chipri Bachao Kruti Samiti; the Maharashtra State Pollution Control Board (MSPCB) sent a show-cause notice to the factory and on May 13, 2002, ordered its closure. On June 7, 2002, barely a month after it had been made to cease production, the unit was allowed to reopen on the condition that it would follow these norms:

- The plant will not discharge, directly or indirectly, any effluents outside the factory premises.
- Scrubbing efficiency will be improved.
- It will upgrade pollution control devices.
- It will use solar evaporation cells for evaporation during monsoon.
- If the surrounding areas were polluted despite these steps, the factory will voluntarily stop all manufacturing activities.

Table 1: Plunging Quality

Parameters	BIS limits	1998	1999
Hardness	300	180	294
Chlorides	250	66	520
Total solids	500	520	5,800

The factory continued to function, but within a year the villagers' woes revisited them. There was a decrease in crop yield and an increase in health-related problems. The water from some of the borewells in the area was tested again in November 2003 and January 2004. The reports said that the samples were not fit for drinking or irrigation. There was a very high content of salts and sulphates, and further tests

Fig. 1. *The Oxalic acid manufacturing unit.*

conducted by others confirmed these findings. Total hardness and total solids were between 1,312 mg/l and 3,104 mg/l and between 634 mg/l and 1,426 mg/l respectively in all the fifteen samples tested. The chloride level was also high. The BIS standard (Bureau of Indian Standards) for water is 350 mg/l for total hardness and 500 mg/l for total solids. The villagers reported the findings to the MSPCB but they did not respond. Then, in July 2004, two girls aged between seven and nine died after bouts of diarrhoea and vomiting. About 1,000 other people suffered from gastroenteritis. Attributing this to the factory, the villagers launched their second agitation against the oxalic acid manufacturing unit. On August 15, 2004, during their *gram sabha* meeting, they decided that the factory should shut down and once again appealed to the MSPCB (Maharashtra State Pollution Contol Board) to come to their rescue. In November 2004, they requested that the license of the company not be renewed after it expired in December that year. The MSPCB remained silent, and two youth from the village began a hunger strike on December 30, 2004 in front of the factory. No official from the government responded though the *tehsildar* did visit once. The callous attitude of the authorities enraged the villagers, and on January 1, 2005, a mob of 4,000 people, including women, ransacked the premises of the factory and destroyed eighteen vehicles; they also entered the factory complex and smashed computers, furniture and other office equipment. The total loss was estimated at Rs. 500 crore. The owner of the Ghodawat group shut down the unit as well as their other factories.

Water pollution over the last three years has forced the villagers to stop using groundwater as a drinking water source. They now get their drinking water from the municipal corporation of Jaysingpur, a nearby

Fig. 2. *The villagers have demanded the closure of the*
Oxalic acid manufacturing unit of the Ghodawat Group.

town. But by now, the ailments had begun to grow and many villagers had developed kidney problems like kidney stones. But their greater worry centres around crop failure. Many farmers lost their *pan* (betel leaf), *jowar*, wheat, banana and tobacco crops in 2004 due to poor water quality. The situation was worse for people whose lands were near the factory. At present, there are about 100 wells in Chipri, but the water cannot be used now. Seven borewells and two open wells have been tested and found to have a hardness of more than 3,000 mg/l. Villagers allege that though the ETP (Effluent Treatment Plant) has been installed, its capacity of 10,000 litres per day is not enough to cope with the effluent generated, which is closer to 15,000 litres per day. They also say that, contrary to its claims, during the monsoon the factory discharged effluents into the rainwater which ultimately leached into the groundwater aquifers. The factory claimed that they evaporate half the effluents through solar cells, but the villagers countered that during the monsoon, in the absence of sunlight effluents are allowed to mix with rainwater. The factory complex has nine borewells, which the villagers say may be used to dump effluents.

Though production has been stalled temporarily, the dispute is far from over. The government is yet to take a decision and cases have been filed against those involved in vandalism. The MSPCB has appointed a committee headed by a villager from Chipri to assess the situation; the report is awaited.

Tense but Hopeful

The villagers are demanding the closure of the oxalic acid manufacturing unit alone and have voiced no complaint against the Ghodawat group's other units. However, the piqued factory owners have now threatened to close all their units in Maharashtra and shift them to neighbouring Goa. There is tension between the factory management and the villagers, and the latter are concerned that the closure of all units of the group may mean loss of jobs for the local people, which is bound to affect the regional economy.

Though the villagers are highly agitated now, they can be persuaded to talk to the factory management since the two have previously enjoyed very cordial relations. The MSPCB has an important role to play here. One option is to shut down the factory, which may lead the Ghodawat group to shut all their units here. This may mollify the local people, but can lead to economic loss for the region, and also for the villagers. The second is the initiation of dialogue between the two parties wherein the MSPCB can play facilitator and regulator. One of the best solutions is to go back to the guidelines of May 2002 and enforce them strictly this time round.

Note

*Table 1 from Jamwal (2004); map and figures by authors.

References

Deulgaonkar, Atul. 2005. 'Too Violent? Was the Attack on Chipri Oxalic Acid Plant Right?', *Down to Earth*, New Delhi: Centre for Science and Environment, January.

Jamwal, Nidhi. 2004. 'Toxic Trap—Village Chipri in Kolhapur: Maharashtra is Fighting Diseases, a Factory and Lax Authorities', *Down to Earth*, New Delhi: Centre for Science and Environment, October.

Unclogging the Khari River in Ahmedabad: Stakeholders Come Together to Halt Pollution

Srinivas Mudrakartha, Jatin Sheth and J. Srinath

Proliferating Industries

Aʜᴍᴇᴅᴀʙᴀᴅ ᴄɪᴛʏ ᴡᴀs ᴏɴᴄᴇ ᴋɴᴏᴡɴ ᴀs ᴛʜᴇ ᴍᴀɴᴄʜᴇsᴛᴇʀ ᴏꜰ ɪɴᴅɪᴀ ʙᴇᴄᴀᴜsᴇ ɪᴛ ʜᴀᴅ ᴛʜᴇ highest number of composite textile mills in the country. This gave rise to a large number of small and medium scale dyes and dye intermediates manufacturing units—most of them located in industrial estates promoted by the Government of Gujarat through the Gujarat Industrial Development Corporation (GIDC). Four such areas are Naroda, Odhav, Vatva and Narol on the eastern periphery of Ahmedabad. These estates were the first to be promoted by the government during the late 1960s and early 70s. At that time, zones for industries according to the type of waste generated, existing environmental considerations, etc. were overlooked, and no provision was made for the safe disposal of industrial effluents.

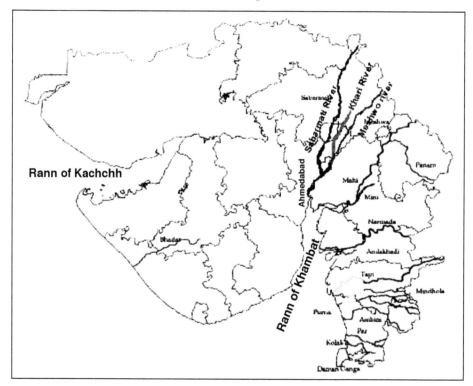

Map 1. *Location of Khari river in Gujarat.*

Most of the factories in the industrial estates are water-intensive and all of them discharge effluents into the nearby Kharicut canal, which flows into the Khari river, a tributary of the Sabarmati. As the canal remains dry throughout the year, the government ignored its (mis)use. The effluents discharged from Naroda and Odhav industrial estates flowed into the canal that connected to Vinzol Vahela, a stream that carries waste from the Vatva Industrial Estate near Vinzol village. The Vatva Industrial Estate houses almost 2,000 industries of which at least 500 fall into the polluting category and let out high volumes of effluent into the stream (Naroda: 3 MLD [million litres per day], Odhav: 1.2 MLD and Vatva: 16 MLD).

Common effluent treatment plants (CETPs) were set up in all three industrial estates after the Gujarat High Court passed orders that they be set up. The CETPs are designed to treat effluent to meet standards related to pH, suspended solids (SS), oil and grease, chemical oxygen demand (COD) and biological oxygen demand (BOD), but they are not designed to meet total dissolved solids (TDS) and heavy metals. Nevertheless, they are useful because the member-industries who also run these CETPs can no longer shrug off their responsibility towards a cleaner environment. As far as TDS and heavy metals are concerned, the solution worked out by the government and the industries involves conveying all the effluents from CETPs, and from factories that have their own secondary treatment facility, into the mega pipeline which joins the Ahmedabad Municipal Corporation (AMC) sewage treatment plant located at Pirana, before being discharged into the Sabarmati. Here, the treated industrial effluent mega pipeline mixes with, almost fifty times greater its volume of treated sewage. At the point where the sewage is finally discharged into the Sabarmati, it reportedly meets all the parameters of the Gujarat Pollution Control Board (GPCB).

Fig. 1. *Vinzol Vahela passing through Vatva Industrial Estate.*

Some small- and medium scale industries continue to discharge effluents into medium and deep aquifers directly through tubewells in a process locally called 'reverse bore technique'. This has resulted in groundwater contamination up to a depth of 183 m. About 110 villages downstream of these three industrial estates (forty on the banks of the Kharicut canal and seventy on the banks of the Khari river) are drinking polluted 'colour' water for twenty years. Representations from the affected villages to the government and pollution control authorities did not yield any results. Finally, the court awarded compensation to the affected villages—the fund is to be used for health and economic development—but nothing came of it because of a disagreement between the villagers and the government on how the money should be spent.

Pristine Beginnings

The Khari river originates from the hills near Nandol, 20 km east of Gandhinagar, bordering Ahmedabad district. This first-order river joins the Meshwo river, a tributary of the Sabarmati near Vautha on the border of Kheda district. Meshwo meets the Sabarmati at Kheda, 50 km down the route.

The 80 km Kharicut canal that begins from Raipur village was constructed more than 100 years ago during the British period for the purpose of providing irrigation support to 10,200 ha in 110 villages, spread over 80 sq km in Daskroi *taluk* of Ahmedabad district and Mahemdavad *taluk* of Kheda district. Today, a total of about five lakh people are affected by air and groundwater pollution, as are thousands of hectacres of agricultural land.

Sick River, Ailing People

In spite of the CETPs, the problem has not been completely resolved. Further, the Khari river and canal have now become a bone of contention between the industries and the farming community, particularly those who live on the banks.

Fig. 2. *Garbage dumped into the Kharicut canal.*

Fig. 3. *Cattle life at risk.*

Owing to the absence of adequate and appropriate drainage systems in these three industrial estates, some of their suburbs let out sewage into the Kharicut canal. While Naroda and Odhav estates manage to dispose of their sewage through soak pits, pipes connected to CETP or nearby AMC drainage pipes meant for domestic sewage, the Vatva industrial estate has an acute problem because of the absence of a sewage disposal network system.

For more than two decades now, people in this belt have been suffering from land and water pollution. Health-related problems include skin diseases, stomach and intestinal ailments, and bronchial problems. The local people say that the mosquito menace is intolerable and that young children often cry right through the night. The strong stench emanating from the polluted water makes it even worse.

Groundwater contamination resulting from the illegal release of untreated effluents into the river during the night or direct injection into tubewells is another serious problem. During 2002–3, when the 183 m deep borewells started yielding contaminated water, the *panchayats* of Chosar, Gamdi, Devdi, Ropda and Vinzol drilled 250 m borewells which also soon started yielding 'colour' water. People are forced to drink this polluted water in the absence of an alternative source. Many families walk long distances to fetch comparatively better quality water from farm tubewells. Livestock casualties and reduction in milk yield have adversely affected the village economy since supplementary income is now denied to the families.

Marriage? No, Thanks

Air and water pollution has also generated a variety of social problems for the people. In Gamdi village, for example, there has been a steady reduction in marriage proposals. The girls' families fear that chronic health problems and difficult daily lives, where women have to walk long distances to fetch drinking water, might make marriage very difficult for them. Also, deteriorating soil fertility has had an impact on agricultural produce which, in turn, has meant reduced income. All this has made the village an unviable place to live in and has led many people to migrate to Ahmedabad city.

Chronology of Events

1978:	People from the affected villages presented their problems to the government.
1988–89:	A group of villagers filed a petition in the Gujarat High Court (Special Civil Application Nos. 7063 of 1989 and 598 of 1989) seeking intervention.
1995:	Two persons from Navagam, Matar *taluk*, district Kheda (one of the Kalambandhi[1] villages) filed a public interest litigation (PIL) in the High Court (Special Civil Application No. 770 of 1995) against government inaction.
1995:	The Gujarat High Court delivered a landmark judgment, 'Polluter Pays', that said: 'Since for the last number of years pollution has adversely affected the 11 Kalambandhi villages of Kheda, as also villages of Lali, Navagam, Bidaj, Sarsa, Aslali, Jetalpur, Bareja, Vinzol and Vatva comprised in Dascroi and Mahemdavad *taluks*, a lump-sum payment should be made by the 756 industrial units, calculated at the rate of 1 per cent of their one year's gross turnover for the year 1993–94 or 1994–95 whichever is more and that amount should be kept separate by the Ministry of Environment and Forests and should be utilised for the works of socio-economic uplift of the aforesaid villages and for the betterment of educational, medical, and veterinary facilities and the betterment of the agriculture and livestock in the said villages.' (Final Order, page 114 – Pt. xii).
1998:	Most of the factories that had been shut down were allowed to reopen after they commissioned primary treatment plants and gave written assurance to the court that they would take the responsibility for secondary treatment either on their own or through the CETP.
1999:	All the three industrial estates had CETPs to take care of secondary treatment for small industries; these are concerns that use less than 25 kilolitres of water per day. Water-intensive industries that consume more than 25 kilolitres of water per day also established their own secondary treatment plants.
2000:	During 2000–1, the AMC laid a mega pipeline with a pro-rata contribution from industries. The mega pipe mainly transports treated industrial effluents from all three GIDC estates; it also carries waste from an Odhav-based estate established by Gujarat Chamber of Commerce & Industries known as Gujarat Vepari Mahamandal Audhyogic Vasahat Ltd., and another estate located at Narol that mostly has textile processing units. The treated effluent was discharged into the Sabarmati at Pirana where the AMC also discharges sewage after secondary treatment. This treated sewage also dilutes the industrial effluent being discharged into the Sabarmati.
2002:	In 2001–2 the government widened and lined the canal to carry more rainwater to avoid flooding; the State also revived it as a live canal by diverting Narmada water through it. Thus, the problem with the Kharicut canal has almost been solved except for the domestic sewage load from the suburbs and occasional leakages from the mega pipeline maintained by the AMC.
2004:	A fresh PIL under SCA No. 4690 of 2004 was filed by the villagers seeking the High Court's intervention for appropriate utilisation of the fund. The High Court has also admitted a *suo moto* (SCA No. 9618 of 2004) application based on a letter by an NGO[2] to the Chief Justice to

address the unresolved grievances of the villagers. The report of the Working Group of the SSF Industry Core Group was one of the main supporting documents.

November
2004:   The Government of Gujarat constituted a high power committee under the chairmanship of the Chief Secretary specifically to address the problem of the Khari river and canal pollution. Apart from the Collector, other members included members from industrial associations, the AMC, Ahmedabad Urban Development Agency (AUDA), GIDC, GPCB, irrigation and industry departments, and the Ministry of Environment and Forests.

NGO to the Rescue

Though the Kharicut canal is now almost free from industrial pollution, the Khari river continues to carry effluents because:

1. Industrial effluents frequently overflow from a few manholes connected to the mega pipeline.
2. A few industries at Vatva Industrial Estate still dump effluents into Vinzol Vahela—they dump the effluents through tankers and at night to save treatment charges in the CETPs.

Fig. 4. *Sabarmati Stakeholders' Forum agricultural group discussing draft water policy.*

A large number of borewells in the villages continue to yield contaminated water. Rainwater harvesting might recharge the groundwater and help dilute the pollutants in a few years' time.

There has been no consensus between the government and the affected villages on the set of activities for socio-economic development. The villages won Rs. 18 crore as compensation as a result of the court's judgment, but that amount, as also the interest on it, remains unutilised as of now.

In 2003, the Vikram Sarabhai Centre for Development Interaction (VIKSAT) studied five villages to understand pollution related problems. In an effort to strengthen the industry sub-group of the Sabarmati Stakeholder's Forum (SSF), a core group[3] was formed in December 2003 with specific focus on four industrial estates: Naroda, Odhav, Vatva and Narol. Regular monthly meetings of the SSF industry core group and discussions on the groundwater, Khari river and Kharicut canal pollution issues facilitated action by the members. Government departments also began to support the cause by providing data, attending meetings and sharing their views. A working group was constituted and it visited the river and canal, covering a distance of about 25 km. Water samples were collected and analysed. The report was tabled at a subsequent core group meeting and then submitted to the Gujarat Pollution Control Board. As a standard practice, the minutes of the monthly meetings are shared among all members including government departments.

The Strife Continues...

The major problems are: First, some factories have refused to cooperate and covertly continue to release untreated effluent into the river; second, there is the issue of leakage from the mega pipeline in the Naroda and Odhav industrial estates; and third, government monitoring and control is not adequate. While serious efforts to resolve the issue should have come from the industries, they largely resort to merely informing the pollution control authorities in case of a problem.

Some Contentious Arguments

Industries:

Industries generate employment and contribute to the economic growth of the state, the country and the people; therefore, the government should take care of factory waste.

Financial impact on the industries: Additional cost incurred for sewage treatment is eating into the companies' profits; this is a difficult situation in a competitive global market.

Villagers:

Damage to the soil and groundwater has resulted in production losses. The household economy has suffered due to problems resulting from pollution; health and social problems have followed. The overall loss of livelihood opportunities is huge, and the long-term consequences are grimmer than loss of employment due to closure of industries.

...But There Is Hope Yet

The response of many of the concerned government departments to the industry core group meetings has been heartening; this positive approach has strengthened the Stakeholders' Forum. Further, the appointment of a high power committee for the Khari river pollution problem by the State government in November 2004 under the chairmanship of no less than the Chief Secretary conveys the right message. As of now, the SSF is not a member of the high power committee. It would be yet another move in the right direction if the two groups were brought together.

The Only Way is Up

Most stakeholders including various government departments certainly appear to be seriously interested in looking for solutions to mitigate the problems of the villagers. Newspapers too are playing their role by helping people air their views. Hopes have been rekindled and it is imperative that the dialogue process moves forward.

Villagers are praying for deliverance from all aspects of the pollution problem. They have certain basic needs:

- Safe and assured drinking water for them and their livestock: This is possible only if all pollution is halted. They hope that the municipal authorities will provide drinking water, including from the Narmada canal. They also demand that contamination of groundwater aquifers should be stopped forthwith through stringent measures.
- Technical support to improve soil fertility.
- Free/subsidised health services in the neighbourhood.
- Special Development Fund for infrastructure development—good roads, schools, drainage system, public toilets, community centre and library.

Light at the End of the Tunnel

A Programme Implementation and Monitoring Committee funded by the award money and supported by NGOs should be instituted to implement the following schemes:

1. Large-scale artificial recharge of groundwater on scientific lines using dry/ abandoned wells, local water bodies such as ponds/ tanks (VIKSAT has demonstrable experience—Mudrakartha 2004).
2. Providing subsidies to farmers for constructing borewells with proper well construction to prevent future rupture of the casing pipe.
3. Suitable compensation packages which include free medical treatment for those who suffered due to pollution.

Because of the Stakeholders' Forum, the people have finally seen a ray of light at the end of the tunnel after twenty-five years of suffering.

Notes

*Map 1 from Irrigation Department, Government of Gujarat; figures from VIKSAT digital photo library.

1. Kalambandhi (bound by the agreement) villages are those that had an agreement with the British government to use Kharicut water for irrigation purpose.
2. Also a member of the SSF Industry Core Group.
3. The Core Group comprises the presidents and secretaries of four industrial associations, representatives of Common Effluent Treatment Plants of Naroda and Vatva, NGOs such as Consumer Education and Research Council, Centre for Environment Education and Centre for Social Justice, representatives of government departments/agencies such as Gujarat Water Resources Development Corporation, Central Groundwater Board, Gujarat Water Resources Department, Gujarat Pollution Control Board, Ahmedabad Municipal Corporation, Ahmedabad Urban Development Authority, etc. In addition, there are also advocates, scientists, architects, engineers, geologists and social scientists from research and academic institutions such as the Physical Research Laboratory and the Ahmedabad Textile and Industry Research Association.

References

Census 2001, Government of India.

'Integrated Plan of Sabarmati River Basin'. Gujarat State Water Resources Planning Group, 1996.

Mudrakartha, S. 'Negotiating a People-Based Governance Institution For Sabarmati Basin Management'. Paper presented at the National Workshop on Deliberative Democracy-Negotiated Development: Prospects For Multi-Stakeholder Platforms (MSPs) in Water Resources Management in India.

———. 2003. 'In Search of an Alternative: A Case of the Sabarmati Stakeholders Forum', Abstract Volume, 13th World Water Symposium, August 13–23.

———. 2004. 'Ensuring Water Security Through Rainwater Harvesting: A Case Study of Sargasan, Gujarat', *Water Nepal*, 11(1): 75–83.

VIKSAT. 2003–04. Annual Reports.

VIKSAT. 2003–04. Minutes of meetings of Sabarmati Stakeholders' Forum.

VIKSAT. 2005. Working Group Report, SSF Industry Core Group. Unpublished.

'Water Scarcity and Pollution Problems in Sabarmati River Basin: A Participatory Approach to Water Management in the Basin'. Project Supported by Gujarat Ecology Commission, Vadodara, 1999.

http://www.gec.gov.in

Bridging the GAP in Kanpur Ganga:
The Failure of Monitoring Agencies

Praveen Singh

An Unholy Mess

THIS CASE CONCERNS ONE OF THE MOST POLLUTED STRETCHES OF THE RIVER GANGA—
specifically the 22 km that fall within the Kanpur city limits. The city covers an area of 1,040 sq km, of
which the municipal area comprises 672.56 sq km.

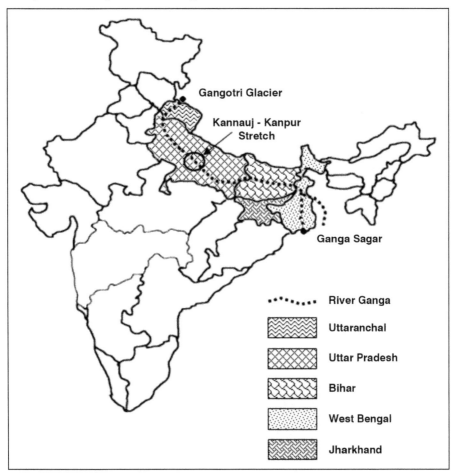

Map 1. *The location of river Ganga.*

Nestled on the banks of the Ganga, Kanpur is one of north India's major industrial centres that has considerable historical, religious and commercial importance. The leather industries (LI) are located in Jajmau, the oldest part of the city.

The geographical area under study is a cluster of villages on the eastern limits of the city, 2 km from Jajmau (henceforth referred to as Jajmau villages). These villages fall within the municipal limits of the city and some of them are situated on the riverbank.

Prior to the implementation of the Ganga Action Plan (GAP) in 1986, the Jajmau villages were being supplied with untreated city sewage mixed with Ganga water (in a 50:50 ratio) for irrigation. This area used to be subject to the vagaries of rainfall because it did not come under any of the command areas of the various canal networks in the region. However, with the introduction of waste water irrigation, the villages prospered and became known for their dairies and floriculture. With the introduction of GAP in Kanpur there was a change in the arrangement; a mix of sewage and tannery effluent was supplied for irrigation after being treated, but without being mixed with Ganga water. This arrangement, ostensibly better because the sewage was treated, is alleged to have completely destroyed agriculture and other means of livelihood in the villages: flower cultivation has declined drastically, groundwater is polluted, stomach and skin-related diseases are reportedly on the rise, as are cattle diseases, and there are fears that poisonous metals and chemicals have entered the foodchain. The villagers claim that this is because both the treatment plants— Sewage Treatment Plant (STPs) and Common Effluent Treatment Plant (CETP)—have failed to treat waste adequately.

There are various conflicts here: between the Jajmau villages and the Uttar Pradesh Jal Nigam (UPJN), which is responsible for operation and maintenance of GAP installations; between the LIs and the UPJN and the Uttar Pradesh Pollution Control Board (UPPCB); and indirectly, between the villages and the LIs. These are apart from another dispute that involves the villagers and Kanpur Nagar Nigam (KNN), the agency responsible for collecting irrigation cess from the villagers.

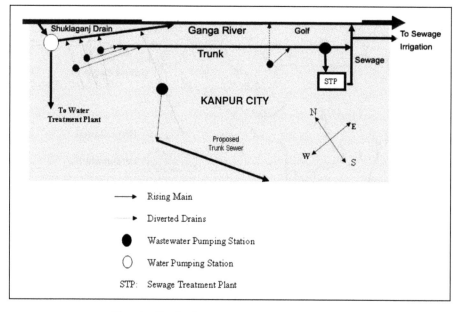

Fig. 1. *Sketch showing GAP I assets in Kanpur.*

The Ball is in the Court

The problem is that the growth of tanneries has not been matched by increasing the capacity of the effluent treatment plant. Of the 220 chrome tanning units in Kanpur, 110 were supposed to install their own chrome recovering plants (CRP), but only eighty-eight have done so until now. And only some of them actually operate these units since running costs are high. Although the proposal to set up a common CRP (CCRP for the smaller units that tan less than fifty hides) has been approved by the NRCD and the funds sanctioned, the scheme is not yet in place. Therefore chromium and other chemicals still find their way into the irrigation channels and the fields.

The National Botanical Research Institute (NBRI) has been carrying out a study on decontamination of chromium from Upflow Anaerobic Sludge Blanket (UASB) treated tannery effluent and contaminated soil in the area adjoining Jajmau by using phytoremediation techniques. Phytoremediation is a cost-effective technique to degrade and manage wastes and environmental contaminants. Some plants are used to capture, sequester, and break down contaminants to reduce their toxicity.

Some effort is now being undertaken to improve the lot of the people in the affected villages. The villages are being electrified, roads are being constructed, work on a half-finished tubewell at Atwan village has been resumed and the tubewell in Sheikhpur village is already in operation.[1] These are deep tubewells that provide safe drinking water to the affected villages.

However, all this happened only after the court intervened. A public interest litigation (PIL) was filed in the Allahabad High Court in 1997 (*R.K. Jaiswal versus State of Uttar Pradesh and others*). It passed several directives from June 1997 to October 1998. Though the court's involvement initially resulted in a significant improvement in the situation, sustained effort on the part of the authorities has been lacking. For instance, the court ordered that there should be uninterrupted supply of electricity for important installations in GAP, but the diesel generation (DG) sets that were provided at these sites for this purpose do not always operate when there is a power breakdown because sufficient diesel is not provided.

In October 1998, the High Court passed an order that stayed the implementation of GAP II till a technical committee nominated by the court was formed (the court also directed the Central government to provide funds for the creation of a River Police Force). The Government of India, through the Ministry of Environment and Forests (MoEF), moved a Special Leave Petition against this order and obtained a stay on it from the Supreme Court. Later, the Sankat Mochan Foundation of Benaras,[2] through two corporators, also intervened in the case in the Supreme Court. The court sent the case back to the High Court where it is still pending.

The government eventually agreed to dump the poisonous tannery sludge in a landfill at Rooma village. The High Court directed the KNN to ensure safe disposal of the waste from the UASB plant, but it continues to be dumped near the villages from where it is later collected and taken to Rooma; and even at Rooma, work on the preparation of the landfill is yet to begin. Meanwhile the dried sludge—all of which is toxic—is sometimes also used by the UPJN to fill breaches in its canal.

The villagers have refused to pay the irrigation cess and the KNN seems to have surrendered. But for the mayor and the corporators the real problem is the media's persistent interest in the case and the pressure exerted by civil society organisations. Politicians and bureaucrats too visit the villages off and on.

Tracing the Conflict

1. The Allahabad High Court reacted to the PIL lodged in 1997. Its directives forced the government to take action and instilled a sense of urgency into its efforts to resolve the problem.
2. In the beginning of 2000, Eco Friends[3] mobilised the Jajmau villagers and several protest demonstrations, sit-ins, *gheraos* (of the mayor) and other campaigns were organised. Memorandums listing the villagers' demands were submitted to the authorities. The issue was highlighted by the media, and the NGO wrote letters and sent reports to relevant parties.
3. An important meeting between all concerned individuals took place in the beginning of 2000 when the installation of CRP was made mandatory. The Chairman, Central Pollution Control Board (CPCB); Director, Central Leather Research Institute (CLRI); a representative of the National River Conservation Directorate (NRCD); the Environment Secretary, Uttar Pradesh government; officials from the Industries Department; representatives from all the tannery associations; and Eco Friends were all present. Prior to this meeting, the Additional Secretary/Project Director of NRCD and secretaries of various departments, including Chief Secretary, Uttar Pradesh, visited the affected villages. A decision was taken to ask all the 110 medium and large chrome tanning units to install their own individual CRPs and the other 110 small units were to set up a CCRP by March 2000. This deadline was extended every three months. As far as the CCRP was concerned, small tanners were requested to contribute to a feasibility study that would be carried out by CLRI. However, they did not do so and the study was ultimately fully funded by the CPCB.
4. Eco Friends called for a boycott of the Uttar Pradesh Assembly elections in 2002. Although the call was later withdrawn, it did its damage. The local politicians who had been neutral to begin with now became antagonistic towards the NGO in particular and the cause in general. There were also signs of disunity among the villagers (divisions based on political affiliations) and the movement lost momentum.
5. Eco Friends commissioned tests to check the quality of water in shallow tubewells, canal water and sludge with the help of the Facility for Ecological and Analytical Testing (FEAT) Laboratory, Indian Institute of Technology, Kanpur, and the Industrial Toxicology Research Centre (ITRC), Lucknow. The test results sent shock waves through the government (CPCB, CLRI)[4] and international agencies like UNIDO and others like the NHRC.[5] The national and international media also highlighted the issue. Eco Friends then organised public hearings in the affected areas, and there was a discernible change in the attitude of the government officials who attended these meetings and listened to the woes of the people.
6. In response to a letter written to G. Thyagarajan, Chairman, Supreme Court Monitoring Committee (SCMC) on Hazardous Waste, and a writ petition filed with SCMC by Eco Friends, SCMC member Claude Alvares paid a preliminary one-day visit to Kanpur on September 18, 2005. He was accompanied by the officials of UPPCB, KNN, Jal Sansthan and UPJN.[6]

Different Stakes, Different Views

The villagers of Jajmau want the earlier arrangement of supplying treated sewage mixed with Ganga water to be restored and the dumping of UASB plant sludge near their village halted forthwith; they want safe drinking water; they also demand that the KNN not badger them with legal notices for not paying the irrigation cess.[7]

The villagers are concerned that they should not be made scapegoats for the failure of GAP.[8] They have already endured hardship and continue to suffer on account of various health problems, decreasing crop productivity, and decline in dairying activity. They hope that studies will be undertaken to examine the long-term effects of the polluted water on their land and health. They have done whatever they can to highlight their plight and now await action to relieve them of their problems.

The LI argue that they have contributed towards the setting up of the CETP (17.5 per cent of the total cost). They believe that the waste water is useful for irrigation and have been flaunting the results of a test conducted at their behest that shows no traces of harmful chemicals in the effluents.[9] The LIs also maintain that since they have all set up CRPs on their premises they are not to blame for the presence of chromium or any other metal in the waste water. They claim that they contribute handsomely to exports and earn valuable foreign exchange for the country apart from providing employment to a large number of people. They believe that the authorities are shifting the blame while the fact is that GAP has failed. They also accuse the monitoring agencies (and local NGOs) of targeting them because they belong to the minority community.[10]

The UPJN takes care of most of GAP's assets and argues that it is constrained by many factors including finances. The agency contends that it does not have the means to replace rusting machinery and passes the buck to the UPPCB, which, it says, does not act against tanneries; the administration fails to collect the O&M dues. Moreover, the machinery is not designed to treat the kind of effluent that is released.

The KNN and the UPPCB, on the other hand, hold the UPJN responsible for the poor maintenance of CETP and the PCB has, in fact, issued a notice to UPJN in this regard. The UPPCB is the agency that monitors and enforces rules, but is considered one of the most corrupt agencies in the State. But under pressure, it has come up with a list of erring industries that have been closed down. The regional officer of the PCB has recently opined that tannery waste should be segregated from sewage and that a mixture of the two ought not to be used as irrigation water.

Being an elected body, KNN could play a crucial role in resolving the conflict. But its role is severely restricted because of limitations of the 73rd and 74th Amendments and the unwillingness of politicians and bureaucrats to devolve powers.

Passing the Buck

Flower cultivation, the main economic activity in Jajmau, has almost completely stopped. Agricultural productivity has declined drastically. Land and groundwater have been poisoned and, if measures are not taken urgently, the damage could be permanent. The impact on the health of the people living in these villages has been considerable: skin lesions, stomach-related diseases and unnatural abortion, both in human beings and livestock, are common. Aquatic ecology in the Ganga has also been adversely affected because of the high levels of biological oxygen demand and chemical oxygen demand. People who live in and around the Jajmau industrial belt are reported to be subject to air pollution as a result of chemical fumes, the stench of the dead carcasses and glue factories. However, because the impact on ecology and society has not yet been scientifically studied, it is difficult to quantify the damage done.

The problem is that although multiple agencies are involved in the matter there is no single department to coordinate efforts or give, decisive direction to the attempt to alleviate pollution. The situation calls for accountability a clear-cut division of responsibilities and a co-ordinating body with corresponding powers.

It seems that the government made a mistake in accepting responsibility for cleaning up the mess created by the tanneries. Rather, the onus for doing this should have been on the polluting industries—collectively or individually. This can be done even now; government agencies could provide necessary technical inputs and also monitor the activity. The leather industry should be encouraged to form a cooperative to effectively undertake the task of cleaning up its mess. A similar experiment is being tried out in Unnao (right across the Ganga) where a group of LIs have set up a CETP which is run by an independent body.[11]

Meanwhile, until a solution is found, the villagers should be provided with an alternate source of safe irrigation water. They should also be given technical and financial assistance to reclaim their lands and to get rid of the poison that has entered the food chain. A scientific study needs to be conducted to reveal the long-term medical effects of exposure to poisonous water.

The Ganga Action Plan was implemented with financial and technical support from the government of the Netherlands. But there does not seem to be any arrangement to make the assets and agencies created under the programme financially self-sufficient. Also, the government has not been allocating enough funds to meet the O & M costs; this is an effort towards which the industries and the general public should also be made to contribute. Agencies like the CLRI and politicians who are trusted by the leather industries should intervene to resolve this conflict. The communal colour sought to be given to the conflict[12] presents a false picture and should be discouraged.

Fig. 2. *Skin lesions caused by polluted water.*

Notes

*Map 1 and figure 2 prepared by Tare *et al.* (2003); figure 1 from *www.ecofriends.org*

1. Although this area falls under the Kanpur municipal limits, infrastructure facilities like roads, electricity and piped water supply had been absent until now.
2. *http://members.tripod.com/sankatmochan/index.htm*
3. For details on Eco Friends visit *http://www.ecofriends.org*
4. Both CPCB and CLRI promised action; CPCB carried out its own test results and recommended the closure of many non-conforming tanneries and started preparing an inventory of chemicals used by the erring industries.
5. National Human Rights Commission (NHRC) lodged cases (Case No. 25499/24/2002-2003/OC, 260/24/2003-2004/OC, 15160/24/2003-2004/UC) on the basis of the reports that were sent to them. It asked the Central and UP Government for an 'Action Taken Report', but it appears that the Commission is not taken seriously by government agencies.
6. See the Visit Report of SCMC at http://www.ecofriends.org/reports/057SCMC3.htm

7. The villagers have stopped payment of the annual irrigation cess since the last four years. They will not pay the cess until safe irrigation water is made available to them. The farmers who have taken land on lease, however, continue to pay the tax.

8. Since there are stricter conditions for discharging wastes into the Ganga and a regular monitoring is done by independent agencies, the authorities instead supply the so called treated waste to the villages.

9. These results have since been refuted by many other reports that have been published by government agencies and Eco Friends-IIT test results.

10. The government made a mistake by agreeing to clean the waste. The error is being repeated in the CCRP case. Tanners now have an easy escape route and an excuse to lay the onus on the government for any failure on their part. Leather factories were required to set up PETP and CRP before the commissioning of the CETP in 1994, but it never happened. So the treatment plants never functioned to capacity. Tanners are also required to share the O&M cost of the plants but they did not do this and were neither held responsible nor, indeed, they felt responsible for the waste.

11. It should be noted that there are problems with this arrangement too, and the treatment process here is not foolproof either.

12. First, because of the religious associations of the Hindus with the Ganga and second, because the leather industry is dominated by Muslims.

References

Tare, Vinod, P. Bose, and S.K. Gupta. 2003. 'Suggestions for a Modified Approach Towards Implementation and Assessment of Ganga Action Plan and Other Similar River Action Plans in India', *Water Quality Research Journal of Canada*, 38(4), pp. 607–26

http://www.ecofriends.org

http://members.tripod.com/sankatmochan/GangaPollution.htm

http://envfor.nic.in/nrcd/limit.html

http://www.auburn.edu/~alleykd/envirolitigators/jaiswaltext.htm

http://www.kanpurganga.com/default.htm

Blank

PART 4

Mining Riparian Health:
A Review

P. B. Sahasranaman

WATER IS AN ESSENTIAL NATURAL RESOURCE FOR LIFE, WHICH, IN TURN, IS PART OF A larger ecological system. Rivers are the main source of water in our country—for drinking as well as for agriculture. Everyone needs drinking water and a very large population depends on agriculture. Rivers are the primary source of freshwater and sustain life and greenery in their basins. There is, therefore, an indisputable need to protect them. What most people do not see is that protecting the river also means protecting the sand that accrues in a river. All efforts to develop, conserve, utilise and manage this important resource of water have to be guided by a national perspective.

The sand that accrues in a river maintains the river's ecology. It stores water and keeps the riparian environment moist long after the river itself dries up. But sand is also a building material and is being extensively used throughout the country today. Indiscriminate sand mining from the river basins in recent years has started creating serious environmental problems. Excessive sand mining leads to lowering of riverbeds and that, in turn, leads to easy subsurface intrusion of saline seawater in coastal areas. The adverse impact of indiscriminate sand mining is felt in many ways: it may pose a threat to the safety of places of worship, important historical monuments, dwelling units and lands in and around riverbeds. Several religious institutions, temples, churches, and mosques are located on the banks of rivers. The famous Sivaraatri festival at Aluva in Kerala is conducted at a temple that is in the middle of the Periyar River. Again, the famous temple at Kalady, the birthplace of Sankaracharya is located on the banks of Periyar, while Sabarimala, for instance, is near the Pampa River. The list is long and such facts are specifically noted in the expert reports. The excess removal of sand leads to land-sliding and may result in these buildings to collapse. It may also cause depletion of water in wells located close to riverbanks because of erosion of the side banks.

Since sand is a mineral available from all the rivers in the country, the framers of the Constitution of India found it necessary to protect it and gave the Central government the exclusive power to legislate on minerals (Entry 54 in the Union List). The Government of India enacted the Mines and Minerals (Development and Regulation) Act, 1957 in public interest, which specified that the Union should take under its control, the regulation of mines and the development of minerals. However, sand being a minor mineral, the State governments have been given rule-making powers over the control of sand mining.

Several States have enacted legislation pertaining to the mining of sand from rivers and riverbeds, but the rules and the law differ from one state to another. What is common is that in almost every State protective measures are flouted or badly implemented. It is extremely difficult to describe how this happens in all the different States because of the varying rules that have been promulgated. In what follows, we shall concentrate on one State, Kerala, with which I am familiar, to illustrate how this happens. Still more research is needed in all States, on how laws and rules interact and what impact they have on riparian health.

Kerala Prepares Sand Mining Regulations

Kerala has forty-four rivers. The extraction of sand from the rivers in the state and its licensing are handled by the *panchayats* of the villages through which the rivers flow. After it came to the notice of the State government that the indiscriminate and uncontrolled removal of sand from major rivers has caused large-scale land-sliding and loss of property, it took control of sand mining regulations and issued a circular on March 6, 1997, imposing the following conditions:

- An expert committee shall be constituted to determine the place and also the quantity of the sand to be mined from there.
- The area to be mined shall be changed every year.
- Every year after November, a re-survey shall be conducted and the quantity of sand available for mining should be identified and determined in May.
- Such a study has to be conducted repeatedly and the sand accrued every year should be monitored.
- Check posts shall be placed on every *kadavu* (place from where sand is permitted to be collected and stored). Vehicles shall not be permitted to be taken into the rivers.
- Sand workers should be given identity cards. The watchman should be given a list of workers engaged in the *kadavus*.
- Auctioning of sand should be stopped immediately.
- Passes should be issued (at the expense of the local *panchayat*) for each load of sand. The amount received (from the sand miners) should be apportioned between the government, the workers and the local bodies.
- A task force of local people shall be constituted.
- There shall be a complete ban on sand mining from June to October.
- No sand mining shall be allowed within a distance of 1 km from constructions.
- Sand shall not be allowed to be used for landfills.
- Substitutes for sand have to be exploited.
- Dumping of human and other wastes in the river shall be strictly prohibited.
- The local administration shall make alternative arrangements for the discharge of waste.
- No agricultural activities shall be allowed to take place on the riverbed.
- Revenue department shall immediately constitute a task force to determine the river boundaries.
- The protection of rivers and river basins shall be entrusted to a separate agency.

The Recommendations of the Expert Committee

The District Collectors constituted a District Level Expert Committee to assess the quantity of sand. It formulated certain measures to regulate sand mining:

- ❏ Sand mining using pole scoop or other mechanised devices should be strictly prohibited.
- ❏ The mining, removal and sale of sand, loading and unloading, storage and all other connected operations should be prohibited within a distance of 1 km on either side of the bridges.
- ❏ No vehicle should be allowed to load/unload sand directly from the river basin. Only the sand stored on the riverbank should be permitted to be loaded into the vehicles.

❑ Mining, removal and sale of sand from the saline water regions of the river should be prohibited until 30 June.

❑ Islands which form/have formed in the river should be protected from erosion due to sand mining.

❑ The vehicles to be loaded with sand should not be allowed to park near the *kadavus*, or along the road to the *kadavus*, beyond the stipulated time of 8 AM to 3 PM.

❑ In order to enforce the recommended regulations and the proposed prohibition at the *kadavus* where extraction has been discontinued, all the shutters may be dismantled and the approach roads fenced off to prevent vehicular traffic. Further to this, all country boats should also be removed.

❑ The crafts, which shall be employed at the permitted *kadavus*, may be registered individually with the *tehsildar* concerned. The registered crafts should exhibit their registration number visibly on all sides.

❑ Arrangements to ensure the supply of sand to domestic and genuine consumers should be made at the local level.

❑ In order to prevent unauthorised mining and malpractices, local-level grievance cells comprising representatives of voluntary organisations and other interested organisations should be constituted. The periodic reports received from such cells should be examined in a timely manner by the district committee.

❑ To prevent unauthorised mining, there should be fully equipped patrols of the river regions during the day as well as night. Extent of time for the removal of sand and the number of crafts to be engaged may be decided by the revenue authorities.

❑ Photo identity cards for the employees should be issued by the *tehsildars,* but only after detailed and thorough verification.

❑ The bathing *ghats* along the river should be preserved and no disturbance should be caused due to sand mining.

❑ Any form of corruption/malpractice noticed in the mining, removal or sale of sand must be dealt with meticulously, and stringent punishment must be imposed on the offenders.

❑ An intertwined locking system should be implemented at the allowed *kadavus* to ensure that the crafts are not used for the mining of sand outside the stipulated time limit.

❑ Taking into consideration the peculiar natural characteristics of the stretch of the river towards the estuarine area, the possibility of allowing mining of sand during the monsoon season may be looked into.

❑ An environment impact assessment (EIA) study should be conducted by the district administration, availing of the services of the Centre for Earth Science Studies (CESS), the Centre for Water Resources Development and Management (CWRDM), and the Soil Division of the Kerala Forest Research Institute.

Had these conditions been strictly followed they would have prevented all illegal sand mining, but that has not been the case. On the basis of suggestions issued by the High Court in public interest, the state has agreed to the reconstitution of the expert committee headed by the Chairman, CWRDM.[1] However, the reports of the committee have been ignored and excess sand mining continues.

The High Court Takes Notice

On the basis of a report by CESS and other materials, the High Court has found that the conditions imposed are not adequate. It issued additional directions on March 26, 2001:

After the period of sand mining holiday from 1ˢᵗ July to 30ᵗʰ September is over , no sand mining shall be allowed by the *panchayats*, unless and until the District Collector, on inspection, is satisfied that kadavus are maintained as per the direction given by this Court and that the *'Karmasamithis'* constituted are functional in each panchayat.[2]

Therefore, it is imperative that the District Collector examines all the *kadavus* and issues a sanction order before the commencement of sand mining each year from October 1 every year.

Despite all these conditions, excessive sand mining continues. Several public interest litigations have been filed before the High Court in this respect. Finally, the court disposed of all such cases by a common judgment, directing the implementation of the present conditions with liberty to legislate.[3] In addition, the court added the following:

❑   No sand mining shall be allowed, unless the conditions incorporated therein are followed, including the obtaining of report from the experts.

❑   Constructions made within rivers (as pointed out by the experts), reported to have an adverse effect on river ecology, should be demolished immediately by the local authority and the expenses incurred on this account should be recovered from the concerned persons.

❑   *Karmasamithis* should be constituted in all *panchayats* and made functional before allowing sand mining. They shall point out all the violation of conditions. All reports submitted by *Karmasamithi* members and the public should be immediately considered by the District Collector, police officials and other authorities, and immediate action should be taken to prevent such violation.

❑   No *panchayat* shall be allowed to extract sand unless all the arrears of royalty and the contribution to the river management fund (RMF) are remitted.

❑   The river management fund should be utilised for protecting the river and the riverbanks.

❑   River sand should be allowed to be taken outside the State only if it is available in excess quantity. (The normal rule prescribed is that river sand from the river should be used in the same district through which the river flows. There are forty-four rivers in Kerala and they flow through most of the districts. Therefore, the authorities can decide how many constructions are planned and how much sand is required.)

❑   Steps should be taken to demarcate the river boundaries and protect them.

❑   The government should consider the constitution of a river protection force and a state river management authority for protecting all the rivers from excessive sand mining.

❑   The government should encourage alternatives to river sand so that construction activities can go on.

❑   All police authorities, including commissioners, should ensure that no illegal sand mining is undertaken in violation of the above conditions.

The 2001 Act: A Step Backward?

The Government of Kerala enacted the Kerala Protection of Rivers and Sand Mining Regulation Act, 2001, to overcome the judgment with these objectives:

•   Regulate the indiscriminate and uncontrolled sand mining from the rivers.
•   Solve the difficulties faced by the construction workers because of the existing administrative regulatory directions.

- Provide the procedures to control sand mining in public interest and for the protection of the river valleys.

The restrictions imposed by this new legislation were:

❏ No sand mining shall be done in any *kadavu* before 6 AM and after 3 PM.
❏ The concerned *panchayat* or municipality shall make arrangements for sand mining in accordance with this Act and the rules made thereunder.
❏ No sand mining shall be allowed in excess of the quantities fixed by the district expert committee.
❏ Sand shall be allowed to be mined only from the riverbed, and no sand mining shall be allowed within 10 m from the boundaries of the river.
❏ No sand mining shall be allowed within 500 m from the bridges and irrigation projects.
❏ No vehicles shall be allowed to be taken to the river. Vehicles shall be kept at a distance of at least 25 m from its banks.
❏ No net or pole scooping or machinery shall be allowed for the purpose of sand mining.
❏ No sand shall be removed where there is likelihood of saline water mixing with river water.
❏ Power is vested in either the State government (at the level of the cabinet, secretariat, etc.) or the District Collector to order closure of sand mining from *kadavus* or river banks.
❏ It is the obligation of the local authorities to maintain the *kadavus* or riverbanks.
❏ No sand shall be removed from a river or any riverbank where the government has expressly prohibited the same by a general or special order.
❏ The *Kadavu* Committee shall fix the price of sand for each *kadavu*, taking into account the availability and accessibility of sand.

The Act empowers the constitution of a River Management Fund (RMF) for meeting all expenses towards the management of the *kadavu* and the riverbanks. Every local authority with a *kadavu* or riverbank shall contribute 50 per cent of the amount collected from the sale of sand towards the RMF maintained by the District Collector. It also contemplates the formation of a river bank development plan to be implemented with the help of the RMF. The Act also contains provisions for the confiscation of vehicles and punishment.

Through this enactment the state has actually reduced the 1 km restriction imposed on the mining of sand from the irrigation projects to 500 m. There is no compulsory sand mining holiday during the monsoons. The practice of auctioning sand has again commenced. *Panchayats* have been given wide discretionary powers. The restriction imposed by the earlier orders on usage of sand for landfills, and the requirement of passes and obtaining of reports have been removed. The expert committees comprises non-professionals and politicians who fix the quantity of sand to be mined on extraneous considerations.

The new law has failed completely in protecting the river ecology. Illegal sand mining continues and the price of sand has increased. Even people living on the banks of the rivers have started digging their land so as to extract sand for sale. This has started adversely affecting the neighbouring lands. The lack of proper provisions in the enactment for confiscation and punishment has given illegal miners an upper hand.

Need for a Comprehensive Approach

What is described here is only an example. The situation in most of the States is similar. Sand is indiscriminately extracted as if it is a free bounty of nature. The regulations imposed on the basis of expert opinion are blatantly violated. The absence of proper legislation for river protection has invited the attention of the court in public interest, calling for intervention in the protection of river ecology. The need for protecting the river ecology and the necessity to have a proper river management authority has been ignored. The authorities empowered to grant licenses are bound to take all precautionary measures to prevent excessive sand mining which destroys river ecology, the only source of drinking water. The UN resolution passed in the Water Conference of 1977 was the following:

> All people, whatever their stage of development and their social and economic conditions, have the right to have access to drinking water in quantum and of a quality equal to their basic needs.

The excessive mining of river sand results in the destruction of aquatic and riparian habitat which has an adverse impact on the ecology. What is important is that the quantity of sand that can be safely mined should first be determined. For example, accumulated deposits and annual accruals can be estimated and the rate of sand mining should be so determined that over a long span of say, twenty-five years, there shall be no appreciable impact of the rate of extraction on riparian health. The quantity of sand allowed to be mined should be fixed on this basis of accrual.

It is essential to provide 'justice' to the rivers through proper legislation and restrictions should be imposed to avoid illegal mining. A river protection authority with ample powers and responsibility should be constituted and entrusted with the control of rivers in the States; control not only of sand mining, but also pollution, use of country boats, construction of riverbanks, irrigation projects, mining, and all other steps required for the protection of the rivers.

Obviously, there will be a gap between the demand of sand for construction and the amount allowed to be mined. There are alternatives to sand, such as crushed stone, quarry dust, glass powder, fly ash, and material derived from wastes. It is important that the government actively promotes these alternatives and plans policies so that they become viable and comparable to sand. At least one of these measures is to fix a high price for sand. At present, it is valued purely at collection cost, and a price so low that no alternative material can stand in competition. The value of sand must incorporate the cost to riparian health that it represents. Unless we do that, we shall be allowing the construction lobby to build houses that are, brick for brick, built at the cost of the health of our rivers.

Notes

1. O.P. (Original Petition) No. 19798 of 1997 dated December 4, 2001.
2. O.P. No. 16272/2000-K dated June 21, 2001. Delivered by Honourable Chief Justice K.K. Usha and Justice Joseph Kurien.
3. For the judgment and other enactments on mining, refer to http://www.elaw.in

Sand Mining in Coastal Tamil Nadu:
A Threat to Local Irrigation Sources

Benjamin Larroquette and Gaspard Appavou

Coastal Dilemma

V ADA AGARAM IS A SMALL COASTAL VILLAGE SITUATED APPROXIMATELY 120 KM SOUTH OF Chennai and 30 km north of Pondicherry in the Marakanam block, Villupuram district, Tamil Nadu. The population is slightly more than 1,000. The soil here is very sandy. The main economic activity is agriculture, practised on 24 ha of land, with paddy being the most important crop. The area is unique in that irrigation is chiefly provided by three natural springs. These springs are in the form of small ponds into which water percolates through the sand.

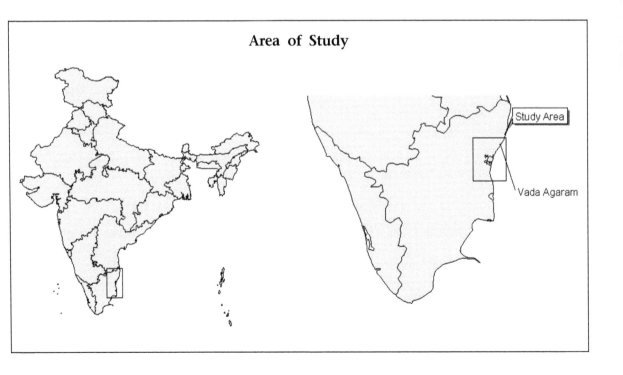

Map 1. *The area of study.*

Sandstorm

The conflict began sometime in 1993 when it was discovered that the sand in some areas of Vada Agaram contained high levels of pure silica, a material used mainly for cement grade testing and glass production. The Mining Department of the Government of Tamil Nadu decided to utilise this resource. The villagers opposed the move claiming that the removal of sand would affect the availability and quality of water in the springs. They insisted that mining would reduce the quantity of water because sand acts as a water retainer. Also, they emphasised that the probability of seawater seeping into the soil and making the water unfit for drinking and irrigation would increase.

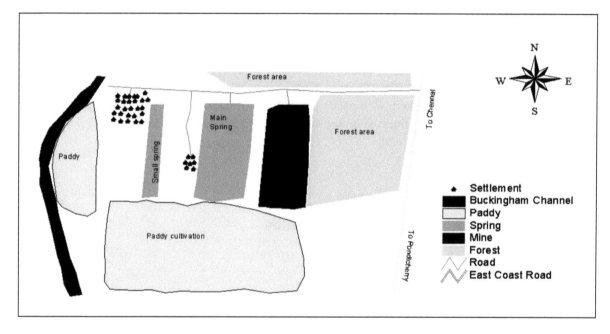

Fig. 1. *This is a sketch of the the Vada Agaram village with the location of the mine, main spring and cultivated areas. The sketch is not to scale and was derived from participatory discussions with the villagers.*

Events

1993: The Tamil Nadu Minerals Ltd. (TAMIN) begins mining activities in Vada Agaram despite local opposition.

1996: The people of the village protest against the mining on East Coast Road. The collector is called in, and he organises a conflict resolution committee to address the issue. He also acknowledges that the government needs the sand and suggests that the villagers take the matter to court.

1996: The Chennai High Court orders a probe by the Pollution Control Board (PCB). Mining activities are stayed. The PCB report of 1997 favours the villagers.

1997: The judge is transferred and his replacement rejects the PCB report saying that the department has no authority to intervene in the matter.

1999: Mining activities resume.

Awaiting Justice

This is an ongoing conflict. Mining continues to take place on a regular basis and the villagers are still opposed to the activity. The matter is now pending with the Supreme Court (petitioned by the villagers), but lack of funds has halted the effort to seek rapid redressal.

Highs and Lows

The protest on East Coast Road was probably the highest point of the conflict. Though the collector was called to address the issue, he made it very clear that it would be difficult to stop the government from mining. His advice was to take the matter to court. The lowest point of the conflict (in the villagers' perception) was when the PCB released a report favouring their claim. At that time, a resolution had seemed within reach.

At Loggerheads

The villagers obviously want the mining to stop because they feel that it will gravely impact the natural springs that are their source of irrigation and drinking water. They feel that mining will reduce the retention capacity of the sand and cause salinity through seepage of seawater into the springs. They claim that the water level in the spring has already receded. The issue was taken to the collector and, subsequently, to the Chennai High Court.

The impact on local ecology and economy, according to the villagers, could be considerable. The main spring irrigates approximately 10 ha of paddy over two seasons. At about 62.5 bags of paddy per ha (assuming there is a change in the quality and quantity of water) the loss will amount to almost Rs. 2,50,000 year. This figure might well double if drinking water considerations are taken into account.

The government, on the other hand, is all too keen to exploit a natural resource that is usually found at a relatively greater depth. The land belongs to it and it denies that mining has any negative effect on ecology. Also, TAMIN claims that it has been replacing the mined sand with seashore sand and the problem of saline water seepage is therefore unlikely to occur. Besides, it has been providing employment to poor women who are paid more than what they are likely to get in the private sector (Rs. 450 per truck instead of 300). The government feels that the villagers just want to get some money out of it.

Moderation Will Pay

A resource as valuable as the sand in Vada Agaram is not going to be sacrificed by the government. As mentioned earlier, the sand—rich in silica—is used mainly to test different grades of cement. In the context of the current scenario in Tamil Nadu, where construction is a vehicle for development, ignoring a resource such as this would be detrimental to their interest. However, mining should be done with great care to avoid damaging the aquifer.

A court battle can be both expensive and protracted and can, therefore, not be considered an ideal method to find solutions to such problems. Various other options can be explored. For example, there exists a federation of all water user associations in the area. This federation comes under the tank rehabilitation project implemented by Palmyra and is called the PALMYRA–ICEF Water User Association Federation. The farmers can appeal to this federation, which has as its members big farmers and other influential members. The federation might find it possible to force the government to ensure safe mining procedures and greater consideration for environmental issues; for, while TAMIN has stated that the mined sand is replaced with sand from the seashore, they must consider that there is a risk of actually increasing the salinity levels by using sand which is already salty.

Another option would be to try and involve the farmers in the process of mining. Increasing the participation of the local population might create a sense of belonging and ownership and thereby reduce apprehension among the locals. Perhaps a percentage of the profit can be allocated to the *panchayat* and earmarked for village development.

Clearly, we are facing a case where tradeoffs need to be made. It will be unfair to jeopardise the village water source but it does not look like the mining can be completely halted. Monitoring systems need to be put in place to regularly test the quality of water, and the company must also put a ceiling on the quantity of sand that can be mined at the location.

Note

*Map 1 and figure by authors.

Sand Mining in Papagani Catchment in Karnataka: Groundwater Depletion in Andhra Pradesh

M. Chandrasekhara Rao

About the Area

THE STUDY AREA IS IN PEDDA TIPPA SAMUDRAM (PTM), *MANDAL* OF CHITTOOR DISTRICT AND Tanakal *mandal* of Anantapur district, Andhra Pradesh, and Bagepalli block in Kolar district in Karnataka. The issue under consideration here has affected a total of fifty villages.

The predominantly affected villages are Kuntapalli, Yerrupalli, Darwalapalli, Chinnabayapalli, Guntipalli, Balireddypalli, Chenraypalli, Chinnapalli, Ramapuram, Yenudala, Anantapuram, Chimanerpu, Jammukanapalli, Kottudam, T. Sadum, Chelur, Puttaparthi and Rekulakuntapalli.

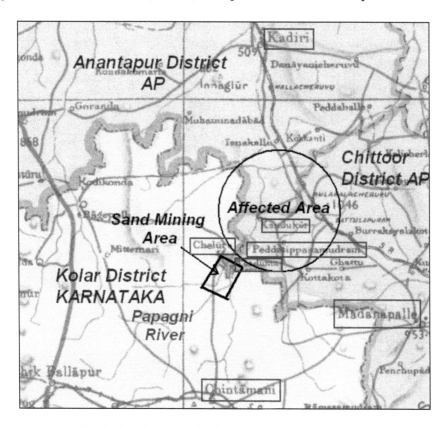

Map 1. *Location of the Sand mining area and the affected area.*

Sand Mining Procedure in Andhra Pradesh

A district-level technical committee comprising a river inspector, and a member of the Groundwater and Geology and Mines Department, inspect the rivers in the district and submit their report to the administrative authority, i.e. the District Collector, District Panchayat Officer, Joint Collector, etc. Based on this feasibility study, an area is auctioned for sand mining.

The Conflict

Illegal sand mining from the Papagani river in Kolar district, Karnataka, has been going on for six or seven years. The Karnataka government initially gave sand mining rights to some contractors, but illegal and excessive mining led to the depletion of groundwater levels in the villages on the banks of the river, causing environmental degradation and problems for the people. This is the border between the two States and the therefore there are implications for both Andhra Pradesh and Karnataka.

It was observed that sand mining has occurred in a haphazard, irregular and unscientific manner. Quarrying has created water stagnation in the riverbed, impairing the natural flow of water which has had a grave impact on agricultural production because of inadequate water for irrigation. The use of machinery like proclains for removal of sand has caused riverbed erosion, collapse of banks, damage to infrastructure like bridges, transmission lines, trees on the bed and the banks, and has also created problems in the drinking water systems. In all these places, the contractors have exceeded the area allotted to them and mined more than the permissible depth. This has resulted in deepening of the riverbed, widening of the river, damage to civil structures like drinking water schemes, culverts and bridges, depletion of the groundwater table, and degradation of groundwater quality.

Fig. 1. *Over-exploitation of sand deposits in Papagani river in Kolar district, Karnataka.*

Chronology of Events

1995: The groundwater levels were satisfactory
1997: Government of Karnataka allotted sand mining rights to contractors
2000: More areas auctioned for sand mining
2003: Farmers and local politicians protest
2003: Farmers lease land for sand mining
2004: An increase in water problems due to over exploitation of sand by illegal methods

Current Status

Wells and borewells are dry, and groundwater has been severely depleted in Tanakal *mandal* in Anantapur district and PTM *mandal* in Chittoor district, Andhra Pradesh.

Sudhakar Reddy is a farmer who belongs to Anantapur village, PTM *mandal*, Chittoor district. He has an open well and a sand bore in the riverbed. His open well (located about 100 m from the riverbank) yielded good water until 1997. But things started changing with the onset of sand mining, and today the open well is completely dry. Reddy recalls that the water level in the well was equal to the water in the sand bed before sand mining began. Sand retains water and helps recharge wells. Apparently, there is no institutional mechanism to support the farmers' cause and they are unable to resist the contractors.

The farmers of Andhra Pradesh feel that the depletion of groundwater levels is mainly to be attributed to sand mining at the riverbed. At present, mining has been temporarily halted. The Government of Andhra Pradesh has also taken measures to control the activity by seizing vehicles that have been ferrying sand from the riverbed. Revenue and police officials have been given instructions to take appropriate action against illegal mining.

There are seventy-two protected water supply schemes in the PTM *mandal*, out of which fifteen had already dried up by January. Seventy-five hand bores have also run dry and only twenty-five bores out of a total of 316 have any water.

About 500 sand wells/filter points have been affected, resulting in heavy losses to farmers on both sides of the river. Each filter point normally irrigates about 2 ha, which means that the total loss stands at 1,000 ha worth of crops. A single crop loss at Rs. 5,000 per ha totals Rs. 25 lakh, and if the area is doublecropped, the annual loss is Rs. 50 lakh.

Highest and Lowest Points

Sand mining was at its peak between August and October 2004. The agitation led by local leaders resulted in a temporary lull. The case was taken to court and the judicial authority promptly slammed a stay order on account of declining water levels. But the contractors' lobby managed to get the stay vacated because the *panchayat* secretary certified that mining was not responsible for decrease in the availability of drinking water. Revenue officials too agreed, and mining took on a frenzied hue. While official figures quote that 1,000 lorry loads of sand are lifted daily from Chelur, the locals claim that close to twice that amount finds its way to Bangalore. The media have certainly done their bit to highlight the problem, and in one instance were even assaulted by miners for their trouble.

The Opposing Stands

The government claims that it has been generous with mining leases because floodwaters have been entering the fields and causing sand-casting problems. The area that is mined is supposed to be earmarked on the basis of depth of sand bed and condition of the embankment. However, there is no monitoring and the riverbed is dug three to ten times the depth mentioned on the license. Often the sand is deposited on private lands. To get the farmers on their side, contractors sublet part of the tank-bed area to them. These farmers in turn lease their land to contractors with the short-term goal of earning quick money. Karnataka farmers felt that if they did not make the most of the opportunity, the profits would go to their counterparts across the border. Also, some local farmers felt that the activity provided them alternate employment in the lean season, as rainfall has been scanty resulting in declining agricultural activity in the last five to seven years.

Scope for Dialogue

The Government of Karnataka should exercise prudence when it comes to leasing out the riverbed for mining activities. The government should demarcate areas clearly and monitor mining through a suitable institutional mechanism. Also, since two States are involved in the conflict, an inter-State coordination committee made up of District Collectors and revenue and police officials should be set up to protect people's interests.

Any exploitation of resources calls for a tempered scientific approach. This will not only ensure good revenue to the government, but will also be in harmony with nature. Sand mining in private lands must be halted forthwith and people should be educated on the ill effects of over-exploitation of a natural resource. NGOs and community-based organisations can play an important role in spreading the message. Lastly, switching to artificial sand like robosand (http://www.robosand.com) may reduce the pressure on river sand.

Role of Non-Governmental Organisations

Jana Jagrithi is a local NGO working in Anantapur district. The organisation works mainly on water resources management and conservation with people's participation. Their involvement in the conflict has been in trying to build awareness in the community about the consequences of over-exploitation of the resource.

Note

*Map 1 from *Survey of India*; figure by author.

References

'Isuka Akrama Taralimpunaku Addukatta—Eenadu Kadhananiki Spandana', *Eenadu,* October 8.
'Nidhulanni Isukapalu–Migilindi Fluoride Saapalu', *Eenadu,* October 4, 2004.

The Baliraja Memorial Dam on Yerala River in Maharashtra: A Case for Sustainable Utilisation of Natural Resources

Drought and Distress

KHANAPUR, WHERE THE STRUGGLE FOR THE BALIRAJA DAM TOOK PLACE, IS ONE OF THE eighty-nine drought-prone *taluks* of Maharashtra. It is also the highest plateau region drained by the Krishna river. According to the Sukhatankar Committee Report of 1973, the average rainfall in this region is about 584 mm and the dependable rainfall is much lower.

The river Yerala, one of the major tributaries of the Krishna, originates in Khatav *taluk* in Satara district in Maharashtra and flows southwards into Khanapur and Tasgaon *taluk* in Sangali district for about 120 km before it meets the Krishna.

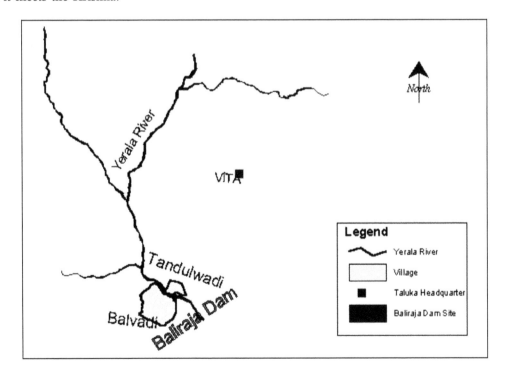

Map 1. *The Yerala river in Khanapur taluk.*

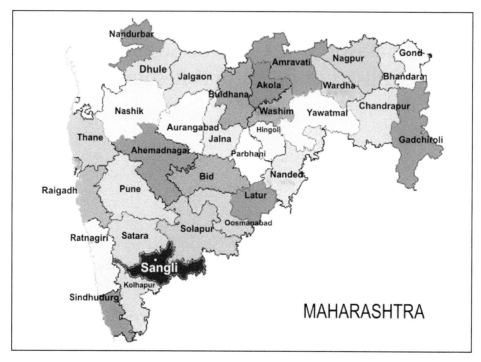

Map 2. *Location of Sangli district.*

The Peasant-King

The villagers have named the dam after King Bali, a mythical peasant-king who offered to be crushed underfoot by the Hindu god Vishnu in his *avatara* as Vamana, the dwarf. Bali is idolised as the perfect king. The farming community recalls his idyllic reign during Dussehra when they pray: *Ida Pida Talo, Baliche Rajya Yevo* (May the sorrows be wiped out, let Bali's kingdom come).

Mukti Sangharsha Chalwal

The movement that led to the building of the Baliraja Dam, Mukti Sangharsha Chalwal, has its roots in the struggle of textile workers who returned home from Mumbai during the great textile mill strike of 1982–83. The movement was founded in October 1983, and initially the focus of the group was drought relief work—they sought fair payment and adequate facilities for the labourers—but gradually their attention shifted to issues that aimed at completely eradicating drought from the area.

Digging for Trouble

Over a period of time, especially after the drought of 1971–72, the Yerala remained dry for more than eight months a year. There were various reasons for this: some medium and small dams had been built at different points on the river. Also, there was increased pumping of water for sugarcane—the crop that had led to the prosperity of a handful of farmers in the area and resulted in income disparity in the community. In the course of the struggle between the government and the villagers, it became obvious that several minor conflicts existed within the farming community that had roots in class difference. There was an unholy alliance between rich farmers who wanted to protect their vested interests and the government which barely gave a thought to the needs of the poor farmers who had neither economic nor political power.

To make matters worse, intensive sand mining was taking place in the riverbed. While a few private contractors made huge profits out of this business, the victims were the local population who had to bear the brunt of its impact on their activities. Mining had also begun to have a disastrous effect on the environment and on the subsurface flows, the major source of water for farmers in the post-monsoon season. Thus, small and marginal farmers enlisted the support of the Mukti Sangharsha Chalwal, and began a concerted struggle to establish their collective right over the natural resources in their area, effectively laying the foundation for the Baliraja struggle that was yet to come.

Fig. 1. *Activists at the time of construction of Baliraja dam.*

Sticking to Their Stand

The locals decided that instead of blindly accepting the government's dictates they would work together to gather information and work out an alternative plan with the help of scientists sympathetic to their cause. Small farmers and landless labourers from Tandulwadi and Balawadi villages had observed that as a result of sand excavation, surface flow was getting depleted. In addition, the pits formed by mining were filling with silt, and this prevented the percolation of water. They organised a *vigyan jatra* in the area and also undertook a study of the villages along the Yerala. The results of the study were published in the form of pamphlets and distributed widely among the villages.

The government has a practice of auctioning sand mining rights to private contractors. It is considered a source of revenue and the authorities do not take serious note of environmental consequences. The movement resolved that it would not allow private contractors to excavate sand from the riverbed. The local people began stopping the trucks from ferrying sand from the site and thus managed to put a stop to intemperate mining. The government was forced to control the excavation and accept the right of the villagers to exercise priority over their resources. They were granted permits to mine the riverbed and sell a limited amount of sand to raise funds for drought-eradication activities. This once again set them thinking. Their need was for an alternative that would help farmers fight drought rather than intensify it. Together, the Tandulwadi and Balawadi villages hit upon the idea of building a dam on the Yerala; capital could be raised by selling a limited amount of sand from the submergence area. Bandu Aappa Pawar, a small farmer, suggested that villagers themselves should mine the area rather than allow private contractors to do so. The dam would also provide them protective irrigation to overcome, at least to some extent, the problem of drought.

One main issue was debated between the government and the people from these villages. Did people have a greater right to local resources? This pertained to both the issue of sand excavation as well as the building of the dam. The villagers believed that they had the greater claim because, unlike the government, they had adopted 'equity' rather than 'convenience of gravity flow' as their rationale for access to water. Since the purpose of building the dam was drought-eradication, their priority was to distribute water on an equitable basis among all the people from Tandulwadi and Balawadi, irrespective of landownership or caste.

The Sand is Ours

The actual process of getting permission to excavate sand and build the dam by themselves was not easy. The government was not convinced about the necessity for a dam in the vicinity since it was intruding into the submergence area of the Wazar dam that was going to be built downstream. But the State realised that it could not guarantee equitable distribution of water or provide the minimum water required for irrigation till the end of the *rabi* season; it was these factors that influenced its decision to grant people the right to control their resources. This was the beginning of the Baliraja memorial dam.

K.R. Datye, a noted engineer from Bombay who is associated with the Mukti Sangharsha Chalwal, worked pro bono and prepared a plan for the 120 m small dam that would distribute a total of 0.68 Mm^3 of water to irrigate almost 360 ha of land.

Incomplete Story

As per the plan, the dam is expected to be 4.5 m high, but the second stage is incomplete. As a consequence, its aim to provide irrigation for 360 ha has not been fulfilled. The water distribution network that is meant to carry water to the fields is yet to be built. This unfinished experiment raises many questions.

The Baliraja memorial dam is a self-reliant, non-exploitative and ecologically healthy model of alternative development. It signals a major shift in the developmental paradigm in India. Though Mukti Sangharsha activists are sceptical about the completion of the project, they are clear that they do not wish to be labelled 'agents' who built the dam by community *shramdan* (voluntary labour). Rather, they want the government to take the work towards completion. It is difficult for them to build *bandharas* Kolhapur-type Weir (KT) bunds based on the Kolhapur pattern which are essential to the project. Since the government has the financial resources, they want it to step in and help them finish the work.

Fig. 2. *The Baliraja memorial dam of today.*

High and Dry

The government did not initially accept the right of people to sell sand to build the dam. They were in favour of private contractors who were beginning to destroy the environmental balance by over-exploitation. The government insisted that the State had the right to decide where and when a dam should be built, but in the case of Baliraja, by exerting their claim, the local people proved that community participation could help conserve their resources and allow them to decide the strategy to eradicate drought. The government actually thought it 'undemocratic' on the part of the people to demand their own dam. In the construction plan of the Baliraja dam, the government prominently marked out silting as a major problem, and officials were questioned as to why there should be another dam when the Wazar dam backwaters could serve most of the farms of Tandulwadi and Balawadi. The villagers argued that they needed a captive source for their area so that they could carry out their innovative experiments.

The Baliraja project has met its political goals, but the point is that it remains unfinished. The experience has taught the activists and local farmers how to make plans that are scientifically oriented to the future. This is a valuable gain, since it also demonstrates that such alternative development is not only more scientific, but it is also something that can be assimilated and put into practice by common people, rather than being the preserve of 'conscious' people who continue to present mere 'alternatives' that are never put to the test of practice.

Equitable distribution of water with the long-term aim of sustainable development involves an alternative path and approach towards agriculture and irrigation as well. While the dam remains incomplete, it is important that the government recognises the concept of people-managed irrigation systems and respects the right of the local population over natural resources in their area; the State, therefore, must provide all necessary financial assistance to complete the Baliraja memorial dam.

Note

*Maps and figure 1 from SOPPECOM, Pune; figure 2 by author.

References

'Baliraja, Pani Ani Rajkaran' (Baliraja, Water and Politics), *Kesari*, June 7, 1987.

Desai, Ramesh. 1998. 'Balipratipadechya Nimittane Hee "Balirajachi Kahani' " (The Story of King Baliraja on the occasion of Balipratipada), *Nisarga Sevak Abhijat*, Diwali Issue.

'Dushkal Nirmulanasathi Lokyojna' (Community Initiated Drought Eradication Programmes), *Kesari*, October 26, 1985.

'Govt Stand on Baliraja Dam Stalls Work', *Lokmat*, May 6, 1987.

Joy, K.J. 1992. 'People-Managed Irrigation Systems: A Case Study of Baliraja Dam'. Paper presented in the National Workshop on 'Farmers Management in Indian Irrigation Systems', held at Administrative Staff College of India, Hyderabad, February 4–6.

Joy, K.J. and Nagmani Rao. 1987. *Degenerated Agriculture and its Effects—A Study of Socio-Economic Transformation in Khanapur Taluka of Southern Maharashtra*. Pune: Shankar Brahme Samajvigyan Granthalaya.

Mukti Sangharsha Chalwal. 1993. 'Baliraja Dharnamage Aamchi Tatvik Bhumika' (The Ideological Stand Behind the Baliraja Dam), *Pudhari*, July 17.

Omvedt, Gail. 1987. 'Of Sand and King Bali', *Economic and Political Weekly*, February 28.

Patankar, Bharat. 1998. 'Balirajadwarey Saman Pani Vatpachya Tatvacha Prasar' (Spreading the Principle of Equitable Water Distribution through the Example of Baliraja Dam), *Sakal*, October 21.

Pawar, Sanjay. 2004. 'Wyatha Baliraja Dharnachi' (Agony of the Baliraja Dam), *Pudhari*, October 26.

Phadake, Roopali. 1998. 'Drought Relief in Maharashtra: A Case Study of the Baliraja Memorial Dam'. Thesis presented to the Faculty of the Graduate School of Cornell University, USA.

'Protests over Govt Bid to Stop Verala River Dam', *Sunday Observer*, April 12, 1987.

Mining and the Nandanvara Dam in Madhya Pradesh: When the State Turns Against its People

Ashim Chowla

A Scenic Setting

TIKAMGARH DISTRICT IS LOCATED IN NORTHERN MADHYA PRADESH. IS PART OF THE BUNDELKHAND plateau in the watershed between the Betwa and Dhasan rivers. Agriculture and related activities are the main occupation of the rural folk. Farmers mostly depend on dug wells and traditional tanks to irrigate their fields. These tanks, many constructed by rulers of the Chandela dynasty, not only enhance the beauty of the landscape but also raise the water table.

The Nandanvara reservoir was built during the Chandela period. It was made by constructing a dam that tapped water from a small river, several local streams and underground springs. The Nandanvara

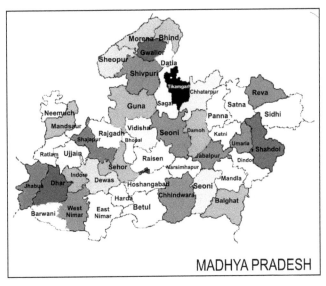

Map 1. *The location of Tikamgarh in Madhya Pradesh.*

Fig. 1. *The dam structure is clearly visible. Mela can be seen on the other side. Quarry Pit No. 1 is seen on the ridge of adjacent hillock to the left abutment of the dam.*

reservoir, the second largest dam reservoir in northern Madhya Pradesh, is flanked by hills on its eastern and western boundaries, an earthen wall in the south and the upstream river on its north. Its catchment area is 221.36 sq km and it stores about 35.97 Mm^3 of water. The dam wall is 249.92 m long and 17.68 m high. This dam was christened Ganesh dam after an idol of Lord Ganesh was placed at the site. After independence, the canal system from the reservoir was extended in order to serve the irrigation requirements of farmers in the region. At present, cultivators from twelve villages derive irrigation benefits from the 16.67 km long right bank canal. The 5.15 km left bank canal serves the needs of farmers in twenty-four villages. In 1990, the total annual irrigation from these canals covered 1,819 ha. The reservoir has sufficient water throughout the year to serve people's needs. As mentioned before, agriculture is the main occupation in this area and the people are heavily dependent on the water from the reservoir. It also plays a major role in recharging the groundwater used by the villagers for drinking water purposes through hand pumps and dug wells. Pisciculture is also carried out in the reservoir.

Dam in Distress

The hills bordering the reservoir are a rich source of the mineral pyrophyllite. In 1984, the Government of Madhya Pradesh granted a twenty-year mining lease to a Jhansi-based private company. The site is now being blasted and excavated, thus weakening the natural barriers that surround the reservoir, making the sides vulnerable to seepage. The blasting site is less than 30 m away from the dam. The farmers are concerned that the blasting has gradually weakened the foundation of the hills which poses a threat to the dam wall, making it vulnerable to breaching.

The growing frequency of blasting started worrying the people of the area. There was a slow build-up of resentment and anger in the community. People's religious sentiments were also hurt since a Shiva temple is part of the area leased to the company. The villagers expressed their concern to the district administration. The Collector of Tikamgarh stayed the excavation work in 1986, and the mining company promptly filed a case against the order in the Jabalpur High Court. In 1992, The High Court revoked the Collector's order and allowed excavation to continue provided:

(a) the company would not use blasting as a method of excavation, and
(b) the actual excavation would be more than 28 m from the reservoir level.

United Front

Mineral extraction continued and although the company authorities said that there was no blasting, the villagers believed that there was clandestine activity. Fears that continuous excavation work would severely impact the dam were compounded by the fact that any breach would immediately inundate three villages downstream. There were concerns about the lack of regulation and monitoring of the mining company's activities. After all, there was no mechanism in place to ensure that the company did not flout the conditions of the High Court order.

The people approached many institutions and individuals to seek assistance in their battle against the mining company. A few State-based NGOs agreed to lend support. They first entered into a dialogue with the authorities and the mining company. They also mobilised people and told them to stand up for their rights. The response from the villagers was overwhelming and this encouraged the NGOs to fight for the cause more earnestly. Almost 600–700 people would participate in the public meetings that were held to discuss the issue. People across all castes and classes, men and women, joined hands in protest. This coming together of a fragmented and hierarchical community was completely unprecedented in the region. The mining contractors tried their best to create rifts between various groups but failed. Representations were made to the state government, Mining Department and District Administration. An appeal was also filed in the High Court. In 1998, perhaps as a result of the mounting pressure from the villagers and NGOs, the High Court instructed that a committee be set up to look into the issue. The report submitted by the special group was rejected by the people and the NGOs since it had representatives from the mining company and was biased in their favour. Yet another committee was set up by the court in response to another appeal. Members of the Dam Security Cell of the Irrigation Department were also included in this new group. However, the report submitted by them was suppressed by the government and never made public. The NGOs were also unable to access the report since the right to information was not mandatory at that time.

Highs and Lows

Around 1997–98, there was a misconception among the NGOs that pyrophyllite was a minor mineral. Taking advantage of the provisions of the 73rd Amendment, which gives the local *panchayat* the right to minor minerals, the NGOs sought to demand the closure of the mine by passing a resolution endorsed by the *gram sabha* and thereafter, all the *panchayats* in the area. Even key members of the *janpad panchayat* and the *zila panchayat* supported the move, but to the dismay of all concerned the Secretary, *Panchayat* Department clarified that pyrophyllite was on the list of major minerals. The resolution was invalidated though the *Panchayat* Department supported it on grounds that the management of natural resources come under the jurisdiction of the *panchayats*. The Department fully intended to take the issue further, but in 1999 the political situation in the state underwent changes. In 2000, the state was bifurcated and the issue was not pursued any further. The secretary was transferred to Chhattisgarh.

Current Status

With the exception of the villagers involved, everyone seems to have lost interest in the issue; the villagers have lost all hope.

Deafening Silence

This conflict has remained unresolved for nearly two-and-a-half decades now. The State, especially the Mining Department, has not heeded the people's plea and has jeopardised the future of the people living in the forty affected villages. The contractors have on occasion gone so far as to try intimidation and inducement.

It is also evident that there has been no inter-departmental coordination. Both the Department of Mining and the Water Resources Department (WRD) have a high stake in this area, though they have been forced to take opposing stands. The WRD is responsible for protecting and maintaining all the water bodies in the state, while the Mining Department is meant to exploit mineral wealth. Despite a conflict of interest, the two departments never bothered to talk to each other at any stage of the dispute. On the other hand, the *Panchayat* Department unfailingly supported the people. Amazingly the three departments—Water Resources, *Panchayats* and Mining—never thought of coming to the negotiating table to try and resolve an issue that had ecological, social and economic ramifications.

The blasting has caused severe ecological damage and ruined the sparse vegetation in the hills. The denuding of the environs has led to silting in the catchment area. There is bound to be a decline in the water level. The Nandanvara dam is an important ecosystem and harbours flora and fauna characteristic of wetlands. The habitat is now under threat. The case of the Nandanvara dam is a classic example of the paradox arising out of having to choose between people's welfare and economic gain. The revenue earned through mining has most likely helped the State's economy but it has also imperilled the livelihood of those who reside in the forty villages dependent on the dam.

Perseverance is the Key to Further Progress

The former Chief Minister of Madhya Pradesh, Digvijay Singh, once remarked that the third world war would be fought over water. Madhya Pradesh today faces a severe water crisis. This particular part of the State has been spared the shortage, mainly due to the presence of ancient water harvesting structures like the Nandanvara dam. It therefore seems ironic that the State should hurt its own interests by creating a water crisis where there was none through its misguided promotion of mining. The indirect social, economic and ecological costs of such blunder will prove a heavy burden to bear in the future. There is a need to create greater awareness among the general public and keep up the pressure on the government to act in the interest of the people it represents. The media could play a big role in this initiative by highlighting the problems of the villages around Nandanvara. National level organisations must be roped in where necessary. Such problems can never be resolved overnight and require the commitment of all parties involved.

Note

*Map 1 from SOPPECOM, Pune; figure from Samarthan Centre for Development Support, Bhopal.

Blank page

PART 5

Micro-level Water Conflicts:
A Review

K.V. Raju

WATER CONFLICTS ARE BREWING BETWEEN AND WITHIN STATES, WITHIN RIVER BASINS, and across competing sectors and uses. The main causes are water scarcity, inequitable access, lack of well-defined property rights, and poor implementation and monitoring of the existing framework for water allocation and distribution. At present, several states and small regions are caught up in water disputes, and unless we move quickly to establish agreements on how to share reservoirs, rivers and underground water aquifers, conflicts over water will continue to be a potential threat. The case studies in this section deal with water conflicts at the micro level, but the issues they capture are not necessarily micro issues; they have a much wider relevance.

The Increasing Demand for Drinking Water

Take, for example, the Dharmasagar case. Initially, the Dharmasagar tank catered to the needs of both irrigation and drinking water. The Public Works Department handed the tank over to the Warangal city municipality for drinking water use. With the increase in population, much more drinking water was required, and very little water was left for irrigation purposes. The Government of Andhra Pradesh then stepped in and came up with the Godavari lift irrigation project that lifted water from the Devadula reservoir, and through a network of tanks and canals ultimately delivered it to the Dharmasagar reservoir. This could ease the situation somewhat, but in the long run the problem will not be resolved until the roles and responsibilities of the Warangal municipality on using neighbouring sources of water are clearly defined.

Conflicting Demands

Similarly, the Morathodu Irrigation Scheme in Chalakudy river basin in Kerala was initiated to divert water from the Chalakudy river to provide both drinking water and irrigation. The main objectives of the scheme were to improve drinking water availability in salinity ingress areas by reducing salinity levels, providing access to irrigation in hilly areas, reducing waterlogging that would be caused in adjacent rice fields by the construction of the regulator, and increasing the cropping intensity of the existing rice fields from two to three crops per year. Instead, the open channel that was dug altered the natural drainage pattern of salt and freshwater in the *panchayat* and led to waterlogging, drinking water scarcity and salinity intrusion into new areas. This was the cause of the growing opposition of farmers to the scheme. This clearly outlines the technical flaws of the scheme.

The scheme was envisaged and granted technical clearance without proper field investigation or a detailed project report which could have highlighted the impact of the scheme once it came into operation. The participation or involvement of the community at various stages—from inception to implementation—

would have resulted in a holistic approach to the design and initiation of this scheme. There has also been no involvement of local political parties in solving this problem. Though there has been opposition to the scheme from various people, most of them have not fully understood the intricacies of the issues and the impact that this scheme will have after it has come into operation. For this reason the conflict has not assumed serious dimensions. Lack of foresight has been a key factor in the failure of this project. In spite of their active involvement, organisations like Janayogam have been unable to resolve the problem.

Sharing Irrigation Water

The conflict between Dhar and Ghelotaon Ka Vaas villages in Rajasthan over the distribution of surface water for irrigation is an example of mismanagement of water resources. The water earlier flowed towards Ghelotaon Ka Vaas, and this flow was later modified due to the construction of the Jiyan Sagar. The State conceived a new drainage system to enhance surface water flow into the lake. Changes in the natural flow initiated the conflict between the two villages. There was a reduction in the quantity of water flowing into the Ghelotaon Ka Vaas, but due to an existing water hole the villagers continued to get some water. Some of the water from the stream entered the water hole and emerged as a natural spring inside the village boundary. Even so, the residents of Ghelotaon Ka Vaas continued to feel deprived.

A dam was constructed to increase water storage to be used for irrigation by both the villages. People in Ghelotaon Ka Vaas considered the dam an impediment to the free flow of the stream, reducing the water entering the water hole and thereby making the spring redundant. On the other hand, Dhar continued to demand an equal share in the water from the stream. Ghelotaon Ka Vaas was in an advantageous position due to the water hole. The locals realised that the dam invariably had surplus and it helped maintain a minimum surface flow for the water hole, which ensured water to Ghelotaon Ka Vaas. In comparison, Dhar lacked any such natural mechanism providing water acceess. So, to bring about equity in sharing the water, Dhar decided to link the reservoir with their farmlands through well-laid-out earthen channels. Farmers from Dhar kept constructing earthen channels and the opposing group kept demolishing them. Due to consecutive droughts, Dhar got another dam sanctioned as part of relief work. But, because of geographical limitations, the work could not be taken up and the villagers decided to augment the capacity of the existing dam by further increasing its height.

The problem here is that the residents of both villages have evolved their own strategy to solve the problem without any consultation. The dispute reaches its peak when there is just sufficient flow to irrigate agricultural lands in both villages. And during excessive and minimal surface flow, the dispute becomes dormant because, in the former case the water is sufficient for everyone, and in the latter it is not adequate for anyone.

Between Two Agencies

The dam in Paschim Midnapur district, West Bengal, has been stalled by a dispute between the Forest Department and the *zila parishad* and is a case of the transgression of an act. The *zila parishad* proposed the Khandarani dam across a perennial stream to enable the adjoining villages to irrigate their fields. The Division Forest Officer opposed the construction of the dam as it violated the provisions of the Conservation of Forest Act, 1980. Two problems encountered due to the construction of the dam were: destruction of the forest due to submergence during monsoons, and the use of forest land for non-forest purposes. The construction of the dam was almost complete before it was brought to a halt. The Divisional Forest Officer was asked to visit the site and submit a report on the matter. The Conservator of Forests recommended that the project be completed.

The major players in this conflict were the Forest Department, the *zila parishad* and the tribal population of the affected villages. The Forest Department and the *zila parishad* had two contrasting views, and the tribals were reduced to mute spectators. While tacitly conceding the benefits of the scheme, the forest officials wanted to abide strictly by the forestry laws and were apprehensive about the destruction of forest lands. The reasoning was that since the dam was being built on forest land, with them rested the final responsibility of its impact. On the other hand, the *zila parishad* was more concerned with the dam's construction. Their argument was that since they had the funds, the expertise, and the development of this difficult area in mind, where was the problem? The tribals, unaware of legal processes and unimpressed by official jargon, only wanted water.

Usufruct Rights in Tank Areas

Sharing of tank usufruct is another issue over which conflicts have arisen. In the Bahoor cluster of irrigation tanks, conflicts arose in Seliamedu village in Pondicherry over the sharing of tank usufruct between the Dalit hamlet (*pet*) and the main village (*ooru*). Earlier, the two different income groups/castes had been sharing the usufructs of the tank in a particular manner. Before 1980 the usufruct income went to the Seliamedu upper caste *ooru* people, but that year a series of events led to its transfer to the Dalits in the *pet*. When the proposal for a tank association to be formed by a common union of the *pet* and the *ooru* people came up, they opposed it on the grounds that all the rights of tank usufruct would be transferred to the tank association. With the intervention of the Centre for Ecology and Rural Development (CERD), the tank association was formed with representatives from both the *pet* and the *ooru* and the tank was rehabilitated by the Public Works Department (PWD) of the Pondicherry government. First, the *pet* people did not like the idea of an all-stakeholder joint initiative as they thought the old rivalry would resurface. Second, they thought that once an all-stakeholder association was formed, their rights to the tank usufructs of trees and fishing rights would be transferred to the tank association.

CERD, which had taken up the task of forming the tank association in Seliamedu village, patiently continued the negotiation with the *pet* and the *ooru* people. As a result, the following arrangements emerged as acceptable to both the groups:

- Most of the usufructs income from the trees on the tank bund shall be retained by the *pet* people and part of it (30 per cent of the auction amount) shall go to the commune *panchayat*.
- Out of the income from fish culture in the tank, 30 per cent shall be paid to the commune *panchayat*, 40 per cent shall go to the tank association, and the remaining 30 per cent shall go to the *pet* people.

Tank-bed cultivation

The dispute around tank-bed cultivation and emptying of the tank in Katgi Shapur village is another case of conflict between two income groups with different stakes in the tank. The Madars, a scheduled caste community (SC), here have been cultivating the tank-bed area for a long time. These families had legal rights (*patta*) for cultivation of the tank bed. Their stake was such that they preferred to release water from the tank so that they could cultivate the tank bed. The command farmers resented this and preferred to store the water so that it could support a second crop. There has been a dispute between the official *patta*-holders and command farmers, the former demanding financial compensation so that they can buy the land.

Some of the villagers feel that the Madars' demand is unjustified and illegal. The Madars argue that earlier when the government decided to launch its programme for development of common property resources, their own lands in the catchment were taken away by the government, whereas the adjacent lands owned by Lingayats were left untouched and were not selected for this purpose. Thus, as vulnerable groups, they have been losing their rights over economic assets.

The Madars do not believe either in traditional or modern (*panchayat*) leadership because both are dominated by upper castes. Madars have been continuing with the cultivation of the tank bed as they own the *patta* document. The case here clearly speaks about the distinction made between upper and lower caste people, which in turn has affected water usage.

Declining Community Control

Failure of community institutions often leads to conflicts around irrigation water. A case that provides insight into water use conflicts on irrigation at the intra- and inter-village levels and failure of community institutions in dealing with them is that of the Barkop village dam. After it was built, a monitoring committee was formed to look into water use issues. A watchman was employed and user charges were levied on irrigation water. There was active participation by the committee members. Incidentally, the watchman quit and as there was no replacement, it resulted in flash floods. A village sub-committee was formed but there were more and more incidents of intra- and inter-village conflicts, several of which also resulted in injuries.

The head-reachers and tail-enders are the main parties to the conflict. Both need water to irrigate their fields in the lean years. The tail-enders have been rallying around benefit-based contribution and argue that everyone who benefits from a resource should pay for it. However, head-reachers have been taking advantage of being in the head-reach and do not pay for the water they draw. This has not gone down well with the tail-enders. The failure of the local village institutions and the committee to curb the head-reachers later prompted them to indulge in the same tactics. This phenomenon is also observed at the inter-village level where the villages upstream try to get more water at the cost of downstream villages, which resist their moves and try to maintain equal water supply to both the villages.

Conflicts between the head-reach and the tail-end farmers on the issue of sharing of water also overlaps with another division—that between two dominant landholding castes/communities in the area. This division between Muslims and Jainas is another example of conflict based on caste/income group. The Muslims were traditionally the largest landholding group in the tail-end but have now acquired more garden land in the head-reach and dry land outside the *atchakat*. Many Jainas, on the contrary, have sold their lands in order to settle in urban areas and have shifted to more profitable economic activities. Earlier the Jainas were the dominant caste, hence all the major responsibilities with respect to tank management were acquired by them. After the tank was taken by the PWD, an irrigation committee was formed to look into matters related to the tank. There are now several conflicts—related to canal cleaning, rotation schedule, rotation for a second crop, etc.—between the two communities.

What is interesting here is that after a protracted conflict spanning over two decades—between the head-reach and the tail-end farmers over a variety of issues pertaining to tank management—the customary rule of water being first supplied to head-reach has been reversed. Water is supplied first to the tail-end; during this time the head- and middle-reach farmers can still irrigate some of their lands with seepage water. Therefore, water is now always supplied first to the tail-end, and the head- and middle-reach farmers do not mind it so much. Initially, there was a lack of understanding on usage between the tail- and headend and this created the problem. But understanding the pattern of water flow and the technical aspects of usage has helped resolve the problem.

Seepage Water is Good for Some

Sometimes conflict over seepage water and competing needs maps on to usage conflicts between different caste or village groups. The Heidyala and Chilakahalli farmers' conflict is one such example. The Heidyala farmers started using seepage water for irrigation. For this purpose, they built bunds across the stream so as to impound water and increase seepage. They pumped water using diesel motors. The result was that no water flowed downstream. The downstream farmers removed all the upstream bunds without warning them. The conflict was further aggravated as there was no department or agency in charge of tank irrigation.

As long as there were good rains in the area, there were no conflicts. But conflicts emerged whenever there was a shortage of rains. Not only is the conflict over seepage water a relatively uncommon phenomenon, some of the issues that have emerged are also rarely seen. First, the conflict is due to the non-functioning of the tank and its poor construction. The conflict over stream water itself is rare and after the construction of the tank, the seepage should not be a major source of water for irrigation. Second, while the conflict appears to be between head-reachers and tail-enders, it is actually not so. In fact, farmers further downstream did not face water problems partly because another stream joins it further down the course. The Chilakahalli farmers are located in the mid-reach of the command area. Third, the conflict is mostly between farmers who use water for irrigation purposes and those households using the same water for domestic and livestock purposes.

Conclusions

These case studies portray water disputes, civil disturbances caused by water shortages, misuse and mismanagement of water resources, and potential regulatory solutions to diffuse water conflicts at several levels. The case studies can be broadly categorised as:

1. Conflicts between two user groups for a common source of water.
2. Sharing of water for different purposes.
3. Inefficient water management policies.
4. Lack of a monitoring or governing body.
5. Practice of conventional methods of water usage.
6. Transgression of law.

A thorough look at the water policy and implementation strategies would help in streamlining water-related crises. The roles and responsibilities of the concerned regulatory agencies—NGOs, governments, corporations, *panchayats*, researchers, etc.—should be clearly outlined. The legal framework should strictly endorse a policy on enforcement of punitive measures for its violation. There have been several conflicts that revolve around the usage of water for both irrigation and drinking water purposes. The sharing of Dharmasagar tank water for irrigation and drinking purposes, the initiation of Morathodu Irrigation Scheme to divert water from the Chalakudy river for drinking water and for irrigation, and the conflict between Dhar and Ghelotaon Ka Vaas villages over the distribution of surface water for irrigation—all these studies focus on conflicts related to the usage of a single source of water for different purposes.

Some of the key points that emerge from these case studies are:

- Any available water source should be clearly demarcated for a specific purpose and should be governed by the concerned regulatory agency. All water-related issues such as usage, conservation, management and sharing should be encompassed by a policy and legal framework.
- In decentralised planning processes, it is essential to take the beneficiaries of the project into confidence and ensure transparency before implementation. Equally important is to study the pros and cons of the scheme before initiation so that it proves beneficial to all users.
- The roles of the various stakeholders have to be outlined clearly, and better coordination between the government agencies, NGOs, the common man and the concerned implementation authority is required. In some cases, the roles of the *zila parishad* and the Forest Department have not been outlined. In the struggle between a government agency and a local body responsible for the initiation and completion of the project, the victims have been poor villagers. By contrast, in the case of the Bahoor cluster of irrigation tanks, the tank association has been instrumental in resolving the issue of sharing of usufructs.
- In some cases, conflicts between the villages are due to individual interests. As new initiatives will jeopardise earlier entrenched interests, they oppose any positive action. In many cases there has been no initiative by the communities to amicably resolve the issue; nor have they sought the cooperation of any external source to mediate and address the issue.
- In some other cases, socially disadvantaged classes have been deprived of their rights. This clearly calls for proper apportionment of water rights and there is a need to make suitable modifications to this effect in acts and rules.
- Lack of an institutional framework and stringent regulation have been the key factors in some conflicts. The amount of water to be used by the head-reaches and tail-reaches should be clearly defined by the concerned authority, as this will help in preventing disputes. Conflict resolution mechanisms will enable better social equity.

Morathodu Irrigation Scheme in the Chalakudy River Basin: A Dream Turns into a Nightmare

M.P. Shajan, C.G. Madhusoodhanan and K.H. Amitha Bachan

Visions of a Better Future

THE PUTHENVELIKKARA *GRAM PANCHAYAT* IS A TYPICAL MIDLAND *PANCHAYAT* IN KERALA, where a majority of the population depends on agriculture. The *panchayat* is situated in the downstream area of the Chalakudy river basin just before the river drains into the Lakshadweep Sea. In fact, it falls in the area between the backwaters and the river. Though the *panchayat* has a cultivable area of 13.97 km², over the years there has been considerable reduction in farmland for various reasons. The average size of a landholding at present is 0.0637 ha and that of agricultural landholding just 0.0545 ha. The Morathodu

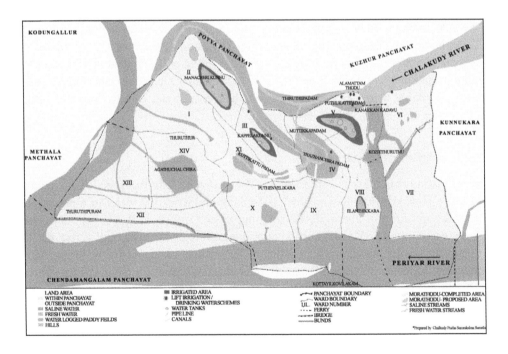

Map 1. *Map showing the Morathodu irrigation project.*

Irrigation Scheme proposal was conceived by the Puthenvelikkara *gram panchayat* in 1997 as part of the democratic decentralisation process of the Ninth Five Year Plan. The authors of the project envisaged that they would implement the scheme in two phases at an estimated cost of Rs. 36 lakh. The scheme was planned to serve 10 km^2 of the 19.87 km^2 falling within the *gram panchayat*.

The *gram panchayat*, although surrounded by water on three sides, faces severe water scarcity due to salinity ingress during summer. The hills too face acute water shortage in the summer. The two major lift irrigation schemes, namely the Pudukad LI (75 hp) and the CCDP Kanakkankadavu LI (300 hp), distribute irrigation water to five wards (2, 3, 4, 5 and 8) in the *gram panchayat*. A major drinking water supply scheme (Elanthikara scheme) provides water to twelve out of the fourteen wards in the *panchayat* at the rate of 8 lakh litres per day from the Chalakudy river. The construction of another drinking water supply scheme, at an estimated cost of Rs. 1.5 crores, is also under progress. Apart from this, there are at least 3,640 private and seventy-eight public open wells in the *panchayat*. But since most of them dry up in summer, these irrigation schemes have been the mainstay of the *panchayat* over the years. Most of the area that will benefit from the Morathodu project is already irrigated by the lift irrigation scheme, and drinking water too is available.

The existing lift irrigation schemes and drinking water schemes are dependent on the Chalakudy river that flows through the east. The fifth largest river in Kerala, the Chalakudy joins the right arm of the Periyar river at Elanthikara before draining into the backwaters at Azhikode. The 144 km long Chalakudy originates from the Anamalai Hills in the Western Ghats and has at least twenty-five *gram panchayats* with 700-odd lift irrigation schemes and thirty-odd drinking water schemes depending on it in downstream areas. Like the Puthenvelikkara *gram panchayat*, all the other *panchayats* too have plenty of lift irrigation projects initiated during the Ninth Five Year Plan dependent on the river. Thus the river is already under severe stress due to over-exploitation of its waters—almost 40 per cent of the water is diverted from tributaries upstream to Tamil Nadu. Thanks to an Inter-State Water Diversion Agreement, 20 per cent is diverted to the adjacent Periyar river basin. Apart from this, water from the river is supplied downstream to a 18,000 ha *ayacut*, a major irrigation scheme. The river flow is so meagre after December that over the years, salinity ingress into the *panchayat* has increased considerably, and has recently started affecting other upstream river basin villages as well.

The Morathodu Irrigation Scheme

The Morathodu Irrigation Scheme was conceived in 1997 to divert water from the Chalakudy river above the Kanakkankadavu regulator (barrage) to provide drinking and irrigation water to a population of over 7,000 out of a total population of about 30,000 covering seven (1, 3, 4, 5, 9, 11, 14) out of the fourteen wards of the *gram panchayat*. The Kanakkankadavu bridge-cum-regulator was built across the Chalakudy river in 1999 to prevent salinity ingress further upstream. The regulator is opened only during high monsoon to release floodwaters. Three wards (1, 9, 14) that are most severely affected by salinity ingress are supposed to be irrigated by this scheme as well.

The main objectives of the scheme are:

1. Improving drinking water availability in the worst affected salinity ingress areas by reducing salinity levels.

2. Providing access to irrigation in hilly areas.
3. Taking care of waterlogging in rice fields adjacent to the Kanakkankadavu regulator.
4. Increasing the existing two-crop pattern to a three-crop one.

The project cost, originally envisaged at Rs. 36 lakh, has already touched Rs. 55 lakh.

Disastrous Consequences

The Ninth Five Year Plan period (1996–2001) in Kerala was considered a path-breaking one—it marked the onset of an era of decentralisation where power was transferred to the people so that the *gram panchayat* had the right to decide how to utilise resources. The people would now plan and implement various development projects along with the *panchayati raj* institutions and government departments. The project proposals prepared by these three groups had to be scrutinised by *gram sabhas*, task forces, working groups, development seminars, *panchayat* committees, and a block level expert committee before being formally submitted for sanction to the District Planning Committee.

Initially, the scheme was welcomed by the people of the area in the hope that it would cater to their drinking water and irrigation needs and solve the problem of salinity ingress. The idea was to open a new channel connecting the river above the Kanakkankadavu regulator into the project area that would carry fresh

Fig. 1. *Kanakkankadavu regulator.*

water inland from the river and replace the saline water all along its route and finally lift fresh water to the hills at suitable points from the channel. Once the digging of the open channel progressed inland, it altered the natural drainage pattern of salt and fresh water in the *panchayat*, leading to waterlogging, drinking water scarcity and salinity intrusion into new areas, resulting in a gradual expression of farmers' opposition to the scheme. The problem was at its peak in 2003–4. While the *gram panchayat* reeled under severe water shortage, the wells and water sources adjacent to the deep channels dug for the project began drying up because water started draining into the channel. Since it was dug deep through the heart of a hill, violating the basic principles of watershed drainage, water that used to drain naturally from Thuruthe Padam rice field to the river during monsoons made inroads into the Thazhanchira padam rice field, creating severe waterlogging in at least 100 ha, thus making rice farming impossible. The farmers, unable to grow paddy in standing water, started selling their fields for reclamation and mining. In other areas, where the rice fields were at a higher elevation from the channel, the problem was the draining away of water from the rice fields into the trench. In 2003 public outcry forced the Assistant Development Commissioner to visit the area to assess the damage. The auditors of the Local Self Government Audit Department (LSGAD) carried out an enquiry into the project's financial anomalies in 2004.

Fig. 2. *This deep open channel dug through the hill has altered the natural drainage pattern and basic watershed principle.*

Questions Raised too Late

A plan fund was set aside for the Morathodu Irrigation Scheme in 1997 by the *gram panchayat*. The implementation has come to a halt following public protest and the funds for the last two years remain unutilised. A local voluntary group, Janayogam, has filed a complaint in the Planning Board demanding an investigation into the technical flaws in the project. The affected people have been raising the issue and its impact on their livelihood at the *gram sabhas*. Development seminars and *dharnas* have been organised in front of the *gram panchayat*. Janayogam argues that the people who sacrificed their land for the project in the hope that they would get water not only lost their lands but also lost their existing drinking water sources and livelihood (farming) due to the draining of water from their wells.

There has been no meaningful response from the *panchayat* governing body or the Planning Board. Meanwhile, local politicians keep promising that the project will be completed soon and that there will be assured water supply for all.

Janayogam demands that, at the very least, there be a thorough study to determine the total water requirement for the project as compared to the water available in the Chalakudy river during the summer when salinity ingress reaches its peak and fresh water availability declines. The group also says that there must be a detailed participatory study on how to make the project viable and more adapted to the local environment. Their third demand is for compensation for those who have lost everything because of the scheme. The people who have borne the brunt want the project to be dropped altogether.

In April 2004, Janayogam along with the Chalakudy Puzha Samrakshana Samiti carried out an independent assessment of the scheme with local participation. The resulting booklet and video were published on January 2, 2005. House-to-house dissemination of information was carried out and the local television channel showed the documentary.

Under pressure from the rice farmers—who experience waterlogging—the *panchayat* has planned to prevent waterlogging in new areas by constructing a temporary sand bund across the channel.

Planning, *Panchayat* and People

In a decentralised planning process, it is essential to take the beneficiaries of the project into confidence to ensure transparency. The Block Level Expert Committee (Parakkadavu Block) reportedly carried out level measurements of the scheme before the commencement of work. However, till date the report has not seen the light of day. The *panchayat* and the people are not even aware of such a study being carried out. The scheme was envisaged and granted technical clearance without proper field investigation, detailed project report or public participation. The local political parties and MLA are maintaining a guarded silence, though opposition parties have publicly criticised the issue. But the conflict has not yet assumed the dimensions of a prolonged movement, the main reason being that the people are not aware of the intricacies of the issue.

Lessons Learnt

The Ninth Five Year Plan witnessed innumerable such projects across the State. These were planned and implemented as people's plans, though in many of the cases there was little sign of public involvement. Therefore numerous such projects are lying in various stages of completion with the State seeking conflict resolution. The overall approach towards decentralised water management under the People's Plan in Kerala itself has come under the scanner. Rather than focus on assessment of the available resource base and the means to use and share it judiciously, most of the *panchayats* used the self-government mechanism to selfishly exploit their resource base to its maximum. Resource over exploitation at local levels can scale up and become overall resource scarcity.

The Morathodu Irrigation Scheme is a typical case of people's plan gone awry. Had it been a People's Plan in the real sense an understanding of the real demands and problems related to water shortage should have preceded it. *Gram sabhas* and working groups should have been used to seek ideas and suggestions from the common people rather than pushing for a sanction. Technical sanction was granted to this project without any detailed study by experts, a master plan or thorough field investigation of the geographical and topographical peculiarities of the area. Since the *gram panchayat* is not equipped with a technical or engineering wing, the scheme implementation was entrusted to the Minor Irrigation Department, whose responsibility was restricted to implementation. The concerned *panchayat* and the engineers who prepared the plan, unfortunately seem to be unaware of the technical and environmental problems that would be created by the project when fully commissioned. Moreover, the bureaucracy and technocracy seem to have the upper hand in planning, decision-making and implementation under the preconceived assumption that people are not capable enough of deciding how to manage their resources. The seeds of conflict are laid if these aspects are overlooked.

Whatever be the scale of the project, it has to be seen against the historical, hydrological and bio-geographical peculiarities of the area. The reduction in the flow of the Chalakudy, owing to several interventions upstream unmindful of the needs of the downstream population, has been taking its toll in the form of increasing salinity; this turn made it imperative to construct a regulator that led to severe water logging in the rice fields upstream of the mechanism. The salinity intrusion proceeded inland despite the regulator. Consequently, the *panchayat* was faced with the dual problem of simultaneously coping with salinity as well as preventing water logging.

Since the problems reached a certain level of complexity, Janayogam was able to act as a bridge between the *panchayat* and the people, make them face the real scenario, help the *panchayat* by providing qualified experts to carry out proper technical and environmental studies on the project and speed up implementation by ensuring minimal damage.

Note

*Map 1 from Chalakhudy Puzha Samrakshana Samithi; figures by authors.

Reference

Visual documentation of the conflict case with a booklet (in Malayalam) by 'Janayogam' and Chalakudy Puzha Samrakshana Samiti.

A Gravity Dam in
Paschim Midnapur District, West Bengal:
Missing the Wood for the Trees

Nandita Singh and Chandan Sinha

On a Downward Slope

Bhulaveda *GRAM PANCHAYAT* IS LOCATED IN BINPUR-II BLOCK IN NORTH-WESTERN PASCHIM Midnapur district, West Bengal, to the east of the Chotanagpur plateau. The terrain is thickly forested and undulating, and the soil red and lateritic.

The 36 per cent tribal population of Binpur-II block is dominated by Santhals and Mundas. There are also a few Oraon villages and a sprinkling of a primitive tribe called Lodhas. Due to the hilly terrain and

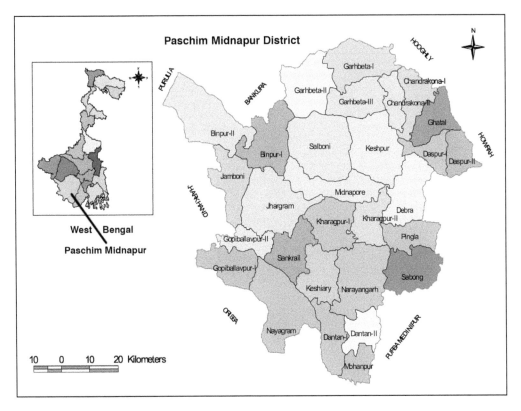

Map 1. *Map of Paschim Midnapur district.*

the extensive *saal* forest, population density is low. The tribals live in small villages comprising 50 to 1,500 people.

The forest has circumscribed the land available for cultivation, and the people depend on the monsoon for irrigation. A single rainfed *kharif* crop is grown on the terraced fields that cover the gentle slopes in these parts, and even that suffers if the second spell of rain does not arrive on time between October and November. The tribals also rely on forest produce such as *saal* leaves, *kendu* leaves, etc. to make a living. The single crop, low-yielding agricultural cycle makes these villages economically vulnerable. What is urgently needed is the creation of irrigation facilities that can render the land more productive.

A Dam for All

In 1995–96, the *zila parishad* proposed to construct a small dam to arrest the flow of Khandarani *khal*—a perennial stream—flowing through the Gadra valley. Although the average rainfall in the area is 1,744 mm, the sloping terrain leaves very little water for irrigation. A small dam on the main stream and check dams further down as well as field channels will enable the adjoining villages to irrigate their fields and increase crop production. The command area of the proposed scheme covers approximately 500 ha.

In 1996, the Midnapur *zila parishad* initiated the construction of the dam, known locally as Khandarani dam, across a rivulet in the forest area at Mouza Amlasole, J.L. No. 83. The proposed dam had a concrete weir 46.20 m in length with an earthen dam on either side of the weir. The total length of the earthen dam was to be 220 m.

Fig. 1. *Khandarani Khal: A perennial stream.*

According to Forest Department estimates, the seasonal submergence is not very large. It would vary between 4 ha in the dry season to 12 ha during the monsoon.

The construction of the dam over the Khandarani *khal* was meant, in varying degrees, to solve the irrigation problems of thirteen villages. Of the 5,803 families that reside in these villages, 155 belong to scheduled castes, 4,618 to scheduled tribes and 1,030 to others. Most of them would benefit from the assured irrigation, which would cover about 500 ha.

The gains of this project may be summarized as follows: If the project is completed, the local population of the Forest Protection Committee (FPC) villages, comprising mainly scheduled tribes who live below the poverty line, will benefit immensely due to: (*a*) assured irrigation both for the late *kharif* season and also for the proposed but currently non-existent *rabi* crop, (*b*) availability of water for domestic use, (*c*) water for livestock, (*d*) proposed pisciculture, and (*e*) improvement of the moisture regime leading to rejuvenation of flora and fauna.

The forest itself, if we consider the 'forest' an independent entity, would gain in many ways by the creation of a reservoir. The permanent water body will help recharge groundwater, thereby improving vegetation. Soil erosion will be checked, reducing surface runoff.

Finally, soil conservation measures—including contour bunding and land development—if taken up as envisaged will lead to improvement of land quality and help increase the groundwater level. As mentioned earlier, the scheme includes setting up a number of small check dams downstream to recharge groundwater in the entire area.

Fig. 2. *The Khandarani dam after most of the work had been completed.*

Bringing out the Rule Book

Keeping in view the requirements and the demands of the local population, the construction of the dam started in 1996 under the supervision of the engineering wing of the Midnapur *zila parishad*. Although inadequate fund flows initially delayed the work, considerable progress had been made by 1997.

Then, in March 1999, when most of the work had been completed—including the 50 m long reinforced concrete dam with supportive lateral earthen embankments—and Rs. 98.91 lakh spent, the work was abruptly halted by the Divisional Forest Officer (DFO), West Midnapur division. Recently transferred to this division he discovered that by constructing the dam the provisions of the Conservation of Forest Act, 1980, had been violated.

Until August 2000, there was no written communication from the DFO to either the *zila parishad* or to his superiors. Construction of the dam was halted by a verbal order and reported verbally by the DFO to the Conservator of Forest. Destruction of the forest by submergence, especially during the monsoon months, was the main reason cited for stopping work. In 1999, the reservoir with its half-finished dam had already submerged some trees, and the Forest Department estimated that about 200–300 trees in all would be damaged when construction was complete. Another concern pertained to legal matters. The land in question had never been transferred to the *zila parishad* which had not submitted a proposal for any such transfer either. Purportedly, several rounds of discussion took place between the *zila parishad* and the forest officials but no records are available.

Between August 2000 and May 2002, further consultations took place between the Forest Department and the *zila parishad*, but the issue remained unresolved. From time to time, helpless villagers who lived close to the reservoir appealed to the Department and the *parishad* to complete the work.

Towards the end of May 2002, the Forest Department's silence prompted the contractor appointed by the *zila parishad* to make another attempt to complete the work, and he moved some equipment to Khandarani for this purpose. The Range Officer, Belpahari, at once lodged a complaint at the local police station. The DFO also met the *Sabhadhipati* and other senior officials of the *zila parishad* and 'explained the legal hurdles of such construction', requesting that the work be stopped immediately.

For another year there was no further progress on the matter. One of the objections raised by the Forest Department was that the construction amounted to the use of forest land for non-forestry purposes and therefore it violated the Forest Conservation Act of 1980. It is interesting to note that the very same Department had attempted to construct an earthen check dam on the very same site presumably for the very same purposes on two earlier occasions in the late 1970s and early 1980s. On both occasions the structures had been washed away.

In May 2003, the *Sabhadhipati* wrote to the minister in charge of the Forest Department, giving him the full details of the project and requesting him to intervene so that the scheme could be completed before monsoons that year. However, there was no response to this communication and the 2003–4 season was also lost.

In September 2004, the *Sabhadhipati*, the *paschim* Midnapur *zila parishad* and the District Magistrate took up the matter with the Forest Department. The *Sabhadhipati* wrote to the DFO, Jhargram, requesting him to take up the responsibility as a forest officer of the *zila parishad* and complete the work on the reservoir at the earliest. The district administration too raised the issue with the government.

Finally, a Resolution

Barring a 10 m section on the southern side of the earthen dam and the proposed lock-gate leading to the narrow canal outlet, most of the work has already been completed.

In November 2004, the DFO was asked by his superiors in the forest directorate to submit a report, which he did. In December 2004, the Conservator of Forest, West Bengal, visited the site along with the DFO. In his recommendation to the Chief Conservator of Forests, the official advised the resumption of work and completion of the project. At the time of writing this piece, the matter was being pursued by the district administration and it is expected that construction will be resumed and that work will be completed fairly soon.

Highs and Lows

The conflict was at its peak when the Forest Ranger, Belpahari range, lodged a complaint against the *zila parishad*. Had the crisis not been defused, things could have taken a turn for the worse. The scheme is now poised to take off from where it has remained unfinished for so many years.

Bullish Stance

The major players in this conflict are the Forest Department, the *zila parishad* and the affected villagers. The Forest Department and the *zila parishad* had contrasting views, and in the tussle for the upper hand the tribals' role was reduced to that of mute spectators. While silently conceding the benefits of the scheme, the forest officials wanted to abide by forestry laws; since the dam was being built on forest land they believed that they would be held responsible for its impact on the surrounding area. On the other hand, the *zila parishad* was more concerned with constructing the dam without getting mired in red tape. Their argument was that since they had the funds and the expertise to develop a difficult region, where was the problem? The tribals, unaware of legal processes and unimpressed by official jargon, only wanted water.

In this struggle between a government agency and local self-government, the victims have been the poor villagers. The Forest Department, by invoking the Forest Conservation Act, 1980, and the *zila parishad*, by not taking prior permission from the concerned authorities, created a situation that delayed a scheme that will yield dramatic gains for this parched region.

Since talks resumed in November–December 2004, things have definitely begun to look up. In March 2005 the Forest Department finally agreed to the completion of the dam along with supplementary soil conservation work in the area. Work on the dam will, in all probability, restart and be completed in time for the Khandarani reservoir to harvest the next monsoon.

Note

*Map 1 and figures by authors.

The Bahoor Cluster of Irrigation Tanks in Pondicherry: Usufruct Sharing between the Dalit Hamlet and the Main Village

T.P. Raghunath and R. Vasanthan

Community Management: Kudimaramathu and Syndicate Agricole

Seliamedu village in the Bahoor commune of Pondicherry, situated about 17 km south of Pondicherry, is a predominantly paddy growing area. The Bahoor tank second sluice (called the Seliamedu sluice) irrigates the lands and also feeds the Seliamedu tank, which has a command area of 27 ha. On the south of Seliamedu is Bahoor village, on the north Aranganur, and Pillayarkuppam on the east. The Bahoor tank lies on the western flank of the village and is the second largest tank in Pondicherry. It has a waterspread area of 455 ha, a command area of 1,747 ha, and feeds seventeen small and medium tanks.

The Bahoor tank was constructed by the Cholas between AD 600 and 800. Tanks were the traditional source of irrigation in this area and a well-defined, community-based irrigation management system known

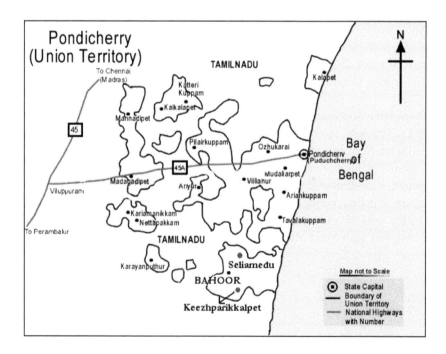

Map 1. *Map showing the location of Keezhparikkalpet and Seliamedu.*

as the Kudimaramathu system was in place. The community had the responsibility of maintaining these tanks and each household had to contribute either in terms of labour or in kind. The Chola kings even had fixed penalties for those who failed to contribute to tank maintenance, and stone inscriptions testifying to this have been found in the Moolanathar temple in Bahoor constructed by Raja Chola, who also built the Chidambaram temple. The French rulers of Pondicherry modified the Kudimaramathu system to what came to be known as the Syndicate Agricole, in which community-based institutions comprising farmers were created and the committees took care of tank maintenance. Each landowner had to pay a land tax, and for every Rs. 5 tax paid, the person had one vote. The French appointed irrigation inspectors known as Chief Brigade to allocate funds for each committee depending on the budget forwarded by them.

Tanks Decline after Independence

When the French left in 1954, the Syndicate Agricole was also abandoned. The early 1960s saw the introduction of the Green Revolution in Pondicherry. New varieties of water and input-intensive paddy were introduced, and electric motors and borewells helped extract groundwater from deeper and deeper aquifers. The cropping pattern changed drastically from one paddy crop a year to three, adding significantly to the demand for groundwater. Incentives and subsidies offered by the State for extracting groundwater led to borewells springing up in large numbers. As a result, aquifers depleted and one has to go as deep as 200 m to strike good quality irrigation water.

Irrigation tanks have been unable to meet this increased demand for water. Farmers began to depend less and less on tanks, and community management practices were abandoned. Tanks were taken over by the Public Works Department's (PWD) irrigation wing, and rights to tank usufructs were handed over to respective commune *panchayats* (who auction the rights, i.e. income from trees, fish culture in tanks, etc. every year). Traditionally, it was the unwritten rule that village leaders would successfully bid a small amount while everyone else refrained from bidding. The village leadership then conducts an informal public auction within the village where the actual higher bidding takes place. The extra amount generated is used by the village leadership for temple expenses, festivals, etc. If there are multiple factions within the village, they come to an understanding and divide the income between themselves. Most of the villages are divided along caste lines and the Dalits live in separate hamlets called *pet*s away from the main village known as *ooru*. The *pet* and the *ooru* have separate leaderships as well as temples. The Dalits are forbidden from going to the temples of upper caste people, though the issue of untouchability, that existed earlier, is not so visible now.

Fig. 1. *Dalit hamlet at Seliamedu.*

Tank Rehabilitation Project-Pondicherry

In 1999, the Pondicherry government started implementing the Tank Rehabilitation Project-Pondicherry (TRPP) through the PWD with financial assistance from the European community. The project aimed at rehabilitating eighty-four irrigation tanks through people's participation and planning. Execution of works and future operation and maintenance of tanks were all to be implemented through an institutional structure called tank association (TA). The TA was envisaged as an all-stakeholder association constituted from the general body which would have two persons—one man and one woman—from each stakeholder household (including the *ooru* and the *pet*). The executive committee was to have 60 per cent command-area farmers, 10 per cent foreshore farmers, and the remaining 30 per cent from landless and Dalit sections. The guidelines for the formation of the executive committee had a provision for one-third representation of women. The project guidelines transferred the tank usufruct rights to the TA so that it would have the resources for annual maintenance of the tank. It was suggested that 30 per cent of the income from tank usufructs be paid to the commune *panchayat* and the rest to the TA.

Who has Usufruct Rights?

Until 1980, the *ooru* people in Seliamedu enjoyed the usufruct income. In 1980, some of the Dalits from Kudiyiruppupalayam took refuge in Seliamedu *pet*, fearing police action due to labour unrest in their village. The police came and threatened the Seliamedu *pet* people with severe action if they continued to protect the offenders. Unrest and chaos followed and the District Collector declared curfew and gave shoot-at-sight orders in Seliamedu, especially targeting the *pet* people. The issue was later resolved, but the Dalits were offended that they did not get support from the upper caste who were influential enough to secure protection from police action. They banded together and, in retaliation, demanded that all the Seliamedu tank usufruct rights be transferred to them. The *ooru* people did not like the idea, but fearing conflict and labour problems, they accepted the terms because they depended on the Dalits for agricultural labour. For many years there was no major friction reported between the *ooru* and the *pet*.

The *Pet* Rejects a Common Tank Association

The Seliamedu tank was selected for rehabilitation in 2002. The NGO, Centre for Ecology and Rural Development (CERD), was entrusted with the job of forming a democratic TA with the participation of all stakeholders as per guidelines. When the CERD studied the village dynamics as part of the preparation for the TA's formation, the past problem of usufruct sharing between the *pet* and the *ooru* came to light. The Dalits categorically refused to be part of the TA along

Fig. 2. *The tree cover around the Saliamedu tank.*

with the *ooru* people. First, they believed that an all-stakeholder joint initiative would revive the old rivalry. Second, they thought that once an all-stakeholder TA was formed the exclusive rights they had enjoyed for two decades on the tank usufructs would be lost.

The *ooru* people welcomed the idea of a TA but were apprehensive that the *pet* would not agree. The problem was complicated because for the past three decades elections had not been held and there were no elected *panchayats* in Pondicherry. Each community and caste had multiple leaders called the *nattaamais* who made up a caste *panchayat* system. But the *nattaamais* in every street were divided over political issues. In spite of a number of meetings, there was no unity on the issue and these leaders finally suggested two separate associations—one exclusively for the *ooru* to look after tank irrigation and one exclusively for the *pet* to mind the usufructs. But the Dalits and their *nattaamais* refused to part with the usufructs, which was the key source of income for the TA for the maintenance of the tank after rehabilitation.

Patient Negotiations Bear Fruit

The CERD continued to negotiate with the *pet nattaamais* for over two months. At one point, the upper caste felt that it would not be possible to form a TA in Seliamedu and the PWD also thought that the tank should be eliminated from the list of tanks to be rehabilitated since the work had to be completed before the onset of the monsoons. The CERD requested for more time, mobilised its full strength and continued negotiations with the *pet* leadership. One of the *ooru* leaders, Sivapragasam, also helped in the negotiations since he was acceptable to both the *ooru* and the *pet*. Slowly the Dalit leadership realised the importance of tank rehabilitation and agreed to participate in the TA, provided their rights over the tank usufructs were retained. The CERD consulted the project management unit and it agreed, as a special case, that usufruct rights of the tank be retained by them on the following conditions:

1. The usufructs income from the trees on the tank bund would be retained by the *pet* people, except 30 per cent of the auction amount which would go to the commune *panchayat*.
2. Out of the income from fish culture in the tank, 30 per cent is to be paid to the commune *panchayat*, 40 per cent to the TA, and the remaining 30 per cent to the *pet* people.

The TA was finally formed with representatives from the *pet* and the *ooru* in the year 2002, almost six months after the CERD started the negotiations. Sivapragasam, who was instrumental in the dialogue process, emerged as the consensus President of the TA. In the days that followed, the Dalits saw for themselves the fruitfulness of the rehabilitation work and cooperated fully with the project.

Improvement in *Ooru* and *Pet* Relations

The TA has helped build a good relationship between the *ooru* and the *pet* people in Seliamedu, thus leaving behind them two decades of enmity. Now the TA feels that it can start a dialogue with the contending parties and convince them to share the usufructs from the trees on the bund with the TA.

Note

*Map 1 and figures by authors.

Conflict over Tank-bed Cultivation in Karnataka: Command Area Farmers and 'Official' Encroachers

M.J. Bhende

Tanks in Karnataka

TRADITIONALLY TANKS WERE CONSTRUCTED TO HARVEST RAINWATER AND USE IT AS supplementary source for irrigation to minimise the hardship of drought in the event of failure of monsoon or insufficient rains. The State of Karnataka has about 40,000 tanks of varying capacity (Raju, et al., 2003). Earlier these tanks were managed and operated by the local communities themselves. However, the ownership as well as the responsibility of operating and maintaining the tanks has now been transferred to various government departments such as the Irrigation Department or *Panchayati Raj* institutions such as the *zila panchayat* or the village/*gram panchayat*. In Karnataka, tanks irrigating less than 4 ha are owned by *gram panchayats*; tanks irrigating more than 4 but less than 40 ha are owned and operated by the *zila panchayats*; tanks irrigating more than 40 ha but less than 200 ha are owned and managed by the Minor Irrigation Department; and tanks with a command area of more than 200 ha come under the jurisdiction of the Major Irrigation Department.

The condition of the tanks has deteriorated over the years because the concerned institutions lack the manpower and funds required for proper maintenance. There are 36,672 tanks with less than 200 ha of command area that serve a potential or designated command area of about 690,000 ha in twenty-seven districts of Karnataka. However, the actual irrigated area is estimated to be not more than 240,000 ha, or only 35 per cent of this potential. In fact, the net area irrigated by tanks has declined from 344,000 ha in 1960–61 to 240,000 ha in 1996–97. The command area farmers who comprise the stakeholder community became indifferent towards the tanks after the management and decision-making powers were taken away from them and transferred to the government. Effectively, farmers' participation has been reduced to merely paying the tax or cess, and over the years this has resulted in an erosion of the sense of ownership that prevailed among them. Moreover, the government also encouraged the development of private sources of irrigation (open and borewells) by providing financial assistance to the

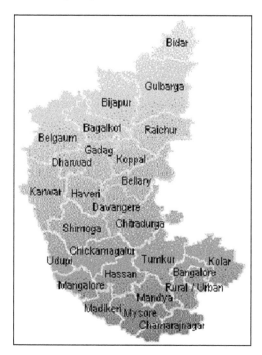

Map 1. *Karnataka.*

farmers. The tanks have gone from bad to worse because of lack of upkeep and from neglect. As the tanks slipped into disrepair, their importance in the rural economy also declined. Nevertheless, there are tanks over which conflicts arise due to different user groups and rights. Somadevi Kere is a case in point.

Somadevi Kere

Somadevi Kere is a tank located in Katagi–Shahpur, 24 km from Yadgir in Yadgir *taluk* of Gulbarga district, and falls in the Bheema sub-basin, part of the Krishna basin. The tank comes under the administrative control of the *zila panchayat*. Official records say that the tank was built in 1956, but the farmers from Katagi–Shahpur say that the Nizam constructed the tank more than 100 years ago. The 27 ha tank has a 4.62 km high bund and a catchment of 5.7 km^2 and is surrounded by hillocks on the west and north-east that are devoid of any significant greenery except for a few shrubs and bushes. The Tank Memoir says that the light black soil *atchkat* area (command area) of the tank is 23 ha, but field visits showed that the actual irrigated area is around 40 ha and this was confirmed by the villagers. The tank helps 100 to 130 farmers irrigate 35 to 40 ha of land. The land owned by the farmers ranges from a minimum of 0.20 ha to a maximum of 6 ha.

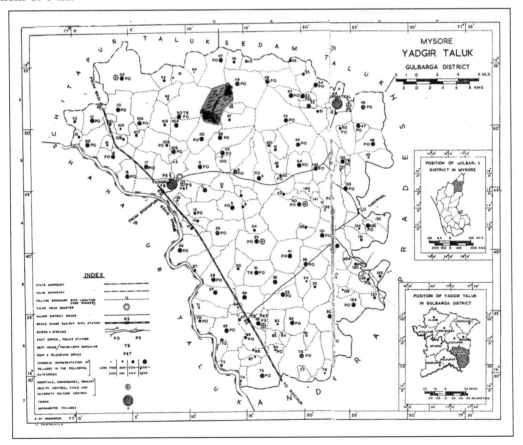

Map 2. *Gulbarga district, Yadgir taluk and Katagi–Shahpur village (shaded).*

The Tank Shrinks

The 27 ha waterspread area of the tank has now been reduced to about 18 ha due to encroachment. In case of shortage, farmers use water from storm water drains which carry runoff water from the village settlement. The main channel was repaired in 1999, but due to the poor quality of work, it has fallen into disrepair. The field channels too have been encroached upon by the farmers and need to be restored. The tank is fed by runoff from the surrounding hills and cultivated land drains into the tank. Between 1994 and 1995, Lamani tribesmen from nearby Hosahalli *tanda* constructed a check dam on one of the feeder channels and created a small tank that they used for irrigation and other purposes. This has reduced runoff into the tank. A sizeable area in the catchment is under cultivation and runoff carries a lot of silt into the tank.

Accumulated silt has reduced the storage capacity considerably. The depth of silt was estimated to be around 1.7 m in 2001 (IN-RIMIT, 2001). Silt from the tank has not been removed in the recent past, except partially under drought relief work in 1972. Though the tank bund was repaired somewhat in the early 1990s with *murram* (soft or weathered rock), it is weak and needs reinforcement, stone pitching and grass turfing. The tank has one sluice and a waste weir, both of which are in need of repair. Part of the tank is infested with *Ipomea fistulosa*, locally known as *Ganeshan kaddi*.

Social Make-up of Katagi–Shahpur

According to the 1991 Census, Katagi–Shahpur had 243 households and a total population of 1,535. The scheduled caste (27.75 per cent) and scheduled tribe (0.72 per cent) population together accounted for about 28.5 per cent of the village population. Madars comprising roughly sixty households were the majority among the scheduled caste community. The Kabbes (fishermen) comprise about eighty households, Lingayats ten, Potters four, the Kurubas, Barbers and Agasas, two each, and there is a lone Brahmin household.

Sociologist M.N. Srinivas had listed three important characteristics of a dominant caste: numerical preponderance, landholdings and economic clout. Like many other villages in the region, the Lingayat community dominates Katagi–Shahpur both politically and economically, even though they are a numerical minority. Half of the Madar households own about 12 ha of dry land but the rest are landless. A few Kabbe households own about 8 ha of land in the command area, but a majority of them are landless. The rest have marginal holdings of about 0.5 to 1 ha. About 30 per cent of the households are listed as cultivators and 30 per cent as agricultural labourers by the 1991 Census. A small number, about 2 per cent, are listed as engaged in livestock/fishing, trade, manufacturing/processing and other services.

About 70 per cent of the land in the command area of the tank is owned by the Lingayats. The lone Brahmin household is an absentee landholder owning about 5 ha of land in the command area. The Lingayat headman of the village owns about 6 to 8 ha of land in the command, and harvested around 450 bags of paddy during the *kharif* season of 2001. A few farmers of Honnagera village also own lands in the command area of the tank.

Head-Reach Lingayats versus Madar *Patta*-holders

Seven Madar households have encroached on the tank bed and have been cultivating it for a long time. These scheduled caste families were bestowed this right after the water was released and the tank bed emptied during the Nizam's rule before the reorganisation of the State in 1956. The families claim legal right (*patta*) for cultivation of the tank bed but other villagers contest its authenticity, though they admit that such *pattas* are prevalent in Bidar and Gulbarga districts which were part of the erstwhile Nizam State. A few households from Chincholi *taluk* of Gulbarga district also cultivate revenue/forest land using similar *pattas*. However, nobody seems to have verified the authenticity of these documents.

These households release the water from the tank soon after the *kharif* harvest so that they can cultivate the tank bed. The farmers in the command area, especially in the head-reach, resent this because they want the water in the tank for a second crop in 15 to 20 ha of land, as do the others who need the tank water for their livestock. The economically better-off Lingayats own most of the land in the command area but feel that since the *patta*-holders are from the scheduled caste and attract political patronage, they cannot stop them from releasing surplus water after the harvest of monsoon (*kharif*) crop. Those in the middle and tail-reach of the command do not vociferously support the head-reach farmers who want to restrict *patta*-holders from emptying the tank so that they can take a second crop. At present, therefore, the conflict is mainly between the *patta*-holders and the Lingayat farmers who own sizeable lands in the upper-reach of the command area of the tank.

With the tank renovation programme, there are chances of the conflict escalating, since the *patta*-holders cannot cultivate the tank bed if the tank is full all the time. They demand compensation in the form of land but also refuse to accept the land the upper castes have identified which, according to them, is forest or wasteland. They expect the government to intervene and provide a suitable alternative. They would prefer financial compensation (current price for this land is estimated to be between Rs. 75,000 to Rs. 125,000 per ha) so that they can either buy land of their choice elsewhere or invest in some other income-generating activity.

Some villagers feel that the Madars' demand is not justified and is illegal, and one cannot sacrifice common interests to satisfy a few. The Madars point out that some years ago, when the government launched its development of common property resources programme, it was their lands in the catchment that were taken away even while adjacent lands were left untouched because those were owned by Lingayats. As vulnerable groups, they have been losing rights even over the economic assets they have built up through hard work, unlike their upper caste counterparts who have simply acquired them as hereditary property.

Moreover, the Madars do not trust either the traditional or the modern *panchayat* leadership since the upper castes dominate both. They accuse the backward Kabbes of joining hands with the Lingayats to exploit the Dalits. The predominantly landless Kabbes depend on upper castes for wage work and a second crop of 20 to 25 ha would augment employment opportunities for them.

Hoping for Government Intervention

So far, none of the villagers has filed any complaint against release of water from the tank. The Madars are firm about continuing cultivation of the tank bed based on their claim to *pattas*. They are also banking on the support of politically influential leaders from the backward community and organisations such as the Dalit Sangharsha Samiti (DSS), which is very active in the district. At present, there seems to be no scope for any dialogue between the command farmers and the *patta*-holders due to local politics and caste equations in the district. However, most of the villagers are hopeful that this conflict will be resolved when the Government of Karnataka intervenes as it has to when it undertakes rehabilitation work on Somadevi Kere.

Note

*Map 1 by author; map 2 from District Census Handbook, Gulbarga District.

References

Census of India 1991, District Census Handbook—Gulbarga District, Part XII-B, Village and Town Primary Census Abstract, Directorate of Census Operations, Karnataka.

'Karnataka Community Based Tank Improvement & Management Project Hydrological Studies'. Report submitted to Jala Samvardhane Yojana Sangha (JSYS). Bangalore: Indian Resources Information & Management Technologies Pvt. Ltd. (IN-RIMT), 2001.

Raju, K. V., G. K. Karanth, M. J. Bhende, D. Rajasekar and K.G. Gayathridevi. 2003. *Rejuvenating Tanks—A Socio-Ecological Approach*. Bangalore: Books for Change.

Head Versus Tail:
Conflict in a Tank Command in Karnataka

Esha Shah

Intense conflicts exist in several tanks between head-reach and tail-end farmers in Karnataka. In the following case, the conflict related to water sharing has not only been going on for a long time, but has also spilled over into other aspects of tank management. The issue is closely related to the cropping pattern followed in the irrigated area, and more importantly, to the landholding pattern and social relations.

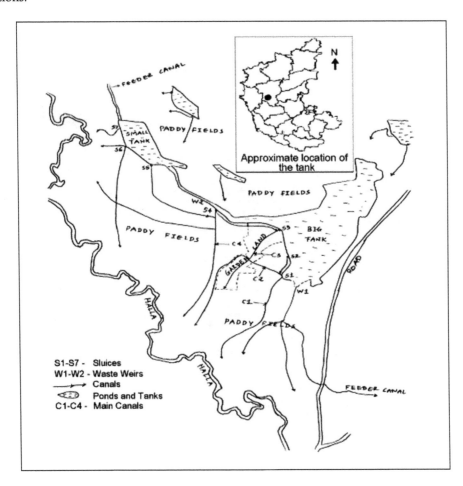

Map 1. *Schematic diagram of the atchakat and the approximate location of the tank.*

The Tank

Kalgudri Tank is located in a village by the same name. It is in Hangal *taluk* of Havery district, which is the semi-Malnad region of western Karnataka, a rain-assured region. The average annual rainfall is above 1,500 mm. Tanks in this region provide irrigation to paddy in the rainy season and to a garden crop of betel nut in the summer. Paddy is cultivated over the entire irrigated area, while the garden crop is cultivated only in a small part of the land located in the head-reach.

The tank under study has a somewhat unique arrangement. As shown in Map 1, it is the bigger of two reservoirs located next to each other and connected by a common waste weir (W2). This means the natural stream that feeds both tanks flows from one reservoir to the other. The embankments of both reservoirs meet in the middle where the waste weir is situated. The tanks in this region are said to have been constructed during the Kalyani Chalukya dynasty between AD 900 and 1100, 800 years ago. Both tanks together irrigate roughly 240 ha, of which 16 to 20 ha is garden land and the rest is paddy land. One crop of broadcasted paddy is grown in the rainy season, followed by pulses. The farmers claim that a part of the garden crop of betel nut is more than five generations old. This crop is considered precious and is prioritised for irrigation, and half the capacity of the tank is always reserved for it.

The bigger tank has a 2.5 km long bund and the smaller one a 2 km bund. The two tanks have 4 and 3 sluices respectively (S1 to S7 in the Map). There are four earthen main canals (C1 to C4) in the *atchakat* (command area) of the big tank, of which three irrigate paddy land but the middle canal, C3, supplied by the deepest sluice S3, exclusively irrigates garden land. Out of roughly 160 ha paddy land irrigated by the big tank, C1 and C2 irrigate around 120 ha and C4 the rest. Around 80 ha get water directly from the main canals and the rest on a field to field basis starting with the ones adjacent to the canals.

Water Availability

The capacity of tanks is smaller than would be required by similar-sized *atchakat* in other regions. Like other tanks in this region, this one too empties after one round of irrigation to the entire *atchakat*. However, it also fills up at least three times during the monsoon and paddy growing season.

It is rare for the tanks in this region to not fill up at all during the rainy season. It has happened only three or four times in the memory of elderly farmers. The tank gets full in July or August and waste weirs usually discharge water until the end of October. November showers are stored in the tank for summer irrigation.

One person is in charge of the tank. He holds the hereditary post of *neerganti*, known in this region as *manegara*. His job is to watch the embankment, open the sluices and ensure that the *atchakat* land is irrigated on a rotation schedule. At present, a *patkari*, appointed by the MID, opens and closes the sluices and watches the structures. Farmers distribute the water themselves.

Tail-end versus Head-reach: Landholding in the *Atchakat*

There are two dominant landholding communities in the *atchakat*—Jainas and Muslims—of whom Muslims are traditionally the larger landholding group in the tail-end. In the last couple of decades, Muslims have acquired more garden land in the head-reach and dry land outside the *atchakat*. A few of them acquired ownership rights to their tenancy land as a result of the tenancy act implemented in the 1970s, though their number is very small. Migration of family members, especially to the coffee-growing areas of Mangalore and Coorg, or to Goa for odd jobs, seems to have enabled many Muslim families not only to retain the ownership of their lands in the *atchakat* but also to expand their economic assets. During the same period, the other dominant landholding caste, the Jainas, lost part of their lands because some of them sold it to settle in urban areas or to shift to more profitable economic activities. At present, 90 per cent of the village population comprise Muslims.

The landholding pattern in the *atchakat* of both big and small tanks highlights the fact that Muslims own a substantial chunk of land not just in the tail-end but also in the garden land. According to the water demand list of 1998, there are a total of 358 landholders in the *atchakat* of both tanks, of which 173 are Muslims (48 per cent), 130 Jainas (36 per cent) and the remaining 55 (15 per cent) are SC, ST and Lingayat. Table 1 gives a rough break-up of Muslim and Jaina landholders in the tail-end, middle and head-reaches based on the water demand list.

Table 1: Number of Landholders in the Tail-End, Middle and Head-Reach of the Tank

	Paddy-tail	Paddy-middle	Paddy-head	Garden	Total
Muslims	43	42	35	53	173
Jainas	13	55	39	23	130
Total	**56**	**97**	**74**	**76**	**303**

Almost all the farmers in the village also have dry land. The cropping pattern on dry land has significantly diversified in the last couple of decades. After the introduction of irrigated varieties of dry crops, a number of crops such as hybrid cotton, hybrid *jowar*, maize, sunflower, mulberry, groundnut, various kinds of vegetables, and even paddy and garden crops like betel nut, betel leaf and coconut are grown on dry land. The economic power earned by Muslims from their garden land and diversification of agriculture on dry land is visible in the form of their collective assertion in the public space of the village. One manifestation of this assertion is the heightened conflict around sharing of water between head- and tail-end.

Conflict, Challenge and Change

Jainas, in this region, are historically a ruling caste. Village officers in the British period were appointed from this community and the practice continued after independence until 1964 when the post was officially abolished. Unofficially, they continued to play an important role in the management of tank resources even after 1964. Until the 1970s, usually one 'powerful' (in the words of the ex-Patel) farmer each from the head-reach, middle and tail-end were selected to form a farmers' committee to take decisions on irrigation matters. This committee and some other influential farmers usually met in the *panchayat* office, especially before the sluices were opened and the canals cleaned. The committee decided all matters related to water distribution and cropping pattern. Although the group had no official mandate before or after 1964, it enjoys a social mandate and legitimacy. After the tank was taken over by the Public Works Department (PWD) in the

1970s, the system continued for another decade or so until the tank was handed over to the MID in the early 1980s. Around 1985 or so, the MID asked landholders to form an irrigation committee separate from the influential farmers' committee. This newly-formed irrigation committee, however, barely functioned because by then there was resistance to the idea of few influential and powerful farmers deciding issues on behalf of all the farmers. This caused enormous conflict in the villages, social arena, as well as in the irrigated area.

The canal cleaning and rotation schedule became the first target of the conflict. Traditionally, the main canals had to be cleaned before the beginning of every irrigation season, failing which almost all the water would be lost through seepage. Conventionally, the influential farmers' committee decided when to open the sluices for irrigation during flowering time after assessing the rainfall and irrigation needs. An announcement was made in the village one or two days in advance of the scheduled day of canal cleaning. All landholders were expected to contribute their labour. However, around fifteen to twenty years ago, some farmers started to object to the fact that the rich farmers sent their labourers in their place. Those who contributed their own labour began to ask questions such as, 'If I have one acre (0.4 ha) of land and I contribute my labour and you have 5 acres (2 ha) of land, how come you send only one labourer?' For a couple of years, the canals were not cleaned because such disputes could not be resolved.

This was also the time when many conventionally accepted norms came to be questioned. If the tank filled up more than half during the month of November, the influential farmers' committee usually gave permission for 16 to 20 ha of land located next to the garden land to be cultivated with a second crop of transplanted paddy. Water was supplied to the second paddy crop around six times. The tail-end farmers began to question this practice on the grounds that the tank would run empty by the end of the summer. In case the rain arrived late the next season, the tank would barely have water during the most crucial phase of flowering for the next paddy crop in the *atchakat*.

This conflict emerged in the context of escalating tensions due to the intensification of paddy cultivation in the *atchakat*. Although the size of the *atchakat* has not been significantly increased, over the last couple of decades every inch of land in the *atchakat* has come to be cultivated. In the last ten years, almost all farmers have also begun to cultivate pulses. As a result of these changes, even though the *atchakat* size has not changed significantly, the demand for water has increased.

The intensification of cultivation also takes other forms. Where previously farmers cultivated a pulse crop using only residual moisture in the fields after the paddy harvest, now farmers demand irrigation even for pulses. The conflict around additional requirement of water for pulses is so intense that during one of the heated discussions, a head-reach farmer who attempted to remind the pulse-growing, tail-end farmers that, 'Your fathers and their fathers never asked for water to grow pulses', received a strong retort that contested sacrosanct tradition: 'Our fathers and their fathers were mad, we are not'.

Tail-enders First

After persistent conflict between the head-reach and the tail-end farmers over a variety of issues pertaining to tank management over two decades spanning the 1980s and 1990s, the customary rule of water being supplied first to the head-reach was reversed. It was reversed as part of the conflict around canal cleaning and tank management, and a novel interpretation by tail-end farmers of the functioning of earthen canals.

There is heavy seepage from the main earthen canals in the *atchakat* because canal bunds have thinned due to encroachments. Rodents have also heavily burrowed through them. As a result, in the words of one of the elderly farmers, 'If the canals irrigate 4 ha, they waste water for another 4 ha'. This was the basis of the argument made by the tail-end farmers at the height of the conflict. The water distribution method, therefore, ought to take advantage of seepage which otherwise would be wasted. They argued that when the canals supply water to the tail-end, they automatically and simultaneously irrigate the head- and middle-reaches due to heavy seepage. Hence, the canals should always supply water first to the tail-end.

It usually takes two days for water to reach the tail-end after the sluices are opened. Meanwhile, head- and middle-reach farmers are allowed to irrigate their lands with seepage water but are not allowed to cut the main canals to take water into their lands. If water is first supplied to the head-reach, 4 ha in the head-reach, 2 ha in the middle-reach and 0.4 ha in the tail-end can be irrigated in one day and by the time water arrives at the tail-end it is merely a trickle. If water is supplied first to the tail-end, the seepage water can irrigate around five to ten *guntha* (100 guntha = 1 ha) in the head- and middle-reaches in one day. Once they are opened, the sluices are kept open continuously for twenty to twenty-five days, including at night, until one round of irrigation is complete. Hence, by the time it is the turn of the head reach, a large part of their lands adjacent to the canals have already been irrigated. Nonetheless, farmers with land away from the canals in the head- and middle-reaches do have to wait until they are allowed to make a cut in the canals and take water to their plots.

Combining Efficiency and Equity

Tail-end farmers justify the 'tail-end first' rule for another reason as well. In their different ways farmers and engineers say the same thing—farmers say 'higher pressure' is needed and when engineers say 'higher discharge' is needed at the beginning of the canal for water to travel a longer distance. Tail-enders argue that in the beginning of the irrigation round when the tank has 'depth' (larger discharge according to engineers) water can travel up to the tail-end easily. Once the water level in the tank is depleted, the pressure/discharge reduces, and the result is a trickle at the tail-end. So it is logical that water should first be supplied to the tail-end to utilise the higher pressure/discharge available at the beginning of the irrigation round when the tank has more water. Such a system is also efficient because it prevents wastage of seepage water. This is how tail-end farmers have interpreted the functioning of earthen canals for the efficient and simultaneous yet equitable distribution of water.

This aspect of the conflict seems to have been resolved at least for now. Today, in this *atchakat* the water is always supplied first to the tail-end.

Note

*Map 1 from Shah (2003).

Reference

Shah, Esha. 2003, *Social Design: Tank Irrigation Technology and Agrarian Transformation in Karnataka, South India*. Hyderabad: Orient Longman.

Kudregundi Halla Tank:
Seepage Water and Competing Needs

Ajit Menon and Iswaragouda Patil

Description

THE KUDREGUNDI HALLA TANK IS LOCATED IN HEDIYALA VILLAGE IN NANJANGUD *TALUK* of Mysore district just north of Bandipur National Park. The village is bound on the north by Chilakahalli village, on the east by Iregaudanahundi, and on the west by HD Kote *taluk*. The tank was constructed with the aim of providing irrigation mainly for crops such as sugarcane, paddy and turmeric in Hediyala, Chilakahalli and Kandagal.

The village lies in a semi-arid zone and receives annual rainfall between 600–800 mm. Agriculture is mostly rainfed and is the mainstay of the local economy. The major crops at the time of the tank construction were *ragi*, cotton, chilli and *jowar* during the *kharif* season, and horsegram and oilseeds in the *rabi* season. People from Hediyala and other neighbouring villages often ventured into what is now Bandipur National Park to collect firewood, carry it by head-load and sell it to the hotels in neighbouring towns during summer.

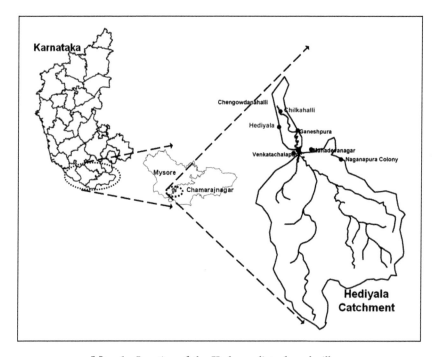

Map 1. *Location of the Kudregundi tank and villages.*

Prominent farmers and leaders in Hediyala and other downstream villages urged the Public Works Department (PWD) to construct a tank across Kudregundi Halla as it was frequently flooded in the rainy season. The PWD eventually acceded to the farmers' request after studying the flow of the stream, the quantum of water available and the geo-hydrological conditions.

Kudregundi Halla Tank consists of a 774 m long earthen dam and a 68 m long spillway. It has a catchment area of 67.35 km^2 and submerges 155 ha of dry land and forest land. The tank has total storage capacity of 3.37 Mm^3 out of which 2.85 Mm^3 is live storage and 0.52 Mm^3 is dead storage. The tank is designed to include a canal network with an *atchakat* (command area) of 816 ha directly benefiting 437 farming families and indirectly a population of 3,436. The tank was constructed in 1982 with assistance from the World Bank.

The tank has, however, rarely filled to capacity since its construction and as a result many lands in the *atchakat* are not sown with irrigated crops. One perception is that there is a technical flaw in the construction and that the stream flow was overestimated to begin with. Also, most of the tank canals are in a state of disrepair or are incomplete, so the sluice gates are rarely opened for irrigation. Tank water seeps through the stream at the dam site as overland flow, and although it has not been measured, it cannot be but a tiny percentage of the live storage. This water is used primarily as drinking water for livestock, for bathing livestock and for washing clothes by the main Hediyala hamlet and Chilakahalli village further downstream. In 1990, four farmers from Hediyala started using this seepage water for irrigation. At the time, no one objected as plenty of water was available in the stream. But soon the number increased from four to more than forty, out of which eighteen were from Hediyala. This is how the conflict over seepage water emerged.

Fig. 1. *Kudregundi Halla tank in Hediyala.*

The Conflict

The socio-economic dynamics in Hediyala and Chilakahalli are important in the context of the conflict and the form it took. Hediyala is composed of Lingayats, Muslims, and scheduled castes and scheduled tribes (SC and ST). Lingayats and Muslims have irrigated and better quality land and grow cash crops such as sugarcane and tobacco. Chilakahalli farmers, though mostly Lingayats, have largely rainfed land and only a few hectares are irrigated. Hediyala, a bigger village, is also a market town and is better represented in the *panchayat*. Hence, economic and political power is concentrated in this village.

The conflict between Hediyala farmers and Chilakahalli farmers over seepage water arose in 1995 when Hediyala farmers, immediately downstream of the tank, started using seepage water to grow water-intensive cash crops such as sugarcane, paddy and turmeric. Though these crops accounted for a small portion of the total gross cropped area of the respective villages, they consumed much more water and were an important source of income for those who grew them. In Hediyala, 10.8 ha of sugarcane, 3.2 ha of turmeric and 2.8 acres of paddy were irrigated in 2003–4. In Chilakahalli, only 3.2 ha sugarcane and 0.8 ha turmeric were watered in the same period. Paddy and turmeric are grown in the rainy season and hence affect water availability in the dry season as well as water availability for sugarcane cultivation. This did not present a problem until 2002 as rainfall was good, water stored in the tank adequate, and seepage water too was plentiful.

In 2003, however, rainfall in the area decreased perceptibly and, consequently, so did seepage from the tank. By now farmers were cultivating a large area of sugarcane and their main aim was to ensure that the crop received enough water. Therefore, eighteen sugarcane farmers, Lingayats and Muslims, built bunds across the stream to impound water that they pumped out with diesel pumps. However, SC/ST farmers who also had land near the stream could not do the same as they did not have the financial resources to invest in pump sets.

The pumping of seepage water by the Hediyala farmers resulted in a decrease in the water flow downstream to Chilakahalli. Chilakahalli farmers had to meet their domestic water needs either by using private pump sets (twelve farmers had private pump sets in Chilakahalli) if they owned any, or public hand pumps. However, farmers were unable to meet the drinking water needs of their livestock. In March 2004, Chilakahalli villagers met in the village temple and decided, without giving the Hediyala farmers any advance notice, to remove all the upstream bunds. Next morning thirty people from Chilakahalli broke the seven existing bunds. Seventeen farmers from Chilakahalli and villages further downstream who pumped water from the stream did not participate. At the eighth bund the Hediyala farmers stopped them. There was a verbal duel and the situation threatened to become violent.

One major reason why this dispute has become a large conflict is that there is no functioning tank irrigation institution. Although the Minor Irrigation Department is the key institution for tank management, including maintenance, repair and constitution of water user committees, they have neglected their duties. Moreover, though a Kudregundi Halla Tank water user committee exists on paper, none of the command area farmers are even aware of its existence. As a result, there are no rules with regard to water usage and no restrictions on cropping choices. Chilakahalli farmers, dependent on seepage water for their domestic needs, have suffered as a result. Not surprisingly, therefore, Chilakahalli farmers lodged an official complaint with the *tehsildar* who sent a revenue inspector to Hediyala to ascertain the facts. When he arrived, there was no water at all in the stream, and he concluded that the Hediyala farmers were not guilty of pumping seepage water and that they had done no wrong.

Current Status

The Chilakahalli farmers were unable to launch another agitation because their demands were not taken seriously by the *panchayat*, which is dominated by Hediyala farmers. During the crisis, one of the farmers who had a private tubewell provided the others with water for livestock and other domestic use. The conflict seems to have been a one-off conflict restricted to a period of acute water scarcity. In 2005, there was no major problem because there was adequate rainfall and the stream had more water despite the bunds the Hediyala farmers have rebuilt since the incidents. But this is a simmering conflict, one that is likely to raise its head again, given the frequency of drought in the area. If bunds continue to be constructed across the stream and the concerns of the downstream users are not addressed, there are likely to be clashes in the future.

Some of the aspects of the conflict are quite unusual. First, the issue has resulted from the availability of water due to the non-functional tank and its poor construction. Second, seepage should not be a major source of irrigation in the first place. Third, while the conflict appears to be between head-enders and tail-enders, farmers further downstream in fact did not face similar problems partly because another stream joins the main rivulet further down the course. The Chilakahalli farmers are located midstream of the command area. Fourth, the conflict is mostly between farmers who use water for irrigation and households that use water (further downstream) for domestic and livestock purposes. Although water for irrigation is a priority in a well-functioning tank system, in the absence of an alternate system, domestic and livestock needs become more important.

Opposing Stands

The 'unusual' dimensions of the conflict complicate its resolution. Chilakahalli farmers insist that the upstream farmers used seepage water illegally. The upstream farmers argue that they were not making use of seepage flow water but seepage pit water. The upstream farmers also consider the seepage water as waste water and claim that using this water is not illegal.

Scope for Dialogue

The ambiguous nature of the conflict and the issue of what, if anything, the Hediyala farmers did illegally, implies that the terms of reference for a dialogue do not exist at present. With hindsight one can say that the revenue inspector's decision was incorrect. Hediyala farmers were tapping the seepage water from the stream by constructing bunds. The fact that there was no water in the stream when the revenue inspector came was because of the earlier pumping by the Hediyala farmers. Bunds should not be constructed on streams and the government should ensure against this, especially in drought years when there is water scarcity.

A number of immediate steps must be taken to ensure against such conflict in the future and to guarantee the rights of all farmers in the *atchakat*. As the stream belongs to the revenue department, they have to prevent the illegal pumping and use of water for irrigation. This includes making two things illegal: the digging of pits and putting bunds inside the stream. Seepage water should be treated as a common property resource to be used by all farmers in the *atchakat*.

In the long run, however, the Minor Irrigation Department should consider repairing the tank so that seepage stops and the canal system is reactivated. Although water leaks from the sluice gates, most of the seepage into the stream is from the earthen tank bund itself. If the tank is repaired, then rules and regulations can be put in place so that all *atchakatdars* are given water. The original design envisaged a command area of 327 ha that would have covered both Hediyala and Chilakahalli. At present, only a few farmers located immediately below the sluice gate are able to lift the sluice gates enough to allow some water to flow through the canal and a few others use the seepage water. A well-functioning farmers' committee can play an important role in managing and maintaining the tank system if water availability improves.

At present, neither the Minor Irrigation Department nor the tank water user committee play any role. It is the job of the Minor Irrigation Department to maintain and manage the tank, repair and renovate sluices, canals, bunds, etc. The *neeraganti*, who is also appointed by the Minor Irrigation Department, is supposed to take water level readings every day, open and close the sluices when water is released for irrigation, and supervise the minor repair work. He too is not performing his role because the tank is in a state of disrepair. If the tank is restored, a rejuvenated tank water user committee will have to take up the role of managing the distribution of water. The biggest incentive for the committee to judiciously play its role is to increase the supply of water.

As we know from other cases, water user associations do not necessarily guarantee equitable distribution of water. Cooperation to ensure that water availability improves downstream of the tank is no guarantee that water will reach all claimants. The head-end–tail-end problem remains a concern, especially in scarcity periods. Also, contested claims to water for different purposes—irrigation and domestic—will continue to be a possible source of conflict. The reactivation of the tank will give priority to irrigation, hence those who depend more on the stream for domestic purposes might be adversely affected.

In the short run, while the tank remains in a state of disrepair, it is necessary that the government enforces rules that will prevent upstream farmers from lifting seepage water. This too is not an easy task. As detailed here, the head-end farmers are better represented in the *panchayat* and thus more likely to influence decision-making processes than the Chilakahalli farmers. In both scenarios, whether the tank functions or not, the possibility for conflict remains.

Note

*Map 1 and figure by authors.

The Dharmasagar Tank in Warangal: Crops Versus People

M. Pandu Ranga Rao and R. Murali

A Prince's Gift

THE ANCIENT DHARMASAGAR TANK IS LOCATED NEAR A VILLAGE BY THE SAME NAME, 18 km from the Kazipet railway station in Warangal, an important town in Telangana, a backward region of Andhra Pradesh.

The tank was constructed in the thirteenth century during the Kakatiya period. In 1919 the tank was extended and renovated by the Nizam. Its three channels—Savaram channel, Kistamma channel and Julu channel—irrigated 674 ha in Dharmasagar, Devanoor, Madipally, Somadevarapally and Anantasagar villages.

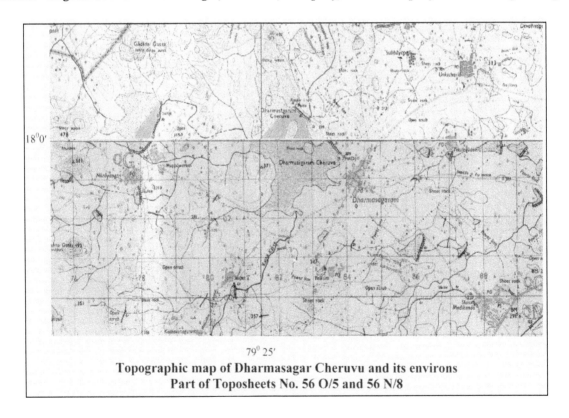

$79^0 25'$

Topographic map of Dharmasagar Cheruvu and its environs
Part of Toposheets No. 56 O/5 and 56 N/8

Map 1. *Location map of Dharmasagar tank.*

Fig. 1. *Opening ceremony of Filter Beds in 1936.*

This photograph from the 1940s shows the Prince and Princess of Berar dedicating the filter
beds to the people of Warangal, thereby commencing the project to
provide drinking water to the people of Warangal municipality.

The Public Works Department (PWD) handed over the tank to the district water works of the local fund department in 1936 to supply drinking water to Warangal municipality. The document that records the handing and taking over of the tank registers that the embankment of the tank had sprung leaks.

Initially, the supply to the town was limited and water was provided for irrigating between 120 ha and 160 ha according to availability. This practice continued until 1963 when the government issued an order that water would no longer be available for irrigation.

The Dharmasagar tank was basically an irrigation tank catering to the agricultural requirements of the people of the village until the 1930s. Having realised the importance of supplying drinking water to Warangal municipality, the Nizam of Hyderabad earmarked the tank waters for use by the municipality through an order. In the early 1930s, the water was used for both purposes and accordingly filter beds with a capacity of 8.18 Mld (million litres per day) were constructed in 1939 to cater to a population of up to 50,000. But with the passage of time and increase in population it became necessary to increase the filtration capacity by 6.82 Mld in 1964 to handle the total requirement from Dharmasagar tank that had risen to 15 Mld by then. The increase in the withdrawal of water from the tank for domestic use meant that the *ryots* could no longer access water for irrigation. Thus arose the conflict.

The Nizam's government had paid compensation to the affected *ryots* while acquiring the tank for the Warangal municipality, but the unhappy farmers filed petitions in the court at Warangal to claim better terms. The matter was temporarily settled when an extra sum was paid to the farmers but things did not end there and the debate over who rightfully needed the water continues till today.

The catchment area of the Dharmasagar has been shrinking owing to encroachments and the inflows into the tank have also reduced substantially, necessitating the need to fill it through external sources. In 1972–73, successive droughts left the tank dry and the government managed the situation only by exploiting the groundwater within the tank bed and in Warangal town.

The depth of the tank at full level is 12.1 m with a full reservoir level (FRL) at 304.7 m above mean sea level (MSL). The total capacity at FRL is 22 Mm^3. The catchment area is about 156 km^2, out of which about 28 km^2 is free while the rest is intercepted. The Malkapur channel starts from a diversion weir constructed 19.2 km upstream of the tank at Lingampally and fills intermediary irrigation tanks before flowing into the Dharmasagar. The capacity of the channel is 9 cumec. There are two tanks—Velairu and Mupparam—whose surplus course also leads to the Dharmasagar tank via the Mupparam channel. The Dharmasagar tank has two surplus weirs—Mondikunta, measuring 306.6 m, and Nakkalathoom, measuring 43 m with a crest level of 305.5 m above MSL. Seepage occurs in the embankment when the tank level climbs 10 m above the sluice sill level.

Warangal continues to grow at an alarming rate. With the present water supply from Dharmasagar rendered inadequate, the authorities decided to tap water from the Kakatiya canal at two places—the Kakatiya University campus and Desaipet. Rudramamba filters, with a capacity of 15.9 Mld, were installed in 1982 at the offtake at Kakatiya University. Another filtration unit of 3.73 Mld was commissioned in 1983 at Desaipet. Thus, a total of 6.82 Mld of protected water was made available to the people of Warangal by 1983 with the Kakatiya canal as the main source. But this canal, being an irrigation canal, was able to cater to drinking water requirements only for a period of eight months while the Dharmasagar took over for the remaining four, and it continues to remain the main summer storage tank. Unfortunately, the vagaries of the monsoon and a highly intercepted catchment have resulted in inadequate inflows into the tank. A decision was therefore taken to fill the Dharmasagar itself by pumping water from the Kakatiya canal through a main that is over 18 km long.

However, the existing one-metre diameter pipeline was not sufficient to fill the tank and the town continued to suffer from drinking water shortage in summer. Repeated requests from the municipality to the state government for infrastructure to help fill the Dharmasagar have not borne fruit. A permanent solution has now been planned at an estimated cost of Rs. 65 crore but this project too is pending approval.

The drinking water crisis is severe despite the fact that Godavari water has been available through the Kakatiya canal since 1983. The failure to fill Dharmasagar is therefore the principal cause for inadequate drinking water supply to Warangal. The tank merely needs to be filled to capacity (22 Mm³) for the shortfall in summer supply to be met. Efforts are now on to fill the tank through a separate pumping main with a two-metre diameter.

On a visit to Dharmasagar at the behest of the local MLA in May 1998, the then Chief Minister made an announcement: 'Supply of water to Dharmasagar tank through Sriramsagar Project (SRSP) canal will provide water for irrigation to Dharmasagar *ryots*'. This created panic among the people of Warangal and they immediately voiced their protest. Political parties too lent their voice to the opposition and asked for a withdrawal of the statement. Professor Pandu Ranga Rao, Advisor to the Municipal Corporation on drinking water and co-author of this article, analysed the issue in a document entitled 'Impact of Dharmasagar Tank on Warangal Water Supply System' and presented it to the Chief Minister. The analysis traced the history of the tank from its inception through its handing over by the Nizam to the people of Warangal right up to its current status. As a result, the government dropped the proposal to allot water for irrigation but the statement had served to bring to the fore the long-buried aspirations of the farmers in the area. They are now looking for an opportunity to raise the issue politically.

Big Plans

The Government of Andhra Pradesh is planning a Godavari lift irrigation project that will ferry water from Devadula through a network of tanks into the Dharmasagar reservoir. The present storage in Dharmasagar reservoir is only 22 Mm³ and the new scheme will help increase the capacity by another 22 Mm³. Thus, a total storage of 50 Mm³ should be available after some improvements are made to the main tank bund and the height increased by two metres, maintaining the FTL at 307 metres. This arrangement is being executed at a prohibitively high cost. The maintenance of this scheme has to be carefully monitored so that it becomes a viable means to increase both the irrigation potential as well as the ability to meet the drinking water requirements of the town. Now that it has become a composite project involving irrigation and drinking water, there might be problems when there is a shortage of water due to technical snags in the project that involves as many as five lifts from Devadula to Dharmasagar. It has to be borne in mind that a large number of lift irrigation projects that had started earlier are now in cold storage for want of maintenance and other ancillary problems. The government should ensure regular non-recurring and recurring grants for the smooth and efficient functioning of the scheme since it involves the drinking water requirements of a very large number of people, not to mention the irrigation requirement of the *ryots*.

There is also a potential for conflict during periods when the water level in the tank is low. At such times the authorities could find it difficult to allot water and might have to prioritise the need of one party over the other. The government must therefore ensure equitable distribution from the very beginning and decide the ratio to be apportioned to the two categories.

Fig. 2. *Outlet structure of the Dharmasagar tank.*

Highest and Lowest Points

The conflict began when the filter beds at Waddepally were inaugurated in the 1940s by the Prince and Princess of Berar, and the Dharmasagar tank was in principle allotted fully to drinking water requirements of Warangal. Subsequently, the filing of compensation claims in the honourable court of Warangal and the subsequent transactions and awards kept the conflict under check. Therefore, the announcement of the Chief Minister in 1998 trying to favour irrigation over drinking water has renewed the conflict.

The present Godavari lift irrigation scheme envisaging the integration of irrigation with drinking water may create a crisis in course of time if the minimum draw-down level (MDDL) norms are not honoured. The government must give a categorical assurance to protect the drinking water rights of the six lakh people of Warangal municipal corporation while it attempts a balancing act to provide water for irrigation.

The Dharmasagar village is close to Warangal, and the impact of urbanisation is being felt in the villages too. Property values have gone up and one hopes that the Godavari lift irrigation scheme will take care of the growing need of the farmers and the town's people. Also, the tank now has an additional storage capacity of 1 TMC (thousand cubic metres).

Fingers Crossed

Though all the problems appear to have been resolved for the moment one must not forget that this tank has been in the eye of the storm on many occasions since the 1930s. Several times permanent solutions seem near but the issue has continued to fester because of the violation of agreements. It is now the responsibility of the stakeholders to ensure that the lift scheme proceeds without glitches so that there is assured water supply for all.

Note

*Map 1 from *Survey of India*; figure 1 from Municipal Corporation, Warangal; figure 2 by authors.

References

Godavari Lift Irrigation Scheme. Technical report. Irrigation Department, Government of Andhra Pradesh.

Rao, M. Pandu Ranga. 1998. *Impact of Dharmasagar Tank on Warangal Water Supply*. Report submitted to the government on December 3.

Unpublished report on Dharmasagar tank—court judgements.

Unpublished report on Dharmasagar tank—official correspondence.

Lava Ka Baas
Traditional Water Harvesting Structure:
The Community behind 'Community'

Prakash Kashwan

THIS CASE STUDY IS ABOUT A TRADITIONAL WATER HARVESTING STRUCTURE CONSTRUCTED on a drain in Lava Ka Baas (LKB) village of Thanagazi block, Alwar district in Rajasthan. Ironically, while the battle was apparently fought to protect community rights over water resources, this study highlights the failure of strengthening community institutions, a must for ensuring equitable distribution of benefits.

Map 1. *Map showing the area of conflict: The Thanagazi block in the state of Rajasthan.*

Lava Ka Baas Builds Its Own Water Harvesting Structure

Tarun Bharat Sangh (TBS), a non-government organisation (NGO), began work in the Thanagazi block in 1985. The Government of Rajasthan had declared four blocks in the region as 'dark zones'.[1] TBS intervened by constructing '*johads*', which are traditional rainwater harvesting structures. They are semi-circular earthen ponds that collect the runoff from tiny streams and rivulets in a micro-catchment. Between 1984 and 2000, some 2,264 *johads* were constructed (Pangare et al., 2003). TBS and several other national and international NGOs supporting its work reported a significant increase in the groundwater table. The area was subsequently declared a 'white zone' by the State government. Five rivers, Bhagani–Teldehe, Arvari, Jahajwali, Sarsa and Ruparel, which had been reduced to seasonal rivers, were reported to have become perennial in 1995, benefiting some 250 villages.

The work on building the 80 m long and 12 m high *johad* at LKB was completed in less than four months beginning March 2001. TBS's Gopal Singh—a *gajdhar*, meaning a rural engineer—with many *johads* to his credit, 'designed' the structure. Noted industrialist P. K. Rajgarhia gave Rs. 5 lakh for the project, and the villagers of LKB and nearby Bhagdoli village pooled the remaining Rs. 3 lakh required for constructing the *johad*.

The Government Objects

Thrilled with its big *johad*, the *gram panchayat* invited then Chief Minister Ashok Gehlot to inaugurate it. During an administrative drill preceding the visit, District Magistrate Tanmaya Kumar Sinha declared that the structure violated a 1910 agreement between the erstwhile princely States of Alwar and Bharatpur that entitled Bharatpur to 55 per cent of Ruparel's water, and that the pond would jeopardise water sharing between the two districts. On June 20, 2001, the Rajasthan Irrigation Department notified TBS that the LKB *johad* was constructed in violation of the Rajasthan Irrigation and Drainage Act of 1954. It asked TBS to stop construction work and demolish the completed portion within seven days or face action under Sections 55 (3) and 58 (2) of the Act that permits the accused to be arrested without a warrant.

Fig. 1. *Lava Ka Baas Johad in October 2001.*

In reply, Rajendra Singh, Secretary, TBS, reminded the Irrigation Department that its own 1998 study had found that the construction of upstream water harvesting structures had not curtailed the downstream flows of nearby Arvari river and that the downstream flows had actually increased. TBS argued that the *johad* would revive the Ruparel and make it a perennial river. In an interview given to the newspaper *Today*, Rajendra Singh declared, 'At least 70,000 people would benefit directly while another two-and-a-half thousand would benefit indirectly from this check dam' (Dasgupta, 2001).

Irrigation Department officials landed in LKB on July 1, 2001 with earthmovers to demolish the structure. Villagers responded by laying siege to the structure. Sensing conflict, district irrigation officials halted the demolition but Rajasthan Irrigation Minister, Kamla Beniwal said, 'Every drop of water that is received through the rains comes under the Irrigation Department and any activity related to the storage of water without any prior permission from the Irrigation Department would not be tolerated' (*Hindustan Times*, July 3, 2001).

Highs and Lows

The incident catapulted LKB and TBS into the limelight in an unprecedented way. The Centre for Science and Environment (CSE), a prominent NGO, launched a concerted media and civil society campaign against the government's attitude. International organisations responded favourably to the online lobbying efforts of the CSE. Incidentally, Rajendra Singh received the 2001 Ramon Magsaysay Award for Community Leadership, though it would be unfair to attribute it to the LKB incident. The CSE also constituted a commission to look into issues of structural safety, violation of law, and the adverse impact of the *johad* on water availability downstream. The commission included top civil society leaders who met the Chief Minister to convince him of the viability of the structure and the approach.

The conflict peaked two years after the controversy, when the LKB *johad* was breached on July 10, 2003 due to torrential monsoon rains. The bureaucracy was quick to ridicule the very idea of building a water harvesting structure based on 'rural engineering' and said that their stand had been vindicated. A TBS team which visited the spot for an assessment attributed the breach to the collapse of six upstream dams—some built by the State government under drought relief work the same year. The TBS informed the media that the villagers wanted to lay pitching on the dam, but the Irrigation Department, the police and the district authorities prevented them from doing so. Post-2003 monsoons, the Tarun Bharat Sangh rarely raised the LKB issue again.

Not Much Scope for Dialogue

The Irrigation Department throughout maintained that the dam was technically weak and that sufficient precautions were not taken to strengthen it. However, the technical committee's report and the notices sent to TBS and Rajendra Singh prior to the enquiry revealed that the structure was built without departmental permission. TBS and CSE maintained that people must be allowed to take control of local natural resources without having to negotiate bureaucratic hurdles. They also pointed out that catchment deforestation had reduced the Ruparel river to a seasonal stream three decades ago. They rejected the validity of the 1910 agreement signed between the two erstwhile feudal States of Alwar and Bharatpur.

The possibilities of a dialogue between the department and TBS seemed remote, owing to ongoing confrontations between the two since 1987 when the department issued a series of notices asking the TBS to stop its work of 'damming' the rivers. The LKB incident was the culmination of a rivalry between the department and an organisation that refused to be cowed. To CSE goes the credit for mobilising top leaders and the national media. It built pressure on the state government, forcing it to ultimately withdraw the demolition orders.

Table 1: Key Institutions and People

S. no.	Key institutions/People involved	Position and impact on the conflict
1.	TBS and Rajendra Singh	Rajendra Singh and TBS promote a vision wherein communities are regarded as being fully capable of dealing with local development and governance issues. Inherent inequities and contradictions in communities do not really figure in their scheme of things. This has not changed despite a few evaluations that have raised this issue very strongly.
2.	People of Lava Ka Baas	Many in the village felt that arrangements should have been made to ensure equitable sharing of benefits. This did not mean that they had any sympathy with the actions of a government that left them to fend for themselves during the severe drought years.
3.	Centre for Science and Environment	CSE has championed the cause of rainwater harvesting at the national and international levels. It mobilised civil society and the national media against the government's decision.
4.	Government and Irrigation Department	The department saw the *johad* and the work of TBS as a challenge to its hegemony over water resources management. State government and politicians generally adopted a hard stand against TBS.
5.	Media (Local/National/International)	Masterly media management by CSE and TBS brought the media to Lava Ka Baas who highlighted the rights of the people.
6.	Civil Society	Some of the best technocrats in the country supported the village community and the NGO.

Another Look at LKB Technology

In order to challenge the dictates of the Irrigation Department, the CSE commissioned Professor M. C. Chaturvedi to undertake a scientific analysis of LKB *johad*. To quote from his report: '*Preliminary studies*'[2] show that the system has much higher capacity than that required for the storage of maximum annual runoff.... A more detailed analysis will be made after obtaining all the information and if any shortcomings are noted they will be rectified well in time.... Measures for upstream and downstream slope protection will be undertaken' (Chaturvedi, 2001).

This indicates that the structure was indeed vulnerable and needed additional work, investment, care and vigilance. In the past too, eminent experts have analysed the indigenous techniques of building water harvesting structures adopted by TBS, notable among them being G.D. Agarwal (retired Professor, Indian Institute of Technology, Kanpur). However, assessing the viability and robustness of an indigenous technique is quite different from assessing the robustness of the actual structures.

Community Effort, Private Benefit?

Reports by Professor Chaturvedi and others suggest that significant support for the structure in question poured in because it was a 'community initiative'; the structure would benefit the poor who had been neglected by the State machinery. It is therefore important to critically analyse the strength of community institutions and to examine how the *johad* benefited various segments within the community. Of the hundreds of media and other reports, there are only a couple that dwell on this issue. Civil society could benefit from carrying out serious introspection on the equity aspects of this work.[3] The analysis that follows derives from evaluation reports by Kumar and Kandpal (2003), and Pangare et al. (2003).

The location of the *johad* was such that to lift water for irrigation, the water had to be brought to a minor ridge (*raadi*) after which it could be taken to the main agriculture fields in the villages of LKB and Bhagdoli. This could be done only by using a submersible pump. As the village transformer had a limit of 25 hp, only three submersible sets (each 7.5 horsepower) were running during February 2003. In addition, eight diesel pumps owned by people from Bhagdoli and three from LKB were lifting water from the bed to irrigate lands lying on the inside of the ridge, where it could be lifted with ease without the use of a submersible pump (Kumar and Kandpal 2003: 51).

The *johad* also led to groundwater recharge. 'However, all groundwater recharge was towards Bhagdoli. Thus 25 borewells were sunk in Bhagdoli of which 20 yielded excellent water at a depth of 400 feet. In contrast, a farmer from LKB dug down till 615 feet without finding any water' (ibid.). On the other hand, as per the reports published by CSE, TBS claimed that the water in the check dam could take care of twelve neighbouring villages and additionally more than 100,000 people would benefit by the 'recharge of thousands of defunct wells located downstream' (*Down to Earth*, online edition, July 31, 2000).

Private investment made for irrigation using the water from *johad* in the two villages of Lava Ka Baas and Bhagdoli during 2003 was worth Rs. 1,657,000 (Kumar and Kandpal, 2003: 51). In this way, rich farmers who owned more than 50 *bighas*[4] (5 ha) were the ones who benefited the most. Of the forty-one families in Lava Ka Baas, thirty had landholdings of less than 10 *bighas* (1 ha) and none of them derived any irrigation benefits at the time of the study. The study also mentions that the *gram sabha* (the local institution promoted by TBS) was more or less defunct (ibid.). This is further supported by Pangare et al. (2003), wherein the evaluation team concluded that a majority of the water harvesting structures constructed by TBS were private assets and not community assets. Even in the case of community structures, no mechanisms were developed at the village level to ensure that benefits were shared systematically and equitably.

Excerpts from an article published in *Manushi* and based on an interview with Rajendra Singh also help us in understanding the development approach adopted by TBS. Rajendra Singh was convinced that 'the Leftist obsession with class struggle, "minimum wage" legislation sought to be implemented through a corrupt and insensitive bureaucracy or propagated through culturally alien, western educated political activists have led to severe fragmentation of the village society. 'Our villages don't need class struggle, but strengthening of their mutual bonds which traditionally knit various caste groups into mutually interdependent and cohesive village communities' (Kishwar, 2001).

While the ideologies of the 'Left' and those of 'western educated political activists' must be debated, no second thoughts can be entertained over the judicious use of public resources for the emancipation of the masses. There is an urgent need to challenge developmental approaches that go overboard in extolling the virtues of greenery without tracing the hands that own the land and harvest the fruits of public money.

Notes

*Map 1 by author, figure from *www.rainwaterharvesting.org*
1.   An area where the groundwater table has receded below recoupable level.
2.   Emphasis added by this author.
3.   For an incisive analysis of this issue readers are advised to refer to Baviskar, Amita (2002). Incidentally, media campaign of CSE on the much celebrated success of Jhabua Watershed Program is at the center of this analysis.
4.   4 bighas = 1 acre

References

Baviskar, Amita. 2002. 'The Dream Machine: The Model Development Project and the Remaking of the State'. Unpublished manuscript, obtained from the author.

Chaturvedi, M.C. 2001. 'Lava Ka Baas johad—A Scientific Analysis', dated July 17. Accessed at *http://www.cseindia.org*

Dasgupta, Kumkum. 2001. 'Jealous Bureaucracy Out to Thwart Village Initiative in Rajasthan'. *Today*, July 28. Circulated over DNRM Listserv. Accessed at *http://www.panchayats.org/discgroup/dnrm_GoR.htm*

'Headline: Who owns the River?', *Down to Earth*, 10(5), July 31, 2001. Accessed at *http://www.downtoearth.org.in/fullprint.asp*

Kishwar, Madhu. 2001. 'Villages in Rajasthan Overcome Sarkari Dependence- Profile of Rajendra Singh and his Work', *Manushi*, 123, March–April. Accessed at *http://www.indiatogether.org/manushi/issue123/rajendra.htm*

Kumar, Pankaj and B. M. Kandpal. 2003. Project on Reviving and Constructing Small Water Harvesting Systems in Rajasthan; SIDA Evaluation 03/40, Department for Asia, Swedish International Development Cooperation Agency.

Pangare, Ganesh, Talha Jamal, Rakshat Hooja. 2003. 'Community-based Rainwater Harvesting- Process Documentation of the Activities and approach of Tarun Bharat Sangh'. Unpublished Report. Mimeograph. Indian Network on Participatory Irrigation Management.

Sharma, Abhishek. 2002. 'Does Water Harvesting Help in Water-scarce Regions? A Case Study of Two Villages in Alwar, Rajasthan'. Iwmi-Tata Water Policy Research Program Annual Partners' Meet 2002, International Water Management Institute.

'Villagers Build Check Dam, the Government Wants to Destroy it', *The Hindustan Times*, July 3, 2001. Accessed online at *http://www.cseindia.org/html/extra/dam/index_news.htm*

Ubeshwarji Ka Nala:
Inter-Village Water Conflict in Rural Rajasthan

Eklavya Prasad

Feuding Villages

DHAR AND GHELOTAON KA WAAS ARE PART OF THE TRIBAL (GAMETHI/BHILS) DOMINATED
Dhar *panchayat* of the Badgaon *Panchayat Samiti* in Girwa *tehsil*, Udaipur district, Rajasthan.

Map 1. *Location of Udaipur where Dhar and Ghelotaon Ka Waas are situated.*

Dhar

Located 18 km west of Udaipur city, the hamlet gets its name from the perennial *dhara* or stream of water, which originates from a natural spring located upstream from the Ubeshwar Mahadev Mandir. This water source is considered a blessing from Lord Shiva.

The village is in the catchment area of Badi Ka Talab or Jiyan Sagar. Dhar has 250 households with a total population of 900, and is predominantly a Gamethi village. It has four settlements—Dhar, Mangri Phala, Pal Khanda and Dar Kheda. The main occupations are agriculture and labour. The landholding within the village varies from 0.2 to 0.6 ha of *hakatbhoomi* (agricultural land) and *padatbhoomi* (wasteland).

Ghelotaon Ka Waas

About 1.5 km north of Dhar, this tribal settlement has a total of seventy households. The name is derived from the Ghelot Rajputs who lived here during the medieval period. The village largely depends on Ubeshwarji Ka Nala, as does Dhar, for water. The other source is the surface runoff from Viyal, a hilly village west of Dhar. Here too, agriculture and labour are the main sources of livelihood. The individual landholding capacity is slightly better than Dhar, varying between 0.2 to 1 ha each of *hakat* and *padat* land. The main produce is maize during *kharif,* the season starting with the onset of the south-west monsoon and ending in October) and wheat during *rabi* (winter cropping) season. A few farmers have diversified into floriculture, horticulture and sugarcane planting. The village has adopted the gravitational flow irrigation technique and there is a mixed network of *pucca* (concrete) and *kutccha* (earthen) channels ensuring equitable distribution of water.

Fig. 1. *Water flow being diverted from Ghelotaon Ka Waas towards Dhar.*

Route Cause: A Historical Perspective

The age-old conflict between the two villages pertains to the distribution of surface water for irrigation. The dispute dates back to a period when Ghelotaon Ka Waas was part of Dhar. At that time Ubeshwarji Ka Nala flowed on the periphery of Dhar before it reached Ghelotaon Ka Waas. From Ghelotaon Ka Waas the stream flowed into Ayad river, which culminates at the Udai Sagar Lake, 15 km east of Udaipur (the lake is on the Berach river and was made in 1565).

With the construction of Jiyan Sagar or Badi Ka Talab between 1652 and 1680 during the reign of Maharana Raj Singh I, the water scenario at Dhar underwent changes. The State conceived a new drainage system to enhance surface water flow from the catchment to the lake. The drainage plan proposed diverting Ubeshwarji Ka Nala to Jiyan Sagar through a water channel connected to the newly constructed check dam. They were to be joined at the junction from where the stream once flowed towards Ghelotaon Ka Waas.

This initiated the feud between the two villages. The quantity of water flowing into Ghelotaon Ka Waas decreased but an existing *khadda* or water hole in the streambed near the village ensured that they continued to get water. Water from the stream entered the *khadda* and emerged as a natural spring inside the village. But in spite of getting this water the locals felt deprived.

After independence an anicut was constructed to enhance water storage, and was to be used for irrigation by both villages. The structure was located downstream from the junction where the spring water resumed its surface flow after falling from the Ubeshwarji Ka Magra (hills) near the fringes of Dhar. People in Ghelotaon Ka Waas considered the anicut an impediment to the free flow of the stream because it had an impact on the quantity of water that entered the water hole. They feared their spring would be made redundant. On the other hand, Dhar persisted with its demand for an equal share of stream water. To them it seemed that the other village enjoyed an advantage because of the water hole. It did not take long for people to realise that the anicut invariably had a surplus. This helped in maintaining a minimum surface flow to the water hole. Since Dhar lacked any such natural mechanism, the people decided to link the reservoir with their farmlands through a well-laid-out channel. This, they felt, would ensure that they got an equal share of the impounded stream water.

During the 1960s and 70s the region witnessed direct confrontation between the villagers. Farmers from Dhar would keep constructing earthen drains that the opposing group wasted no time in demolishing. In the late 1970s, Seva Mandir, a voluntary organisation—in an endeavour to promote the 'Food for Work' scheme—renovated the government anicut with the help of workers from the local village community. The work was accomplished without any dispute from either village but the feud resumed once the construction was over. Although the organisation has made several attempts to resolve the issue, both parties have consistently refused to cooperate.

The region reeled under drought in 2002, and the Dhar *panchayat* sanctioned the construction of an anicut as part of the relief work. The then *sarpanch*, Savalibai Navaram Kalava of Ghelotaon Ka Waas, was keen to construct it in her village but geographical limitations did not permit this. The villagers therefore decided to augment the capacity of the existing anicut by further increasing its height. However, people from Dhar noticed that the top portion of the anicut tilted towards the other village to permit water to overflow from the reservoir. Not to be outdone, they in turn constructed a temporary drain to siphon water from the anicut into their fields.

To counter the growing antagonism between the two villages, Savali bai later got two additional projects sanctioned. Both hamlets were to construct a concrete drainage system from the anicut. However, even this initiative flopped. While Dhar managed to build its drain, Ghelotaon Ka Waas could not do so because a flawed estimate resulted in insufficient funds. Also, Dhar clearly indicated that it would not allow the impounded water to be carried to Ghelotaon Ka Waas through the drain. They preferred that the *khadda* remain the only medium for the other village's water supply.

Status Quo

The dispute between the two villages has reached a dead end. Both the villages have independently evolved their own strategies to address the issue. Needless to say, all solutions are unacceptable to the opposite group. The problem is being further aggravated because of the ad-hoc measures that have been adopted to gain access and control over the stream water. At present, things are quiet because of the lack of sufficient surface flow in the stream, but hostility is perceptible in both the villages. By now the history of the dispute is well entrenched in the conscience of all the villagers though the quarrel has acquired a seasonal quality. The fierce arguments peak when there is just enough water to irrigate the agricultural lands in both villages. On the other hand, when there is either too much or too little water both sides see no point in fighting.

Bitter Memories

This incident took place thirty-five years ago but continues to infuriate both the communities. Devangaba Gamethi of Ghelotaon Ka Waas and Gangaram Vadera of Dhar village got involved in a bitter quarrel over the water issue leading to legal action. It so happened that farmers from Dhar kept constructing earthen drains while villagers from Ghelotaon Ka Waas consistently demolished them overnight. This led to a fight between the two men and the case was referred to the *panchayat*. When this proved unsuccessful, Ghelotaon Ka Waas took the matter to the *patwari* (representative of the revenue department at the village level) and the revenue inspector. The revenue officials filed a case against Gangaram and fifteen others at the *nyay panchayat* in Thoor. They believe that people from Dhar used their contacts and resources to influence the presiding officer to get the verdict in their favour. The *nyay panchayat* declared that Ghelotaon Ka Waas did not have any right to the stream water, but the setback did not in anyway deter them from continuing to voice their demand for their share.

The Great Divide

Dhar

The people of Dhar are willing to share the stream water with Ghelotaon Ka Waas on the condition that they limit their access to the overflow that enters the water hole and not ask for the stored water. In any case they believe that Ghelotaon Ka Waas gets far more than Dhar receives, and therefore an additional supply scheme would create inequity. They also think that even during the lean season there is just enough surface water in the stream to feed the *khadda*. The water might not surface at Ghelotaon Ka Waas through the spring but it recharges the groundwater reservoir, which indirectly benefits the farmers. Dhar claims that the rich store of groundwater has prompted farmers of Ghelotaon Ka Waas to grow sugarcane which would not have been possible in a water-deficient region.

Ghelotaon Ka Waas

Limiting their access to stream water under the pretext that the existing natural subsurface drainage is enough to meet their needs is an argument that is unacceptable to Ghelotaon Ka Waas. Their options to resolve these age-old conflicts are:

1. Equitable distribution of impounded water.
2. For ten days a month, release water to the village from the anicut so that irrigation needs are met.
3. Review the extent of land being irrigated in both the villages and accordingly allocate stream water.

The general conception in Ghelotaon Ka Waas is that Dhar has the right contacts and therefore manages to influence the *nyay panchayat*. The villagers also claim that Dhar has been projecting inflated figures of potential beneficiaries to account for the increased water requirement. According to Dhar, the village has fifty farmers who utilise stream water for irrigation, but Ghelotaon Ka Waas maintains that the figures are manufactured.

There is no denying that the long-standing dispute has created an environment of uncertainty among the stakeholders in both villages. During the peak season, despite there being sufficient water in the stream, farmers from both villages tend to be ambiguous about their ability to access water because it could lead to hostile confrontations. There is tremendous pressure on all those involved, especially during the leaner periods when everybody's attention is fixed on the farmlands on the one hand, and the limited water supply on the other. Ironically, the conflicting groups belong to the same caste. The water war in these two villages continues to intensify and defy justification or logic.

Time to Talk

The conflict has its roots in the perception that individual interests might be jeopardised. Therefore all suggestions and initiatives are coloured by prejudice and lack the commitment to explore viable options for the collective good of the community. Neither community has taken the initiative to sincerely try and resolve the issue, nor have they cooperated with the external sources which have sought to mediate.

Dialogue as an option has never been considered in all these years. Localised intervention in the form of direct communication between stakeholders might be the most effective and sustainable approach. Both communities need to involve local voluntary organisations. The Dhar *panchayat* has the distinction of having elected a new *sarpanch* unopposed for the second time. This indicates consensus within the Gamethi community (they are the dominant group in the region). A similar effort needs to be undertaken to evolve some kind of an agreement to resolve this dispute. Legal action should be the last resort as it might further entangle an already complicated issue, leading to further crisis. A self-regulated, localised framework of action is the only option that the communities should pursue.

Note

*Map 1 from SOPPECOM, Pune; figure by author.

Shapin River Basin in Jharkhand: The Failure of Community Institutions

Pankaj Lal, Kamaldeo Singh and Kapildeo Prasad

Shapin: The Serpentine River Without an Outlet

Rainfed agriculture is the primary means of sustenance in rural India, and Godda district in Jharkhand is no exception. More than 96 per cent of its population is rural and only 14.21 per cent of land is under irrigation. The district depends mostly on the seasonal river Shapin, a name that literally means snakelike or serpentine, and farmers of the Pathargama block depend on it for irrigation.

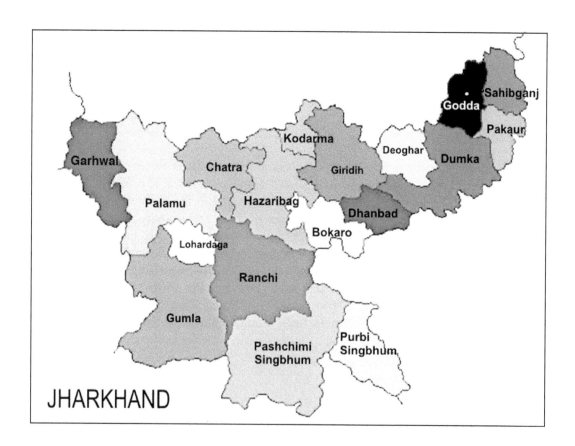

Map 1. *Location of Godda in Jharkhand.*

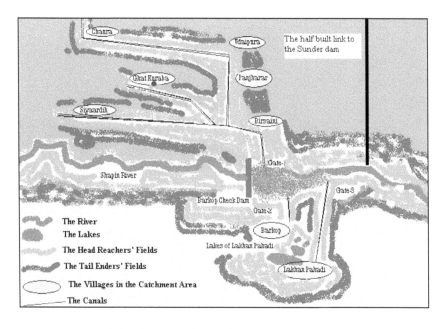

Map 2. *Detailed map of the study area (not to scale).*

Shapin has two specialities: first, it has no outlet, which means that its flow is consumed within the district; second, the riverbed is at a height greater or at the same level as the fields on its banks, which means the water does not need to be lifted.

This case study focuses on a group of eight villages in the Pathargama block of Godda district, on water use conflicts between and within villages, and failure of the community institutions in dealing with them. These villages lie in the command area of a dam built in 1971 on the Shapin in Barkop village with the help of an NGO named Gadadhar Mishra Smarak Khadi Gramodyog.[1] The dam is situated at a distance of 4 km from Pathargama, benefits about 11,350 people, and irrigates 2,400 ha of land (Table 1).

Table 1: Villages Benefiting from Barkop Dam

Village	Cultivable land (ha)
Barkop	360
Lakhanpahadi	480
Birsaini	60
Siyardih	180
Ghatkuraba	520
Panjharar	280
Udaipura	216
Chaura	240

Fig. 1. *Barkop dam, its iron gates as well as the dam watchers-room, can be observed.*
These iron gates have to be opened in case of heavy downpour to avoid flash floods.

Eighteen kilometres further east is a large dam on the Sundar river. The villagers feel that if the Barkop dam is linked to the Sundar dam, irrigation potential can increase to 6,000 ha. The State government tried to link these dams but a one km canal stretch could not be completed. The command area of the dam has a slight slope from the north-east to the south-west and, as mentioned earlier, the riverbed lies higher than the fields it irrigates.

Two unlined irrigation canals—Ghatkuraba canal upstream and Siyaardih downstream—supply water to fields lying on both sides of the command area. The farmers make their own arrangements to take water from the canals to their fields. The command area has highly fertile black soil. Sixty per cent of the working population are cultivators and 24 per cent are agricultural labourers. There are two cropping seasons—*kharif* and *rabi*. In kharif, paddy is the major crop, cultivated in 95 percent of the area, while wheat, maize, pulses, peas and mustard are cultivated in *rabi* season. The productivity of paddy varies greatly from Rs. 6,250 to Rs. 37,500 per ha.

The Committee is Formed, the Committee Declines

In 1971, at an NGO's initiative, an advisory committee consisting of five villagers and two ex-officio NGO representatives was formed for site selection and for facilitating the construction of the Barkop dam. The committee members comprised farmers with big landholdings of more than 4 ha and none of the Musahars (a scheduled caste community) are members of this committee. After construction of the dam, the same advisory committee was converted into a user group's management committee. It was responsible for maintenance of the dam and unlined canals, and decided on access, use, contribution, etc. An annual charge per *bigha* (a local measure of land equal to about 0.27 ha) was levied for water and stands presently at about Rs. 94 per ha. Marginal farmers (less than one *bigha*, i.e. 0.27 ha) were supposed to contribute in terms of labour. A watchman employed to take care of the dam was paid from the collections. Other expenses incurred were for dam maintenance and upkeep of canal streams. Committee meetings were not held at regular intervals, but only when there was a need to take a decision or arbitrate in conflicts.

The NGO members in the committee did not have much to say from the beginning. During the first ten years, everything went smoothly as all community committee members made decisions collectively and took care of whatever needed to be done. In the following decade, the decisions and responsibilities were limited to just two members—Deendayal Roy and Kapildeo Prasad. The watchman left in 1991 and a replacement could not be found. Managing the dam and the committee's work became a tedious task. Roy and Prasad could not devote as much time and energy as they had earlier. Defiance was building up. To address the situation, a sub-committee—drawn from the youth of the area—was formed.

Neither the committee nor the sub-committee members were compensated for their time. Becoming a member was considered a matter of prestige and respect. However, collecting money from reluctant people, using the funds effectively and ensuring equity was a tiresome and thankless job, and soon most of the new sub-committee members quit, leaving one individual, Arun Kumar, to manage it. By now, the committee and the sub-committee had almost become defunct. People started making individual efforts to carry water to their fields. At the same time, problems of free-loading were on the rise while user charge collections dipped. One of the reasons was that maintenance through contributions was virtually impossible as collections were small and insufficient for proper upkeep. The contributions started dipping after 1991, and last year were a paltry Rs. 5,000. In short, what was left was a weakened institution and insufficient funds. Conflicts surfaced both within and between villages.

Chronology of Events

1971a:	The Barkop village dam was built and the monitoring committee formed.
1971b:	A watchman was employed and user charges on irrigation water levied.
1971–1981:	Active participation by all committee members in all the activities of the committee.
1981–1991:	Deendayal Roy and Kapildeo Prasad were appointed to look after work.
1991a:	The watchman quit, and there was no replacement resulting in frequent flash floods.
1991b:	Village sub-committee of youths was formed
1991 onwards:	There was an increase in incidences of intra- and inter-village conflicts.
2000:	Flash floods, sand fill fifty acres of land.
2001-2003:	Around thirty intra-village conflicts in Ghatkuraba village occur every year.
2003:	Conflict between the villages Ghatkuraba and Siyaardih result in three casualties.
2004:	Good monsoons. No intra- or inter-village conflicts.

Conflict between Villages

The villages are located at different heights. The better-located village not only has to spend less money and labour to get water to its fields, but also gets it more easily. Conflict usually arises when there is scarcity, because the village that puts in more time and effort ends up not getting enough water, while those who spend less get enough. As mentioned earlier, the Shapin has no outlets and is prone to a sudden rise in water level when there is heavy downpour. If the dam gates are not opened on time, flash floods occur. With no watchman to open the gates, frequent flash floods began to occur.

In fact, the dam gates were not even maintained, and now they are completely unusable. No wonder then that the largest flash flood in the area occurred in 2000 and submerged all nearby villages, and 50 *bighas* (13.5 ha) of fertile land were covered with sand.

Conflicts Within Villages

Two distinct user groups can be identified in the command area. The first are the head-reachers who have land adjacent to the canals. All field channels have to pass through their fields first, and unless there is ample water in their own fields they don't allow it to pass on to others. The second are the tail-enders who are much more active in the upkeep of the canal and the channels because irrigation in their fields directly depends on quality water management.

Proper dam and canal maintenance means increased water availability; it translates into water for tail-enders. Around 70 per cent of the tail-enders paid water charges before 1991, but now only 30 per cent do so. Only 25 per cent of the head-reachers have been regular contributors since 1991. In the lean years, all head-reachers are able to irrigate their land but few tail-enders can do so. The resentment is caused by the fact that the head-reacher who has not contributed can irrigate his land, while the person with a field just behind is unable to irrigate his land in spite of contributing.

Rising Conflicts

The head-reachers and tail-enders both need water to irrigate their fields in the lean years. The tail-enders have been arguing for contribution-based benefit—everyone who is using a resource should pay. The failure to curb free-riding by head-reachers has prompted them to indulge in free-riding themselves. This phenomenon is also observed between villages, where the upstream villages try to get more water at the cost of those downstream.

Within the villages too there was a sharp increase in conflicts. Between 2001 and 2003, there were around thirty conflicts in Ghatkuraba village alone. Villagers report that such tussles increased substantially after 1991. Earlier there were about five to ten water-based conflicts within a village, whereas now these have increased to about twenty to thirty such cases. Conflicts between villages were a rarity before 1991, but now there are about two every year.

The potential for conflict within and between villages is at an all-time high. In years of water scarcity, the conflict is at its ugliest, but when there is enough rainfall the year passes virtually without any problems. For example, there was good rainfall in 2004, and no major quarrels were reported. The best years were between 1971 and 1981 when the committee was active. The sub-committee is not only inefficient but also lacks widespread acceptability. The free-loading tendency has increased. The flash floods are synonymous with heavy rainfall; the dam gates have holes in them and are out of order. Conflicts flare up in an instant and spread like wildfire, mainly due to the lack of a controlling institution.

Elders Resolve a Clash

Ghatkuraba village is at a higher altitude than Siyaardih, but the former has been given a larger and the latter a smaller opening. Normally, therefore, the two canal streams are such that both villages get an equal share of water. The years 2002 and 2003 were two consecutive lean years and there was inadequate water for irrigation. The farmers had begun to get restive, fearing poor harvest in the coming years. One night in 2003, the farmers from Siyaardih cleared and enlarged the opening of the canal whereby the entire supply was drawn into their canal. When Ghatkuraba farmers came to know of this in the morning, 150 farmers marched to Siyaardih. There were angry discussions, scuffles broke out and people drew out guns and rifles. About twenty persons were injured and three needed urgent medical attention. The situation was brought under control by the village elders who decided to settle the issue amicably. Three representatives from each village were selected for arbitration and they judged that the Siyaardih farmers were at fault. The old system was restored at once, the defaulters fined and the money put in Ghatkuraba's village development fund.

Informal institutions such as village elders, have been used to resolve conflicts both between and within villages. Both parties involved select their representatives (normally two or three for each party) and the decision taken by this group is binding. A mutually accepted group of village elders are often entrusted with the responsibility of arbitration in a situation where the committee is significantly weakened. In conflicts between villages too, the village elders sit together and seek resolution. Till date these arbitrations have been respected. The decisions have been unbiased on the whole and therefore this system has worked. But lately, defiance is mounting and the village elders have begun to distance themselves.

The government machinery here favours local resolution of the conflict and facilitates the process. Only when informal efforts fail do they resort to formal institutions such as police, courts or other state machinery. Formal institutions like *panchayats* and state government representatives such as block development officers have a potentially positive role to play, especially when informal arrangements are seen to be weakening and a need is felt for formal institutions to step in. Strong local institutions, and elected representatives of the *panchayats* or the State legislature can establish a structure of equitable contributions and benefits.

Impact of Conflicts

1. Decline in the resource in the absence of proper upkeep of canals and dams resulting in leakages and less water for irrigation.
2. Falling contributions and growing tendency to free-load.
3. Flash floods because of defunct dam gates in the year 2000. There was a loss of Rs. 5 lakh worth of paddy crop.
4. Fragmentation and weakening of traditional institutions such as committee of village elders and arbitrators.
5. Growing levels of defiance resulting in withdrawal of the village elders, and bitterness and acrimony between villages.

Better Institutions are the Key

The situation here has not reached a point where conflicts have become irresolvable. A two-pronged approach needs to be followed. First, the people must strive to increase the resource base, and second, they must strengthen institutions. The first can be achieved by completing the 1 km stretch of canal to join the Barkop and Sundar dams, controlling seepage loss, and repairing the dam gates. Institutional strengthening can be accomplished by implementing a strict central regime. The sub-committee should be dissolved and a new committee formed, which should plan and utilise the resources efficiently. The involvement of *panchayat* or state representatives can give the new committee legal standing while participation of members from all social groups will give it acceptability. It is important to co-opt formal and informal institutions within the available legal space. Pressure needs to be mounted against free-loaders and the committee must establish and justify regular contribution for upkeep and maintenance.

A tiered contribution structure can be planned, based on the 'beneficiary pays' principle. Graded pricing mechanisms can also be followed with the downstream compensating the upstream. Charging high usage rates in the lean years should discourage the head-reachers from monopolising irrigation water for a second crop. In the case of inter-village conflicts based on upstream-downstream divisions, people should explore tradable water rights. The disincentive or penalty should be graduated, and based on the frequency and intensity of damage.

Note

*Map 1 from SOPPECOM, Pune; map 2 and figure by authors.

1. The genesis of the Barkop dam can be traced to the efforts of Rameshwar Thakur (presently the Governor of Orissa) who was behind many watershed projects in the district. He, along with an NGO, Gadadhar Mishra Smarak Khadi Gramodyog, built this dam in Barkop village in 1971 with an aid of Rs. 3 lakh.

PART 6

Dams and Displacement:
A Review

Bharat Patankar and Anant Phadke

THE CASE STUDIES IN THIS SECTION DEAL WITH SOME OF THE IMPORTANT STRUGGLES BY Project-Affected People (PAP) during the last two decades in various dam-related projects in India. They also contain a critical analysis of the concerned projects, and one of them focuses on an alternative design of the Sardar Sarovar Project (SSP). The conflicts they cover range from conflicts around smaller projects like the Haribad Minor Irrigation Project affecting thirty-five families, to the Sardar Sarovar mega-project affecting thousands of families. However, despite the great variation between these projects and the conflicts around them, certain common issues can be identified and there is a need to evolve a pro-people consensus around them.

The Decision-makers and their Processes

In all the projects discussed here, the PAP were not taken into confidence and had no say in the overall decision-making or design of the project. The elected representatives of the people are supposed to be party to the decisions about developmental projects, but there is no proper mechanism to involve the PAP in the decision-making process.

The conflicts depicted in this section show that at stake is not only the question of adequate compensation or rehabilitation of the PAP, but the very desirability and design of such developmental projects can also be in question. It is therefore necessary to evolve an open, transparent decision-making process that provides adequate scope to the PAP and their protagonists to critically examine such projects. Yet, as all these case studies show, whenever the PAP sought information about the details of the project from the authorities, they were met with a negative if not hostile response. This attitude has to change fundamentally. All the details of the project plan should be available publicly and there has to be a mechanism to invite suggestions, including alternatives, and to discuss these in a transparent manner.

Even in the case of a mega-project like the SSP, when the decision to go ahead with this gigantic project was taken, neither the detailed studies of its environmental impact nor a realistic cost-benefit analysis or a proper master rehabilitation plan were ready. Even the basic assumption about the availability of water flow in the Narmada river was mistaken. In the case of the Tehri dam, seismologists have pointed out that in the next 100 years the dam has a 60 per cent chance of experiencing a major earthquake measuring more than 8.0 on the Richter scale. On this ground Professor Vinod Goud has questioned the design of the Tehri dam, as mentioned in the case study on Tehri dam in this section, but this warning has been sidetracked.

There is a need to build consensus on how to arrive at a final decision about any project that harnesses a natural resource in a given area. What would be the role of the local community in such a decision-making process? Can, for example, a local community claim royalty for the use of water in its area for the benefit of the society at large? There are many such issues, and they need to be debated and settled.

Rehabilitation Policy

All dams, big or small, create problems of displacement for farming communities, agricultural labour and artisans of the submerged area. This is not just an economic issue that can be resolved simply by creating new rehabilitation villages, or settlements, or the provision of land. There is also the problem of people being uprooted from the original habitat where they have lived for generations: *adivasis* are rehabilitated in non-*adivasi* irrigated areas; people from the mountains forced to settle in the plains; people habituated to rainfed farming and small irrigation systems suddenly transported to large irrigation systems. In short, there are also the psycho-social problems of being uprooted and being forced to find roots in a new and alien socio-cultural and geopolitical milieu. These additional concerns—those that drive many protest movements of the displaced—are often overlooked.

However, it should also be noted that in some drought-prone areas, for example in many areas of Maharashtra, millions of people are uprooted due to drought and the lack of supplementary irrigation from dams. They migrate to large cities like Mumbai and live in slums in the most degraded conditions. Further, the assumption that people do not want to move but remain where they have their roots, also contradicts historical experience—people have been migrating for ages across landmasses, seas and continents. It is untenable if people have to leave only to lead a degraded, miserable life where they are forced to go. But if people are rehabilitated in a humane way and if they can become a part of a developed society in a manner that preserves their self-respect and their capacities for their own development, then this is acceptable displacement. We are not just making a theoretical point; this is our experience. For example, many people living in the Sayhadri ranges along the backwaters of the Koyna reservoir would gladly accept 'developmental displacement' as a boon when compared to their present situation.

Once any particular developmental project is considered necessary, some amount of rehabilitation would generally be involved, especially in larger projects. While proper design could reduce the need for rehabilitation (for example, the Paranjape–Joy redesign of the Sardar Sarovar Project could reduce displacement by 70 per cent), some rehabilitation would be necessary. The real problem is that there is no national policy or Act(s) to ensure just rehabilitation and to ensure that improvement in the conditions of the oustees is no less than the improvement in the conditions of the beneficiaries. The movement of PAP in Maharashtra has demonstrated that the needs of rehabilitation must take precedence over the progress of the project, emphasised by the slogan, *'adhi vikasansheel punarvasan, mag prakalp'* (development-oriented rehabilitation first, only then the project). None of the projects discussed in this section even had a plan that could do justice to this demand. Despite a spate of dam-oustees' movements, government policy in different States has not changed.

It is difficult for movements to approach the courts every time. The studies show that even in the case of the SSP and the Tehri dam—where the battle on the legal front was also waged vigorously—the outcome has not been very positive. We need a law and a change in the policy and approach of the government towards the dam oustees. In the absence of these, rehabilitation is only a cruel joke. The following description is not an exception; it could apply to any other dam as much as it does to the Tehri dam:

The rate of compensation has varied from village to village and sometimes even within a village. Rural markets are not part of the plan. Innumerable mistakes and inadequate land settlement have meant that many of those affected are not even being considered 'project-affected persons'.

There has been a failure to compile proper eligibility lists in the displaced villages; in several cases either people have not been awarded land at all or, if allotted, have been unable to take possession for some reason or the other; housing plots have not been awarded; agricultural land has not been considered; improper compensation for housing and failure to take shops into account—the list is endless. While the administration is in possession of all sorts of records it is the displaced who are asked to produce records, to run from pillar to post.[1]

Surprisingly, while there are specific directives and government resolutions in respect of rehabilitation, only Maharashtra seems to have a comprehensive rehabilitation law at the State level. Moreover, it appears from many reports that the demands made by movements and the activists involved have been limited mainly to particular rehabilitation problems related to particular dams. Barring Maharashtra, there seems to be no movement in other States that consistently pursues a State-wide comprehensive rehabilitation act. In Maharashtra, under pressure from the movement, the first comprehensive act for rehabilitation came into force in 1976, was revised in 1986 and, more recently, in 1991. It is time for movements in different States to come together and pursue a national-level comprehensive enactment covering rehabilitation policy, especially in view of the fact that the draft policy the Central government has declared is more backward than even the existing Maharashtra rehabilitation act.

Small versus Big Dams?

The track record of big dams in rehabilitating the people displaced by them has been dismal. Yet the papers in this volume clearly show that the issue of rehabilitation can be as bad in the case of small dams. Secondly, as the case of the Uchangi dam shows, the scope for reducing submergence without reducing benefits is the same in small dams. Despite the problems associated with big dams, it may be necessary to provide water from large sources (big and medium reservoirs) to severely drought-prone regions where a minimum prosperous life for *every* rural family cannot be ensured only on the bais of small, local water sources. We need a perspective that does not counterpose small to big dams, but rather provides for a combination of minor, medium and large irrigation projects in such a way that it assures everyone adequate access to land, water and forest.

The 'small versus large' dichotomy also tends to gloss over the fact that Indian society, whether in the hills or plains, forests or agricultural land or deserts, is divided along class, caste and gender lines. The impact of dams on the downtrodden is not merely a function of its size. Local percolation tanks victimise Dalits and other oppressed sections; almost 80 per cent of the percolation tanks built in drought-prone areas submerge the lands of Dalits and other downtrodden castes, while the wells that benefit from the percolation tanks belong disproportionately to large farmers from the upper castes. On many small dams, submerged land belongs to small farmers and the downstream beneficiaries are big farmers. In Maharashtra, more and more medium and small dams release water into the stream without canal systems, thus benefiting only those farmers with large landholdings who can afford to lift it and who later sell water at onerous rates. Small systems are as liable to be controlled by capitalist landowners and become obstacles in equitable water distribution as large systems. It is important to realise that merely giving control of natural resources to 'villages' will not solve this problem.

In many drought-prone areas in Maharashtra, Gujarat, Andhra Pradesh, Karnataka and Madhya Pradesh, given the amount of rain, the nature of soil, topography, population density, etc., local watershed development will not provide adequate water for livelihood. Hence exogenous water from larger dams would be essential to supplement and strengthen local harvesting of water. In light of this, there is a need to forge a broader perspective as illustrated by the suggested redesign of the SSP by Paranjape–Joy.

The Drought-affected Versus the Dam-affected?

The case study entitled *Valley Rises* points out that the design of the SSP had major flaws. However, the drought-prone areas in Gujarat undoubtedly need 'outside' to supplement local watershed development if livelihoods are to be assured. This gave the establishment the handle to pit the drought-affected people in Gujarat against the dam-affected people. There is a need to overcome this division between the two groups who comprise the deprived sections of the toiling people; we need a strategy that unites rather than divides them. A broad outline of how this could have been done by redesigning the SSP has been indicated in the case study on the Sardar Sarovar Project by Paranjape and Joy. If we are able to do so, it will be possible to achieve a humane rehabilitation policy as well as a policy of equitable water distribution. It requires dialogue, interaction and coordinated activity at a *mass* level among these apparently opposed sections of people. The Haribad Minor Irrigation project study is a good example of how the conflict between the dam-affected people and the project beneficiaries can become explosive even in a small project displacing only thirty-five families. There is a need to forge a common perspective to overcome such an impasse.

The experience of the south Maharashtra movement shows that equitable distribution of water is a common issue that can help build such an alliance. Here, dam oustees have supported the demand of the drought-affected for equitable distribution of water because, as PAPs, equitable distribution can get them water for their plot of land in the command area. For example, in the case of the Uchangi dam in the Kolhapur district of Maharashtra, people in the command area and the catchment area came together to develop an alternative to the proposed dam and its distribution system.

It is also necessary to take note of the larger struggle that is being waged unitedly by the dam-affected and drought-affected people in four districts of south-western Maharashtra over the last ten to fifteen years. It has demonstrated that it is possible to force the government to institute a policy of implementing rehabilitation in new village settlements with specified modern amenities as well as alternative land in the command area before actual submergence takes place. This united movement of the drought-affected and dam-affected, the Pani Sangharsha Chalwal, has forced the Government of Maharashtra to adopt a policy of equitable water distribution regarding many dams on the Krishna and its tributaries, such as Urmodi, Wang–Marathwadi, Karli or Uttar–Mand dams. The movement has also succeeded in getting the Government of Maharashtra to sanction a water allowance of Rs. 600 per month to the rehabilitated farmer families until such time when the systems actually supply water to irrigate the land they have been allotted. These successes were made possible only because the movement forged a unity of the drought-affected and the dam-affected people, making it a model for struggles elsewhere.

Viable Alternatives

Many social movements start as protest movements with purely negative connotations. But if they are to move ahead, they must elaborate their own perspective, forge their own alternatives, and rally large masses of people around such alternatives. One of the important reasons why the existing State models of development retain legitimacy and continue to dominate people's minds is that often people's movements have no coherent alternative that is convincingly practicable and realistic. However, there is now an increasing, and welcome, trend of trying to evolve such alternatives. Case studies in this section show that people's movements have put forth viable alternatives even as they opposed official plans. Such examples help us come out of the 'There Is No Alternative' (TINA) syndrome. The challenge is to work out alternatives that aim at equitable, sustainable prosperity and that are viable—socially, technically, economically and ecologically. Though movements cannot be expected to work out alternatives in great detail, they do need to establish the basic features in sufficient and convincing detail.

Another small but important example is the success story of the alternative provided by the various fishing cooperatives in respect of the Tawa dam in Madhya Pradesh. As the study notes,

> The TMS's (Tawa Visthapit Adivasi Matsya Utpadan Evam Vipanan Sahkari Sangh) performance in terms of productivity was exceptional. The average annual fish production rose to about 327.63 tonnes, an increase of 274 per cent over the corporation's performance; the per hectare fish productivity in the reservoir rose from an average of 7.3 kg/ha/yr to 26.95 kg/ha/yr; the average annual income per worker was almost three times what it had been earlier; the number of people employed in organised fishing in the Tawa reservoir grew from an average of 137 to 422; among other factors, one reason for the involvement of a larger number of people was the improved access to nets and boats.
>
> In addition, the TMS took a number of other initiatives. These included greater focus on local markets, more fishing days to ensure livelihood security for the workers, use of smaller nets as against the dragnets promoted by the corporation, worker welfare policies like subsidised loans, etc.

The example of the Tawa cooperatives shows that when there is genuine people's control, both productivity as well as sustainability can be ensured along with increase in employment and income for the toilers. Such struggles need to be linked into State-level, even national-level movements that put forward alternative irrigation and rehabilitation policies.

The Uchangi dam case study briefly describes a viable alternative put forth by a people's movement. With the help of pro-people experts and with the active participation of the local people, a viable alternative plan was suggested that mainly consisted of: (*a*) constructing three supplementary storage dams instead of one large reservoir at Uchangi, allowing for a reduction in the height of the Uchangi dam and consequently the submergence area; (*b*) combining this with watershed development work to augment water; and (*c*) equitable distribution of water so that more area gets irrigated per unit of dam water. The important thing is that the alternative suggested better rehabilitation options and means of reducing submergence without compromising on the area irrigated. This plan was partially accepted after a lot of pressure was built up.

One of the issues in evolving a viable alternative is the availability of eco-friendly sources of power. If displacement is kept at a minimum through proper site selection, its social consequences minimised and properly remedied, hydro-electrical projects can play an important role by providing sustainable and cheaper electrical power. To be sure, there is a need to avoid indiscriminate, excessive and unnecessary use of electricity that the dominant paradigm of development incorporates. Nevertheless, if all citizens are to attain sustainable prosperity and get an opportunity to unleash their creative, human potential, the per capita energy consumption in India will have to increase significantly. Hence it is important to tap hydropower potential wherever it is possible to do so without sacrificing the interests of the local population. The Bhilangana project is one such project as the impasse it faces is entirely resolvable. It seems that it is only the lethargy and lack of will on the part of the political establishment that is holding it up.

The papers in this section have thrown up some very important issues and we hope that in the coming years society at large and the next generation of movements will address these issues ably.

Note

1. *Tehri Dam Project: A Saga of Shattered Dreams* by Vimal Bhai.

The Dam That Was Never Built:
The Pulichintala Project on the Krishna River

R.V. Rama Mohan

A Well-meaning Project

The Pulichintala project was planned across the Krishna river at Pulichintala village in Guntur district. The east-flowing Krishna originates from the Mahadev range of the Western Ghats near Mahabaleswar in Maharashtra and enters Andhra Pradesh in Mehbubnagar district. In Andhra Pradesh, the Krishna traverses the border of Nalgonda and Guntur districts for a distance of about 200 km before it branches out into a delta region. The project location is 85 km upstream of Prakasam barrage at Vijayawada and 115 km downstream of Nagarjunasagar.

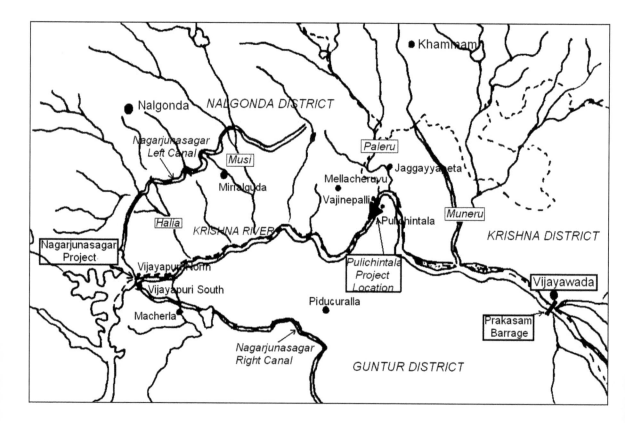

Map 1. *The location of the Pulichintala project.*

Pulichintala reservoir was planned for a total storage capacity of about 1,300 Mm³ and a live storage of about 1,025 Mm³ by harvesting the runoff from the 20,028 km² catchment between the Nagarjunasagar project and the proposed project location. The scheme is aimed at ensuring irrigation water to 5.23 lakh ha of land in the Krishna delta as per the Krishna Water Disputes Tribunal (KWDT) award of 1976. The project will not feed any irrigation canal, but regulate flow to the Prakasam barrage to ensure timely irrigation water for the Krishna delta.

The following are the various features of the Pulichintala Project[1]

Top of bund level (TBL)	58.24 m
Full reservoir level (FRL)	53.24 m
Lowest bed level of river	21.58 m
Height of dam	36.66 m
Power potential (4 units of 30 MW each)	120 MW
Total project cost	**1,100 crore**
Maximum flood discharge	57,674 cumecs
No. of spillway gates	33
No. of river sluices	4

Ten villages in Nalgonda district and eight in Guntur district will be affected by submergence: 3,361 families (15,465 people as per 1991 Census) will be displaced. It will submerge 12,884 ha (967 ha of forest, 626 ha of the Nagarjunasagar *ayacut*, 8,576 ha of private cultivable land, 1,769 ha river, and 946 ha *vaagu* [rivulet] & miscellaneous). A few quarries that supply limestone to surrounding cement factories will also be submerged.

A Century-old Plan

The Pulichintala project was conceived way back in 1903 but was shelved for a long time due to political and financial constraints (such as the First World War in 1914; the economic crisis in the country between 1930 and 1933; and persistent shortage of financial resources for the project). The Krishna delta region has been irrigated from the delta canal system of Vijayawada barrage (later named the Prakasam barrage) since 1855. Realising the need for storage to efficiently utilise Krishna waters, Andhra Pradesh planned and built two reservoirs—Srisailam and Nagarjunasagar (upstream of the present Pulichintala project location) between 1950 and 1980.

The Krishna Water Disputes Tribunal Award or the Bachawat Award that was finalised in 1976 gave a second lease of life to the Pulichintala project proposal. The award allocated 5,135 Mm³ water to the Krishna delta, which includes 2,868 Mm³ water yield from the Nagarjunasagar reservoir and the Pulichintala project site; another 2,267 Mm³ was allotted from the Polavaram project. In other words, the Award envisaged delinking the Krishna delta from the Nagarjunasagar reservoir and making the delta self-sufficient through efficient utilisation of runoff below Nagarjunasagar. Delta farmers demanded the project, and the government too was keen to start work on it. Once constructed, the Pulichintala project could reduce the

Krishna delta's need for Nagarjunasagar water, and provide that water to the upper areas in Telangana and Rayalaseema. Also, carry-over storage in Pulichintala would facilitate early release of water to the Krishna delta during the months of June and July when the inflows into the Krishna are minimal.

But since the 1980s, the project has been delayed due to stiff opposition from submergence villages, financial constraints, and lack of clearance from relevant central government ministries and agencies.

Chronology of Events

November 1988:	N.T. Rama Rao, then Chief Minister of Andhra Pradesh, laid the foundation stone for the project in 1986 and committed to a three-year deadline.
1989–96:	Project stalled due to stiff resistance from people on the submergence issue and lack of clearances from the Central government.
1996:	Environment Protection Training and Research Institute (EPTRI), Hyderabad, prepares and submits Environmental Impact Assessment, Management, and Rehabilitation Plans.
1998:	Government of Andhra Pradesh under N. Chandra Babu Naidu again attempts to initiate construction. People from affected villages and People's War Group (PWG) break the foundation stone.
1999 and 2001:	Cases filed in High Court against the project, citing lack of clearance from concerned Central agencies.
August–September 2004:	Government under Y.S. Rajasekhara Reddy begins construction at the site.
September–October 2004:	Protestors object to construction without clearance from Environment Ministry.
October 15, 2004:	Y.S. Rajasekhara Reddy lays fresh foundation stone at Vajinepalli amidst commotion at project site. People damage equipment and burn temporary shelters at the site.
November 2004:	Responding to a petition by V. Venkata Reddy, the High Court stays construction work and asks the government to obtain environmental clearance. District Collector of Nalgonda conducts public hearing—a mandatory exercise to obtain clearance from the central government. The event is widely criticised for denying people and environmentalists an opportunity to express their thoughts.
December 2004:	District administration convinces people and negotiates a rehabilitation package that it promises to implement at the earliest.
January–February 2005:	Special MRO and teams conduct household and land surveys in all submergence villages and draw up an estimate of loss to property.
June 2005:	The project finally gets environmental clearance from the Central government and construction resumes at the site.
July 2005:	A divisional bench of the High Court admits a writ petition challenging the construction without necessary clearance and permission from other Central agencies.

A History of Opposition

Though the project was conceived a century ago, the conflict arose in 1988 when the then Chief Minister of Andhra Pradesh laid the foundation stone and promised to complete the project as soon as possible. But the government had to put off its plans for reasons already mentioned. The issue remained dormant until 1998 when once more an attempt was made to restart work on the construction. However, that time too things did not take off, and in August 2004 the present government finally started work against the backdrop of violence. The conflict has three major dimensions:

Fig. 1. *Construction stalled since October 2004.*

Technical Feasibility

Eminent engineer, T. Hanumanta Rao (former Engineer-in-Chief, Department of Irrigation and CAD, Andhra Pradesh), perceived that Pulichintala would not get sufficient inflows if constructed at the present site. Many engineers and activists also take the same view. The inflows in two upstream rivulets—Halia and Musi—have reduced considerably over the last thirty years due to various changes in the catchment area. Therefore, it is unlikely that Pulichintala will harvest additional water; on the contrary, it will have to depend on the release of water from the Nagarjunasagar project, especially since rivulets like Paleru and Munneru have good flows during the rainy season. As an alternative, Rao suggested five smaller barrages on the Krishna at the confluence of these rivulets. These barrages could be designed to harvest the same quantity of water (about 990 Mm³) but without causing any submergence upstream. But until now the government has not heeded these suggestions from experts.

Environmental and Other Clearances from the Central Government

The Government of Andhra Pradesh allowed the contractor to begin construction in September 2004 without getting environmental clearance from the Ministry of Environment and Forests. There was great opposition to this move and a writ was filed in the High Court, and the government was directed to complete necessary formalities with the concerned ministry before starting construction.

Between November 2004 and March 2005, the government conducted public hearings in the project submergence villages and submitted its

Fig. 2. *Eastern Ghats abutting reservoir area.*

environmental management and rehabilitation plans to the central government, anticipating an early clearance from the Ministry of Environment and Forests. Permission was finally obtained on June 9, 2005, clearing one of the major hurdles. However, before the Andhra Pradesh government could heave a sigh of relief another writ petition made its way to the court in July 2005, this time to do with clearance from the Central Water Commission and the Planning Commission.

Doubts about Rehabilitation Efforts

People from the submergence area have steadfastly opposed the project for the last twenty years. A few months ago the government finally managed to convince them and negotiated a rehabilitation package which included compensation for privately owned land between 1,50,000 and 2,87,500 Rs./ha depending on the type of land and 220 sq m of housing site for each displaced family. The District Collector 'verbally' agreed to this deal and assured speedy resettlement at a place near Kodad, about 35 km from the project site. But the local people as well as the activists involved are suspicious about the sincerity of the government. Meagre budget allocations during 2005–6 (about Rs. 130 crore for the project[2]) raise doubts about whether the government will be able to stick to its commitment, especially since the compensation for loss of agricultural lands alone may amount to about Rs. 150 crore. Moreover, social activists and environmentalists insist that the government's estimate on submergence area is faulty, and that the affected area is actually much larger.

Highs and Lows

The people from the submergence villages have consistently opposed the project for the past twenty years, but the conflict remained at a low level since there was no move to actually start work on the scheme. The problem escalated when the new government in Andhra Pradesh promised to complete the project and took up construction. The highest point of the conflict was in September–October 2004 when work began without clearance from the Central government. The agitation took a violent turn and resulted in damage to machinery and dismantling of a few temporary constructions at the project site.

Need for Transparency

Construction of a storage reservoir at Pulichintala is relevant and appropriate in the context of the Bachawat Award, which apportioned Krishna waters below Nagarjunasagar project to the Krishna delta. Flood water and runoff downstream from the Nagarjunasagar reservoir can be efficiently utilised by this project, which is presently not possible due to the lack of storage facility at the Prakasam barrage. At the same time it makes sense for the government to keep an open mind, seriously consider expert advice and explore better alternatives.

The government also needs to adopt a systematic and transparent approach if it is to make any headway in completing this project. Initiating work before necessary clearance and consent is obtained will only complicate the issue. Therefore, the government should communicate with the people and address their concerns on various aspects of environmental management and rehabilitation.

It is also necessary to establish the exact area of submergence and offer people good compensatory packages. Most importantly, the government must allocate sufficient funds for the project and prioritise rehabilitation over construction.

Notes

*Map 1 and figures by author.
1. Comprehensive Environmental Impact Assessment and Management Plan on Pulichintala Project. 1996. EPTRI & SROSS, Hyderabad.
2   'Higher Allocations for Irrigation Projects', *The Hindu,* February 20, 2005.

References

CAD Annual Report. Andhra Pradesh: Irrigation and CAD Department, 2003–04.

Comprehensive Environmental Impact Assessment and Management Plan on Pulichintala Project. Hyderabad: Environment Protection, Training and Research Institute (EPTRI) and Society for Research Operations and Social Services (SROSS), 1996.

Hanumantha Rao, T. 2004. 'A Note on the Lecture on "Technical Alternatives to Pulichintala Project"', Presented at the Institute of Engineers (India), Andhra Pradesh State Centre, Hyderabad, February 13.

'Higher Allocations for Irrigation Projects', *The Hindu,* February 20, 2005.

'Projectuto Chintalundav', *Eenadu* (Telugu daily newspaper), November 11, 2004.

Rehabilitation & Resettlement Plan for the Pulichintala Multipurpose Project. Hyderabad: Environment Protection, Training and Research Institute (EPTRI) and Society for Research Operations and Social Services (SROSS), 1996.

Shiva, Vandana. 1991. *Ecology and the Politics of Survival: Conflicts over Natural Resources in India.* New Delhi: Sage Publications.

Sivaramakrishna, P. 2004. 'Taatalanati Kalalu—Eenati Tala Tikkalu', *Nadustunna Charitra* (in Telugu), December.

Water Resources and Utilisation in Andhra Pradesh. Hyderabad: Dr. K. Sriramakrishnaiah Smarama Seva Samithi (DKSSSS), 2004.

The Polavaram Project on River Godavari: Major Losses, Minor Gains

R.V. Rama Mohan

THE POLAVARAM PROJECT IS A MAJOR MULTI-PURPOSE IRRIGATION PROJECT ACROSS THE Godavari river at Ramaiahpet village in Polavaram *mandal*, west Godavari district. The project has long been in the pipeline and the proposed location is about 42 km upstream of the existing Sir Arthur Cotton barrage at Dowleswaram and 34 km upstream of Rajamundry.

The Godavari originates in the Sahyadri hills in Triambakeswar, Maharashtra. Pravara, Poorna, Manjeera, Maneru, Pranahita, Indravati and Sabari are important tributaries of the Godavari. Pranahita, Indravati and Sabari contribute about 70 per cent of the river flow. After the Sabari meets it at Kunavaram in Khammam district, the river traverses a narrow 250 to 275 m wide pass in the Paapi hills before it widens to 3.5 km at Rajamundry.

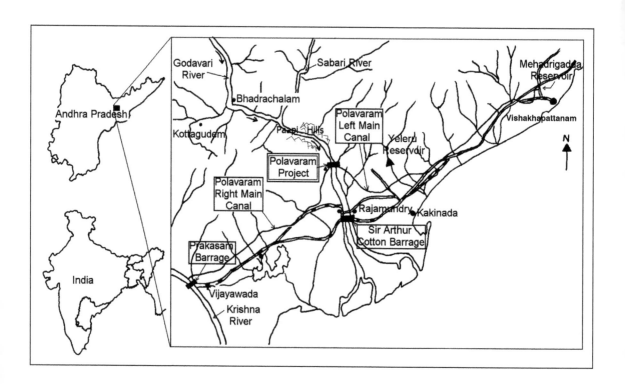

Map 1. *Map of the Polavaram project.*

Sri Ram Sagar Project (SRSP) at Pochampadu and Cotton barrage at Dowleswaram are two existing major projects on the Godavari in Andhra Pradesh. The Polavaram project is intended to utilise about 8,500 Mm^3 of water out of a dependable yield of 85,000 Mm^3 (including surplus waters) in the river. Of approximately 42,000 Mm^3 that is Andhra Pradesh's share, about 19,000 Mm^3 is diverted to various projects while about 23,000 Mm^3 is let off into the sea.

The project is supposed to have a full reservoir level (FRL) of 45.72 m and top of bund level (TBL) of 53.32 m. The head works include construction of an earth-cum-rockfill dam 2.31 km in length; a powerhouse with 960 MW capacity on the left flank; a left main canal 181.5 km long; and a right main canal of 174 km.[1]

The Major Benefits

1. Irrigation water for 1.6 lakh ha in east Godavari and Visakhapatnam districts through the left main canal and for 1.28 lakh ha in Krishna and west Godavari districts through the right main canal.
2. Generation of 960 MW hydelpower.
3. Transfer of 2,200 Mm^3 water to Krishna basin at Prakasam barrage, Vijayawada.
4. Availability of 653 Mm^3 water to meet the domestic and industrial needs of Visakhapatnam.
5. Drinking water supply for all the villages that lie on the route of the left and right main canals.

The project also involved the submergence of 42,000 ha of land in its foreshore (37,743 ha in Andhra Pradesh; 1,618 ha in Chhattisgarh; 2,786 ha in Orissa) and displacement of 30,607 families (1,28,913 people as per the 1991 Census) in 292 villages, according to the designs made during the 1980s. These were later modified, and the 2003 designs are expected to reduce the submergence to 276 villages (1,17,034 people as per the 2001 Census) in seven *mandals* of Khammam and one *mandal* each of east and west Godavari districts in Andhra Pradesh. There will be no submergence in Chhattisgarh and Orissa.

The Classification of the Submergence Area in Andhra Pradesh[2]

Wet	Dry	*Poramboke**	Forest	Total
398 ha	22,218 ha	11,941 ha	3,186 ha	37,743 ha

*village wasteland

A total of 46,929 ha (wetlands 11,782 ha; dry lands 32,667 ha; garden lands 2,481 ha) need to be acquired on account of submergence/canal construction. The original project was estimated to cost Rs. 12,234 crore, going by the rates in 2001–2. The revised proposal is estimated at 8,198 crore according to 2003–4 rates, with a project cost-benefit ratio of 2.4:1. The modified design lowers the crest level of the spillway for peak discharge in order to avoid inter-State submergence.

In August 2004, the government proposed extending the left canal by 100 km to irrigate an additional 3.2 lakh ha in Vizianagaram and Srikakulam districts.[3]

A Jinxed Project

The Polavaram project was conceived way back in the 1940s. Initially it was called the Rama Pada Sagar Project and was originally supposed to be constructed across a narrow river section at Paapi hills. Later, the project was shelved because of the huge costs involved in laying the foundation in the absence of shallow hard rocks at the proposed dam site.

Fresh investigations about the viability of the project began in 1977, and inter-State agreements were signed by 1980. By 1985, detailed proposals had been prepared to construct a dam at Polavaram, but the plan could not be implemented since clearance from CWC and various ministries of the Central government was not obtained.

Chronology of Events

1941:	Conception of the Polavaram project.
1951:	Detailed project proposals prepared for the Rama Pada Sagar Project (as it was first christened), but the scheme is shelved due to technical shortcomings.
1980:	Andhra Pradesh enters into inter-State agreements with Madhya Pradesh and Orissa.
1981:	Shri T. Anjaiah, Chief Minister of Andhra Pradesh, lays foundation stone.
1985–90:	Detailed project proposals and two Environmental Impact Assessment (EIA) reports—one in 1985 and the other in 1990—are prepared by the Irrigation & CAD Department, Government of Andhra Pradesh.
1991:	About fifty NGOs organise a Manya Praanta Chaitanya Yaatra lasting fifty-two days to raise public awareness on the social, economic and environmental pitfalls of the Polavaram dam.
1993–95:	The Godavari Basin Action Group (GBAG), a network of NGOs working in the Godavari basin, conducts a field study to show loss of resources and livelihood in the nine submergence *mandals* of Khammam district.
1995–96:	The Government of Andhra Pradesh identifies 3,327 ha of wasteland in Visakhapatnam, east and west Godavari districts for compensatory afforestation.
1996:	Environment Protection Training and Research Institute (EPTRI), Hyderabad is assigned the job of developing EIA and Environmental Management Plan (EMP).
1996:	Centre for Economic and Social Studies (CESS), Hyderabad, prepares and submits detailed rehabilitation plans costing Rs. 1,511 crore. The report was revised twice to follow AP-III guidelines on rehabilitation.
July 2000:	The Government of India constitutes a high-level committee chaired by a member-CWC to resolve inter-State issues.
May 2001:	Chhattisgarh categorically objects to submergence in its territory. Orissa response awaited.
November 2003:	EPTRI submits interim EMP report that places the estimated cost at Rs. 2,788 crore. Final report incorporating the observations of the Ministry of Environment and Forests is awaited.
November 2003:	Andhra Pradesh modifies designs to limit the submergence to natural levels in Chattisgarh and Orissa, and submits fresh proposal to CWC for examination.

Going Under

The problem is the magnitude of human displacement and submergence. The project displaces 1,17,034 people in 276 villages besides submerging 37,743 ha of land—farmland, forests and wasteland. The livelihood and habitation of many tribal and poor people are at risk.

The major dimensions of this conflict are:

1. Construction of the dam will lead to the displacement of an enormous number of people. Rehabilitation is a herculean task because of large-scale submergence.
2. Clearance from CWC, Ministry of Environment and Forests, Ministry of Tribal Welfare, and the Planning Commission will be difficult because of the controversy around the vast submergence of forest and farmlands, a low benefit–cost ratio, and the loss of livelihood.
3. Environmentalists and social activists are sceptical about the sincerity of the State government. Andhra Pradesh has a poor record in implementing rehabilitation programmes in projects such as Yeleru, Kovvada, Surampalem, Nagarjunasagar and Sri Ram Sagar.
4. There is bound to be a loss of biodiversity and wild life, and submergence of graphite mines leading to water pollution and loss of aquatic life, not to mention the submergence of heritage.
5. Sedimentation and silt deposition in the Godavari delta is reportedly high. The life-span of the project itself is in question if silt deposition continues at the current rate.
6. Though the Government of Andhra Pradesh came out with a Resettlement and Rehabilitation Policy in April 2005, no consultations were held with the people from the submergence area. To add to the scepticism the annual budget allocation of 300 crore rupees for the year 2004–5 does not explicitly mention rehabilitation activities.[4]
7. Andhra Pradesh has to part with 992 Mm³ water—intended for Karnataka and Maharashtra—in lieu of transferring 22,267 Mm³ water from the Polavaram project to the Krishna river as per the Bachawat Award of 1976. Therefore, the net gain from the inter-basin transfer is only 1,275 Mm³.
8. *Gram panchayats* and local *mandal parishads* do not have any information about the extent of submergence or what government plans have in store for them. People assume that submergence levels will match the 1986 floods.
9. People have been protesting against the dam and want to know why their life and livelihood should be sacrificed to benefit the people of coastal regions, who are economically better-off.

Fig. 1. *A submergence village on the shore of Godavari.*

Pressing Ahead

Polavaram came into the limelight owing to the Andhra Pradesh government's renewed interest in the project in early 2004. Despite the lack of clearance from central government agencies and ministries, the Chief

Fig. 2. *A glimpse of Chaitanya yatra in 1991.*

Fig. 3. *An All-party committee meets in V.R. Puram mandal, Khammam district, in November 2004.*

Minister went ahead with the foundation-stone ceremony for the construction of the right main canal on November 8, 2004, and the endeavour was renamed the Indira Sagar project. The State government has been negotiating with the Austrian government for financial support for the project. The Polavaram project has top priority, second only to the Telugu Ganga and Devadula projects, as far as budget allocation goes, and Rs. 300 crore were set aside for the year 2004–5.

The people of the submergence villages are deeply anguished at the recent developments. But they have neither managed to get themselves organised nor do they understand the larger implications of displacement. They only wish for a good rehabilitation package in case submergence is inevitable.

Disaster Preparedness Network (a group of twenty-seven local NGOs), inspired by Manya Praanta Chaitanya Yaatra of 1991, has been fighting for the cause of the project-affected people in seven *mandals* of Khammam district since August 2004. The network constituted all-party committees at the *mandal* level and chalked out action programmmes to raise awareness among the people on the impact of the project on their lives and environment.

The network helped all the *panchayats* in the nine submergence mandals of Khammam district to pass resolutions against the construction of the project, and also conducted a field survey on loss of resources and livelihood in three villages. The group is now making efforts to negotiate with the government on behalf of the villagers for effective rehabilitation prior to construction.

Different Strokes

The government and technical experts argue that:

1. There is plenty of surplus water in the Godavari that can be harnessed effectively by constructing projects such as Devadula, Dummugudem and Polavaram to increase irrigation potential and serve the drought-prone regions in the State.
2. The Polavaram project will not only provide irrigation water to about 2.88 lakh ha, but will also offer multiple benefits in terms of electricity generation, and industrial and domestic water supply to villages and cities along the canals.
3. The transfer of 2,267 Mm^3 water to the Krishna basin at Prakasam barrage will enable diversion of water from Srisailam and Nagarjunasagar to the drought-affected Rayalaseema region.

Social activists and concerned citizens are of the view that:

1. Other alternatives should be explored in light of the huge environmental and social damage due to submergence.
2. The government should plan a project that involves minimum submergence.
3. Alternatively, the government should have an effective rehabilitation package in place and implement it before the construction of the dam.

Fig. 4. *Panchayat representatives chalking out future strategies in June 2005.*

Stop, Think Again

The State government's intention to effectively use the surplus water of the Godavari is laudable. Indeed, water needs to be efficiently harnessed by planning appropriate irrigation projects.

At the same time one cannot deny that the Polavaram project, as it stands today, involves human displacement and environmental destruction on an unprecedented scale. About 276 villages in nine *mandals*, predominantly populated by poor and neglected people, stand to be submerged; all of this is meant to benefit the relatively better-off regions of coastal Andhra Pradesh where it will irrigate 2.88 lakh ha of land. The only exception is the backward area of Rayalaseema that will definitely benefit from the transfer of 2.267 Mm^3 water (net gain 1,275 Mm^3) to the Krishna basin.

Therefore, the government really needs to explore alternative means to achieve these ends in a manner that is people-friendly and environmentally sound. The mid-Godavari basin, for example, not only has a good water yield but is also suitable for locating projects that will benefit the backward Telangana region. The government should keep in mind two essential concerns:

(1) any alternative plan will have to prioritise minimum submergence; and
(2) such a project should benefit the poor and backward regions of the State.

Notes

*Map 1 and figures by author.
1.   Annual Report 2003–04, Irrigation and CAD Department, Andhra Pradesh.
2.   *Polavaraaniki Punaadi – Girijana Prantaaniki Jalasamaadhi, SAKTI, Hyderabad,* October, 2004.
3.   Polavaram–Marintha Vistharana, *Andhra Bhoomi*, August 24, 2004.
4.   'Higher Allocations for Irrigation Projects', *The Hindu*, February 20, 2005.

References

Annual Report. Hyderabad: Irrigation and CAD Department, Andhra Pradesh, 2003-04.

'Anumati Leni Padhakalaku Inta Aarbhatama!', *Vaartha* (Telugu daily newspaper), November 9, 2004.

Bhushan, M. Bharat and R. Murali. 1992. 'One More Peril—Socio-cultural and Environmental Implications of Proposed Polavaram Dam'.

'Higher Allocations for Irrigation Projects', *The Hindu*, February 20, 2005.

Manya Praanta Chaitanya Yaatra (Part 1 & 2). Hyderabad: SAKTI, 1991–1993.

Polavaraaniki Punaadi–Girijana Prantaaniki Jalasamaadhi. Hyderabad: SAKTI, October 2004.

'Polavaram-Marintha Vistharana', *Andhra Bhoomi*, August 24, 2004.

Socio-economic Study on People of Polavaram Submergence Areas in nine Mandals of Andhra Pradesh. Reports of Godavari Basin Action Group (GBAG), 1993–1995.

Sri Sarampalli Mallareddi. 'Godavari Jalalu—Viniyogam'. Kadalika, Anantapur: Democratic Youth Forum, January–March 2003.

Water Resources and Utilization in Andhra Pradesh. Hyderabad: Dr. K. Sriramakrishnaiah Smarama Seva Samithi, March 2004.

Kolhapur's Uchangi Dam:
Disputes over Height and Alternatives

Raju Adagale and Ravi Pomane

Need for a Dam

THE UCHANGI DAM IS SITUATED IN AJARA *TALUK* OF KOLHAPUR DISTRICT, MAHARASHTRA, one of the few States in India with a high per capita income. However, as per the socio-economic abstract of 1995–96, 34 per cent households are below the poverty line. In Kolhapur district, one of the more prosperous districts, 63 per cent of the people are dependent on agriculture for their livelihood. Rural poverty in Kolhapur is inextricably linked to lack of access to water. While gravity canal irrigation is the common solution to address water scarcity problems in the plains, hilly and steep regions like Kolhapur rely on lift irrigation to pump water to fields that are located at elevations much higher than the riverbed.

In 1985, the regional irrigation agency, the Maharashtra Krishna Valley Development Corporation (MKVDC), proposed a dam in village Uchangi. The site is on the banks of the Tarhol stream, downstream from Chafawade and Jeur villages. The submergence area of the Uchangi dam has a fertile landscape of cashew groves and lush sugarcane crops. There are some cashewnut processing units and a sugar factory in Ajara town. The major crops of the region are paddy, wheat and sugarcane.

As per the original design of the MKVDC, the Uchangi dam has no canals, and farmers will be allowed to sink pumps in the river at pick-up weir sites for lifting water to their fields. The estimated submergence is 222.10 ha. This mainly affects Chafawade and Jeur, and another village, Chitale, partially. The original project design serves ten villages in the command which fall in Ajara and Gadhinglaj *taluks*.

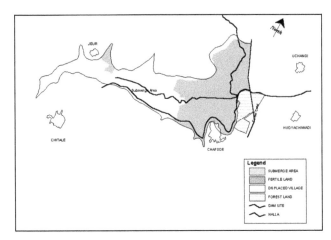

Map 1. *Location of the Uchangi dam in Kolhapur district.*

Opposition and Conflict

One of the main reasons cited for the MKVDC to proceed with the construction of the Uchangi dam is the Bachawat Award,[1] according to which Maharashtra had to utilise its full share of the Krishna waters by May 2000. Although the dam was approved in 1985, construction could not proceed for over fifteen years because

of resistance from the local people who will be displaced by the reservoir. Villagers from Chafawade and Jeur began with minor protests in 1986. The opposition became much more dramatic after 1997 when irrigation officials began plotting markers for the submergence zone of the reservoir. In the same year, leaders from the project-affected villages approached a local political organisation known as Shramik Mukti Dal (SMD) to seek support for their cause. The Project-affected People (PAPs) got together under the banner of the SMD and articulated their opposition to the project, criticising its poor technical design, the large submergence area, limited benefits and lack of disclosure of project details.

Fig. 1. *Fertile land of the submergence area.*

However, they were keen to suggest an alternative to the proposed reservoir and hence approached a Pune-based organisation called Society for Promoting Participative Ecosystem Management (SOPPECOM) for technical help. SOPPECOM, along with Bharat Gyan Vigyan Samiti (BGVS), undertook a Participatory Resource Mapping (PRM) exercise in the two villages to gather detailed information on the local ecosystem and land use patterns. The PAPs argued that the 222 ha that were to be submerged were part of the most fertile land in their valley. They decided to submit an integrated alternative proposal for watershed development to the MKVDC, the main features of which were:

(*i*)   constructing three supplementary storage dams at Dhamanshet, Khetoba and Cherlakatta, instead of one large reservoir at Uchangi;

(*ii*)   bringing down the Uchangi dam height by 5 m to reduce the submergence area;

(*iii*)   building eleven additional weirs between Khetoba and Chafawade through a watershed approach.

Needless to say, these demands created a conflict between the PAPs and MKVDC that broadly involved two issues:

(*i*)   right to information on all technical details and design plan of the proposed dam in order that a viable alternative be found;

(*ii*)   an alternative that promised reduced submergence and better rehabilitation options without compromising on the area under irrigation.

But the MKVDC was extremely reluctant to provide systematic information. Not only was there a delay in accessing data, but available information too was peppered with several inadequacies. It was clear that the movement could not proceed beyond its anti-dam stand unless people could access technical and other details. This led to tension, although not an open conflict, between the potential beneficiaries and the PAPs; the former thought the SMD was sympathetic to the PAPs and would therefore pursue an anti-dam stand anyway. Several dialogues, rallies, *morchas* and 'sit-ins' later, the MKVDC finally parted with the details.

The second major conflict arose when the proposed alternative was rejected by the MKVDC on the grounds that they were restricted by their mandate. The proposal for three alternative sites was not accepted in its entirety. Of the three sites suggested in the alternative plan only Khetoba found acceptance. The corporation also relented somewhat on the issue of reduction in dam height, agreeing to reduce it by 2 m.

But matters did not end here and the conflict reached its peak when, despite having agreed to the terms, the MKVDC went ahead with the construction of the dam without holding any consultation with the PAPs. It was only after a prolonged struggle that the corporation conceded to some of the demands of the SMD.

Chronology of Events

1985:	The Uchangi dam is approved by the State government.
1986 onwards:	Minor protests by villages in the submergence area.
1997 onwards:	Increased opposition by PAPs.
April 1999:	The Uchangi Dharan Kruti Samiti, representing the project beneficiaries, demands that construction of the dam be started at once; on the other hand, SMD activists organise a *morcha* (*Tarun Bharat*, March 15, 1999).
June 1999:	A large *dharna* organised by the SMD creates a law and order problem; the Superintendent of Police arranges for the protestors to meet the District Collector.
October 1999:	PAP bar entry of an MKVDC official attempting to carry out a survey; MKVDC agrees to reduce the height of the Uchangi dam by 2 m and to build a small dam at Khetoba.
December 1999:	The government gives a written promise to hold a meeting at a higher level (*Sakal*, December 31, 1999).
January 2000:	Activists of the SMD oppose this decision and again demand reduction of height by 5 m (*Tarun Bharat*, January 22, 2000). The SMD decides not to compromise on its demand for 'rehabilitation first and then construction'. The government is forced to stop the construction of the dam.
April 2004:	A Government Resolution gives the MKVDC powers to construct the dam as per original specifications.
2005:	The deadlock continues.

Refusing to Look at Alternatives

The MKVDC has agreed to:

(*i*) Reduction in height of dam by 2 m.
(*ii*) Additional storage site at Khetoba.
(*iii*) An 85 per cent subsidy on a lift irrigation scheme for Chafawade and Jeur.
(*iv*) Bimonthly joint meetings between the PAPs and district collector to discuss the rehabilitation plan.

Fig. 2. PAPs gathering at the Uchangi dam site.

However, the protestors were still not satisfied with the reduction of the height by 2 m because the submerged area would still be considerably large. The land in the command area that is required for rehabilitation is close to 242 ha whereas the available land is not more than 28 ha. The movement refuses to compromise on the rehabilitation issue, and categorically states that it has to be 'first rehabilitation only then dam construction'.

In 2000, the SMD-led movement forced the MKVDC to stop construction and there the matter rests now. According to Sampat Desai, an SMD activist, 'The government will not be allowed to construct the dam if they cannot fulfil the demand of prior and proper rehabilitation.'

SMD activists are part of the Maharashtra Rajya Dharangrasta va Prakalpagrasta Shetkari Parishad is an organisation of the dam-affected, and are involved in bringing about larger policy changes at the State level. The Parishad has been stressing on the need for progressive rehabilitation for all dams and this has implications for the Uchangi issue too.

According to R.B. Vengurlekar, Deputy Engineer MKVDC, they have chosen the best site and have planned a good design based on topological, geological and hydrological assessments. The cost of the Uchangi dam was estimated at Rs. 40 crore. However, this has been going up steadily. When the government agreed to reduce the height of the dam the farmers allowed the construction of the dam to proceed. About 30–35 per cent of construction has now been completed. However, for the last three years construction has come to a halt due to lack of funds and an inadequate plan for rehabilitation of PAPs.

The MKVDC has, however, through a Government Resolution of April 2004, decided to go ahead with the building of the dam according to the original design and height.

Project Beneficiaries

The Uchangi Dharan Kruti Samiti represents the interests of the project beneficiaries. Albert D'souza, the leader, feels that the dam is necessary to irrigate the parched land in the area. But he too agrees that rehabilitation is a priority. The Samiti has identified approximately 640 ha of land that belong to the Swami Shankaracharya Trust. But the entire land is outside the command and scope of any irrigation scheme, and is therefore unsuitable for rehabilitation. The Samiti is prepared to support the SMD alternative and stand by it if in return it promises to irrigate the same amount of area. Despite their differences, the Samiti has never really come into acute conflict with the SMD.

Highest and Lowest Points

The highest point of the conflict came in October 1999. Despite a series of protests, the MKVDC paid no heed to the demands of the PAPs and inaugurated the construction site. It initiated surveys of the command area and the area of submergence. The SMD refused to allow MKVDC officials to carry out the survey on the dam site. Thousands of men and women marched with their cattle to the dam site to oppose the dam. A huge police force was stationed there to beat back the protestors. However, the Superintendent of police realised the gravity of the situation and arranged a meeting with high-level officials of the MKVDC and the District Collector. The meeting ended with an understanding that construction would not proceed until decisions are taken at the policy level on issues of rehabilitation and the alternative proposal was duly considered. The conflict is still unresolved. It is latent and likely to resurface and gather momentum if the MKVDC proceeds with the construction without complying with the demands of the SMD.

Butting Heads

As mentioned, the conflict is dormant at this point but is sure to be aggravated if the MKVDC decides to abide by the April 2004 resolution. Although the fact that the SMD managed to bring construction to a standstill on the issue of rehabilitation, and that reduction of the height of the dam has been a victory of sorts, things are now uncertain because of the resolution that gives the MKVDC the powers to proceed with the building of the structure despite being aware that such an act does not respect the spirit of the negotiations between the two parties.

There is urgent need for dialogue between the corporation, the project beneficiaries as well as the PAPs on the issue of rehabilitation, as well as an examination of the alternate proposal.

Instead of adopting an adversarial stance the MKVDC could play a proactive role in (*a*) evolving a concrete rehabilitation plan jointly with PAPs, and (*b*) studying the proposed alternative of height reduction by 5 m as well as the promised additional storage structure at Khetoba on the Tarhol stream.

Thus the MKVDC needs to (*a*) examine the implications of the height reduction for the command and submerged area, and (*b*) identify land in the command area to rehabilitate the PAPs. Both measures are very much within the present legal framework and the mandate of the MKVDC/State government. What is lacking at present is the will, the accountability and transparency in sharing information, as well as financial commitment. In this context mobilising support from both project beneficiaries as well as PAPs becomes crucial.

The second level of conflict, which arose as a result of demand for integrated watershed-based planning perhaps cannot be handled by the MKVDC, and therefore needs to be on the agenda of a much wider movement that addresses broader concerns in the water sector.

Note

*Map 1 from SOPPECOM, Pune; figure 1 from MSP Research-IWE 2004, figure 2 by authors.

1. It was set up in 1975 and specified that if the quota meant for the concerned state is not utilised by May 2000, then the unutilised share will come under consideration during the next round of negotiations.

References

Alternative Proposal to the Proposed Uchangi Dam in Ajara Taluka. Kolhapur: MRDPSP, BGVS and SOPPECOM, July 1999.

Kesari, April 21, 2004, Pune.

Lokmat, May 12, 1999, Kolhapur.

Maharashtra State Water Policy. Government of Maharashtra Irrigation Department, July 2003.

Patankar, Bharat. 1999. *Uchangi Dharnala—Paryay*, SMD, Ajara–Kolhapur.

Phadke, Anant. 1994. 'Anti-Drought Movement in Sangli District', *Economic and Political Weekly*, November 26, 1994.

———. 2000. '"Dam-Oustees" Movement in South Maharashtra', *Economic and Political Weekly*, November 18.

———. 2002. 'Anti-Drought Movement in Sangli District', *Wasteland News*, May–July 2002.

Phadke, Roopali. 2002. 'People's Science in Action: Participatory Watershed Mapping in India'. University of California, Santa Cruz: Department of Environment Studies.

Pudhari, June 18, 1999, Kolhapur.

Sakal, June 20, 999, Kolhapur.

Sakal, December 31, 1999, Kolhapur.

Sakal, March 13, 1999, Kolhapur.

Salient Features of Uchangi Minor Irrigation Tank. Pune: Maharashtra Krishna Valley Development Corporation (MKVDC); Kolhapur Irrigation Circle, Minor Irrigation Department, Kolhapur.

'Socio-Economic Abstract of Kolhapur District'. Mumbai: Directorate of Economics and Statistics, Government of Maharashtra, 1995-96.

Tarun Bharat, March 15, 1999, Kolhapur.

Tarun Bharat, June 19, 1999, Kolhapur.

Tarun Bharat, January 22, 2000, Kolhapur.

Tarun Bharat, March 5, 1999, Kolhapur.

People's Struggle in the Narmada Valley: Quest for Just and Sustainable Development

Sanjay Sangvai

Prelude

THE SARDAR SAROVAR PROJECT (SSP) IS AN INTER-STATE PROJECT SPANNING GUJARAT, MADHYA Pradesh and Maharashtra with a 110.64 m high dam situated in Narmada district, Gujarat. It is part of the Narmada Valley Development Plan (NVDP) consisting of thirty large, 135 medium and 3,000 small dams on the Narmada and its forty-one tributaries. The SSP and the Narmada Sagar Project (NSP) near Khandwa in Madhya Pradesh are the two mega-dams in the plan.

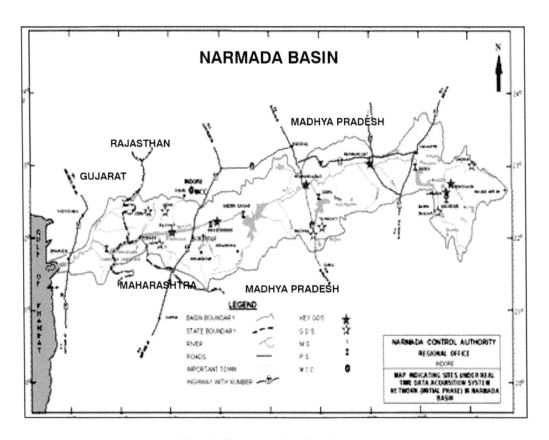

Map 1. *The Narmada valley development.*

After a series of disputes, the Narmada Water Dispute Tribunal (NWDT) was formed in 1969. In 1979, the NWDT gave its award and decided that the top of the dam would be 138.8 m above mean sea level. There was opposition to the high dam on the Narmada in Madhya Pradesh, and in 1980 sporadic protests like the Narmada Bachao Andolan (NBA) erupted. By 1990, five large dams in the NVDP had already been completed, viz Tawa, Sukta, Kolar, Barna and Bargi, all in Madhya Pradesh. However, their record of resettlement, rehabilitation and even benefits is miserable.

SSP: Some Features

(*i*)   Nine million acre feet (MAF) water storage capacity.
(*ii*) Irrigation: 18 lakh (1.8 million) ha in Gujarat; 75,000 km canal network.
(*iii*) Power: 1,450 MW of installed capacity.
(*iv*) Drinking water to over 8,000 villages and 131 cities.
(*v*)   Water for industries in Gujarat.
(*vi*) Flood control, tourism, etc.

The NWDT has offered a liberal policy for rehabilitation of displaced people including land-for-land and community resettlement—at least a year before submergence—and linkage between resettlement and the height of the SSP dam. The loan agreement with the World Bank (US $ 450 m in 1985) was signed with guidelines on resettlement and rehabilitation (R and R), none of which were followed.

The Government of India, through the Ministry of Environment and Forests (MoEF) and the Planning Commission, gave conditional clearances to the project in 1987–88. The MoEF clearance was withheld for a while because important studies pertaining to the SSP and NSP were incomplete. But the World Bank decided to fund the SSP in 1985 itself.

In 1987 the SSP-affected tribal villagers in Maharashtra belonging to Akkalkuwa and Akrani *tehsils* (Nandurbar district) organised the Narmada Dharangrasta Samiti. They demanded full information, questioned the displacement and did not take it as a *fait accompli*. They asserted their right over the land, forest and waters, asked for a master plan of community-based resettlement and guarantee of land and other natural resources, and questioned the public purpose of the displacement and the dam.

Very soon the Narmada Asargrasta Sangharsha Samiti in Gujarat and the Narmada Ghati Navnirman Samiti in Madhya Pradesh were also launched. However, their attempts to get information and answers to displacement-resettlement or cost-benefit at various levels did not get any response. A united NBA was then formed and it declared its opposition to the displacement brought about by the projects. It called for a total review in August 1988.

Intense Struggle

Thereafter, it has been a constant struggle against the three State governments, the Union government, the police and the administration, the World Bank and other multilateral donors; against direct or indirect attempts to displace people, increase the height of the dam and thereby make it irreversible. There were mass actions at both the local and national level, indefinite fasts by the main activists, and innumerable attempts to involve the State and Central governments in dialogue—the NBA spared no effort to utilise all possible democratic means and forums to look for a just solution. The Independent Review by the World Bank (*Morse*

Report, 1992) dismissed all tall claims relating to benefits, resettlement and legalities and it finally withdrew from the project in March 1993.

Even while the government made suspect claims about the benefits from the project, several parameters of the dam were changed and though the people did not move out of their villages, the authorities continued to raise the height of the dam. People resisted the politics of intimidation non-violently through *satyagraha—doobenge par nahin hatenge* (we will drown but not move out)—and stood defiant against forcible submergence and displacement. Between 1995 and 2002, the *satyagrahis* stood firm during many monsoons, neck-deep in submergence water.

The Union Water Resources Ministry appointed a five-member group to review all the aspects of the project on August 5, 1994. The Supreme Court also took note of the Group's recommendations. The work on the dam came to a halt as a result of the people's struggle and in December 1994 the court disallowed resumption of work for five years (until 1999). The dam stood still at 80.3 m.

The NBA had moved the Supreme Court in April 1994. Though the court initially showed willingness to reopen the NWDT and protect human rights, there was a change in the bench in 1998 and it also brought about a change in its attitude. In its final order in October 2000, the Supreme Court allowed construction to proceed, making unsubstantiated generalised claims that large dams were beneficial to the nation, or that all oustees were resettled—without any evidence or hearing. Justice Bharucha, however, asked for fresh environmental clearance.

In its judgment in March 2005, the Court told the State governments to adhere to the NWDT orders on land-based rehabilitation, including land for 'encroachers'. People's protest forced the Maharashtra government to appoint a task force to review the status of rehabilitation and establish the Joint Rehabilitation Review and Planning Committees. The NBA meanwhile continues to fight for resettlement in the three states and is trying to stop the dam from rising beyond the present 110 m.

Just Count the Losses

Decision-making

There has been no consultation with the people, nor has information been made available to them till date about the magnitude and extent of submergence, the availability of land and other resources, master plan for resettlement, etc. The conditional clearances by the MoEF and Planning Commission have been violated because studies and plans pertaining to rehabilitation, command and catchment area treatment, flora-fauna, seismicity, carrying capacity, compensatory afforestation and health-related issues are still incomplete. The original plan envisaged that the amount of water in the Narmada would be 27.2 MAF but it is only 22 MAF.

Mirage of Benefits

The NBA has always maintained that the water issue in Gujarat will continue to remain grave despite the SSP.

(*i*) The ultimate justification for the SSP is that it will irrigate parched Kutch, Saurashtra and North Gujarat. But according to the plans only 9.4 per cent of total cultivable land in Saurashtra may see canals by 2020 while 1.6 per cent of cultivable land in Kutch will get Narmada water by 2025. Kutch and

Saurashtra have been given a 2 per cent share of the total 9 MAF water. Plans to start at least twenty-five sugar factories in the early reaches of the canal will further erode their meagre share, not to mention the plans to sell water to cities and industries.

Though the height of the dam was raised to 110 m, water has not reached even the limited area that was to be served by this increase in height. The 2002–3 report of the Comptroller and Auditor General (CAG) refutes claims that the drinking water crisis in Ahmedabad and Saurashtra has been solved by the Saurashtra Pipeline Project.

(*ii*) *Electricity*: Though 1,450 MW is the installed capacity, firm power is 425 MW in stage I and 50 MW in stage II.

Unprecedented Displacement

The SSP reservoir will displace over 45,000 tribal and peasant families from 245 villages in three States. Other project-related construction like colony (six villages), canals (25,000 landholders), sanctuary (4,500 people), downstream effect, power sub-station, secondary displacement, involves the displacement of thousands of people in Gujarat alone. However, all these figures are not being taken into account and only those directly affected by reservoir submergence are counted as 'dam-affected people'. Thousands of tribals are termed 'encroachers' on their own land and there is no final estimate of project-affected families (PAF) or persons (PAPs). All this when, till date, even those below 80 m have not been resettled as per NWDT provisions.

Madhya Pradesh (33,000 PAFs) has refused to give land-for-land despite assuring the Supreme Court that it would; Maharashtra finds it difficult to resettle 3,000 PAFs. In Gujarat, hundreds of 'resettled' families have returned to their original villages while countless other 'resettled' PAFs lead a life of destitution and poverty.

Financial Crunch

The cost-benefit of the dam has always been lopsided. From the original 4,240 crore rupees (1985) the cost has risen to 25,000 crore. The actual costs might be anywhere near Rs. 50,000 crore. About 85 to 90 per cent of the irrigation budget of Gujarat is devoured by the SSP alone, leaving little for other projects. The State government resorted to market borrowings, taking out money from relief funds for the earthquake and riot-affected people in the form of bonds. Now it finds it difficult even to defray the 17 per cent interest (Rs. 8,000 crore every year) on the bonds.

Environmental Aspects

The dissenting Supreme Court judgment by Justice Bharucha pointed out the violation of environmental norms and asked for fresh clearance. The SSP will submerge 13,744 ha of forests, an equal area of fertile land, rare flora-fauna as well as the breeding grounds of the Hilsa fish. There has been no progress on catchment area treatment and compensatory afforestation. The downstream impact is likely to affect 10,000 fisherfolk, adversely affect water availability for hundreds of villages, and exacerbate pollution and the sea water ingress. The dam is situated in a seismically sensitive gorge. There were earthquakes in Jabalpur (1996) and Kutch (2001), and frequent tremors have been felt in Bhavnagar and Khandwa as also at the dam site.

Fig. 1. *The present SSP.*

The SSP and Narmada project will also submerge, without fully exploring its potential, a region that is considered a treasure trove of Indian archaeology. It is the only place in the subcontinent where historians have been able to find uninterrupted links from the pre-historic era.

Political

The SSP threatens the peasants' and tribals' right to life and resources. There is no attempt to consult, inform, invite participation or seek consent which amounts to a violation of democratic rights and brutal repression. The displacement has been carried out by compulsion, deception and atrocity; it has fragmented and disempowered the communities.

The Struggle Continues

The Sardar Sarovar dam has reached 110(+3) m, having submerged the houses, farms and villages of over 12,000 tribal and peasant families. Of them not even 10 per cent have been resettled according to NWDT norms. The benefits have proved illusory, far fewer than anticipated. The continuing people's struggle in the Narmada valley has for the following objectives:

(*i*) Full and just resettlement of all those affected due to height of 110(+) m in accordance with the NWDT and court orders; restoration of the rights of PAFs, particularly tribal PAFs.

(*ii*) Comprehensive and participatory review of the dam and no further increase in height beyond 110 m in view of the large-scale violation of conditions and human rights.

(*iii*) Implementation of decentralised, people-centred and sustainable water harvesting schemes in Kutch and Saurashtra.

(*iv*) A new policy for development projects, displacement and resettlement.

(*v*) A people-centred water policy.

Dialogues and Alternatives

The NBA has always sought dialogue with the government to find a just solution to the issue that is based on people's rights, entitlement, equality and sustainability. It has convened several meetings that have involved socio-political workers, experts and government representatives.

The Andolan presented a full plan for a participatory review in May 1990 and again in January 1991. The five-member group formed by the union government has also conducted an independent analysis and suggested a review of the project—with particular reference to the hydrological, environmental, seismological and displacement–resettlement aspects. The NBA held official discussions with the Union government on June 30, 1994.

There are many proposals to reduce the height of the dam to an RL of 117, 133, 94 or 64 m and alternative plans for water and land development. In 1990, water expert Vijay Paranjape, and in 1994, Suhas Paranjape and K.J. Joy, presented alternative blueprints on water and irrigation. The NBA convened meetings of officials and activists to discuss these plans. The Madhya Pradesh government and the NBA conducted a workshop on Review of the Narmada Valley Projects and Alternatives in December 1995. A task force was appointed and the members recommended decentralised, people-controlled, supplyside-oriented sustainable alternatives. In Maharashtra, the State government instituted a State-level joint resettlement review, and in September 2001, appointed planning and overview committees, as also an expert committee to examine the cost-benefit from the dam at the insistence of the NBA.

The NBA and concerned citizens' groups in Gujarat pointed out that the real solution to the water problem in the State lies in locally harnessing rainwater, recharging wells, ponds, tanks, opting for different crop patterns and rural-urban and intra-sectoral equality in water distribution. However, bureaucracy and lack of political will put paid to any hope of finding solutions.

The NBA has also undertaken watershed development, alternative schools or health workshops, and *panchayat* training in the valley. The villagers themselves have built two microhydel power plants in Domkhedi (2000) and Bilgaon (2003).

Twenty years on ...

1. The Narmada struggle has raised major issues that are relevant to all present water-related projects and the water policy; it is trying to evolve a new policy and paradigm with like-minded organisations and striving to underline the importance of communities themselves to take up development projects with a rights-based approach to displacement and rehabilitation. All this becomes important in an era when national and multinational companies have their eye on the resources of the common people. A new 'Draft National Policy on Development Projects, Displacement and Rehabilitation' has been prepared and conventions and national and State level action-programmes have been undertaken by the National Alliance of People's Movements (NAPM).

The government claims that about 10,000 PAFs have been resettled and that they have both land and livelihood. Even though the NBA considers this a partial resettlement, the fact is that resettlement on such a scale is unprecedented in India and is the outcome of the struggle against displacement.

2. As mentioned earlier, the NBA has been trying to evolve a new people-centred and people-based equitable and sustainable water policy. To this end a large convention was held in the Narmada valley in 1996 at Kasaravad, and also at other places in collaboration with NAPM. The Andolan is also a part of the campaign opposing the Interlinking Rivers' Project.

It has been the NBA's endeavour that issues pertaining to large dams, displacement and new water and land policies or sustainable development are not just seen as 'environmental' or 'humanitarian' issues, but as part of a larger politics of an egalitarian, just and sustainable socio-political and ethical system—a politics of technology, capital, socio-political and cultural relations and sane development. The NBA's struggle is a fight for a just society and polity.

Chronology of events

1979:	The NWDT Award; sporadic protests in Madhya Pradesh.
1985:	World Bank loan for the Sardar Sarovar Project.
1987:	Conditional Clearance by the Government of India.
1987:	Formation of the Narmada Bachao Andolan.
1988–90:	Mass actions within and outside Narmada valley.
1990 May:	Prime Minister V.P. Singh orders review of the dam.
1990–91:	Long march by thousands of oustees; twenty-two-day fast.
1991, June:	Manibeli Satyagraha against submergence; Jeevanshalas start.
1993:	World Bank withdraws from the SSP after the Morse report.
1993 August:	The union government starts reviewing the SSP and forms the five-member group.
1994:	NBA files case in the Supreme Court. First submergence in the SSP.
1995–99:	NCA stops work on the SSP due to NBA action.
2000 October	Supreme Court orders resumption of work.
2002, Monsoon:	Satyagrahis protest by standing neck-deep in submergence waters.
2002 September:	Mumbai: Eleven-day fast; task force for review of R&R.
2003 January:	Villagers build second micro-hydel power plant in Bilgaon.
2004 May–July:	Land satyagraha; Mumbai fast for R&R, no rise in height of dam.
2004 June:	Harsud submerged by Narmada Sagar.

Note

*Map 1 from Narmada Control Authority; figure from Sardar Sarovar Nigam.

Alternative Restructuring of the Sardar Sarovar Project: Breaking the Deadlock

Suhas Paranjape and K.J. Joy

The Issue

THE SARDAR SAROVAR PROJECT (SSP) IN GUJARAT IS A HUGE PROJECT WITH A PLANNED service area of 1.8 Mha or 18,000 km². It has a submergence area of 360 km², spread largely between Madhya Pradesh, Maharashtra and Gujarat, and has been the focus of a conflict from its very inception. The main parties representing the two extreme positions can be clubbed as the official establishment comprising the Gujarat government, and experts and anti-dam activists who view big dams generally, and the SSP in particular, as unmitigated environmental, social and economic disasters. In some ways there is also an underlying conflict between the people of the drought-prone areas of Gujarat who think the Narmada is their 'lifeline' and the *adivasis* in the upstream region who stand to lose their land and way of life for an abstract 'common good'.

We believe that there is a third way; an alternative that can effectively reconcile the opposing stands. This alternative (the SJ plan) was put forth in the book *Sustainable Technology: Making the Sardar Sarovar Project Viable*, published in 1995. Ten years later, the SJ plan itself became less relevant as the State continues

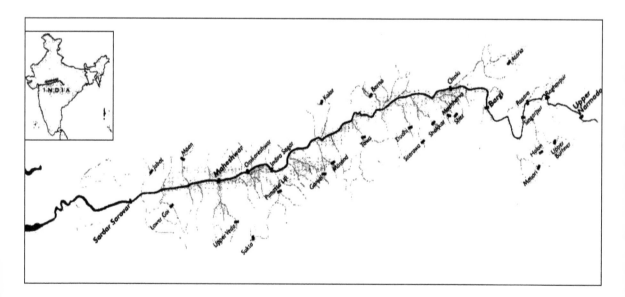

Map 1. *Proposed big dams in the Narmada valley.*

to inexorably raise the height of the dam. The principles behind it, however, are still pertinent. In what follows we revisit these principles in the context of the alternative restructuring of the SSP. We have concentrated on a few of the central precepts rather than on the details and, more importantly, have left out of this exposition the energy component, not because it is irrelevant—in fact, it informs the entire argument—but because it needs to be dealt with separately.

A Summary of the Alternative

The alternative proposal ensures a delivery of 9 MAF of Narmada waters to Gujarat, but with a much smaller height, and brings down the submergence by over two-thirds of the current level. It retains the power benefit at levels very close to that of the present SSP plan and within the same order of cost. It greatly increases the share of Narmada waters delivered to the drought-prone regions of Gujarat without a substantial reduction in the incremental irrigation benefit to the rest of Gujarat, and extends the total service area from 18 Mha to 41 Mha. It provides for equitable and sustainable water use in the service area and for permanent tree cover over a little less than one-third of the service area of the project (for a comparison of salient features, see Table 1). This is made possible by a radical change in approach that is presented here.

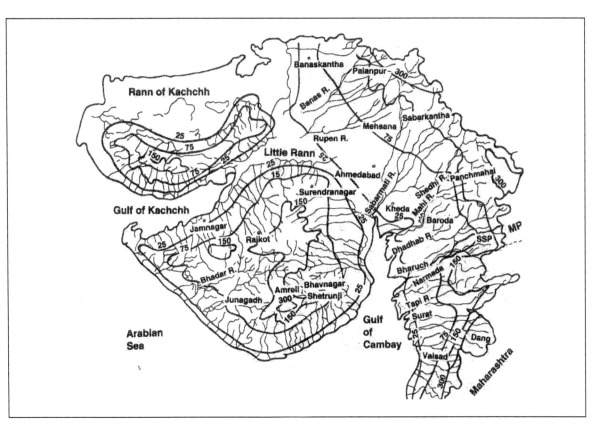

Map 2. *Alternative canal alignment for Kachchh and Saurashtra.*

Table 1: Comparison of the Alternative (SJ) Plan and the Current Plan

No.	Item	Alternative (SJ) plan	Current plan
1.	Storage level at SS dam	107 m (90 m baseline level)	140 m
2.	Total Submergence	10,800 ha	36,000 ha
3.	Displacement	Drastic reduction in displacement	1.5 lakh people displaced
4.	Rehabilitation	Within the same area with assured share of Narmada water	Uprooted, rehabilitation in new area
5.	Upstream Service Area	More than 1 lakh ha	Nil
6.	Gujarat Service Area (of which)	41 lakh ha	18 lakh ha
	Saurashtra	13.1 lakh ha (32 per cent)	3.9 lakh ha (22 per cent)
	Kutch	4.0 lakh ha (10 per cent)	0.4 lakh ha (2 per cent)
	North Gujarat	14.7 lakh ha (36 per cent)	3.1 lakh ha (17 per cent)
	Rest of Gujarat	8.9 lakh ha (22 per cent)	10.6 lakh ha (59 per cent)
7.	New electricity generation	850 MW 2,600 MU	1,400 MW3, 600 MU
	Consumed in the project	1,646 MU	1,138 MU
	Peak load capacity	1,200 MW	1,400 MW
	Gas-solar hybrid generation out of saving	200 MW1, 750 MU	Nil
8.	Surplus energy	At least 4,410 MU (26.3 MT) produced as biomass	Not planned
9.	Equitable water distribution and sustainable development	Basic issue	Not planned
10.	Total cost	12,920 crore	13,000 crore
11.	Expenses on local employment and services	3,620 crore	Negligible
12.	Cost recovery	Based on distinction between basic and economic service	No such plan
13.	Gujarat's total share of Narmada water	9 MAF	9 MAF
14.	Loss of forest	3,000 ha by submergence and 10,000 ha low grade forest for rehabilitation	13,700 ha substantial prime quality forest
15.	Permanent vegetative cover in service area	11 lakh ha (23,000 ha in upstream contiguous to forest area)	No provision

Integration of Large and Small

Dispersal of Storage

The alternative does not see the conflict as one of large versus small, but as a question of the relation between the two. When planned and used properly, large sources can support smaller and local systems and are important in increasing their reliability as well as sustainability. Most large systems today are planned with behind-the-dam storage comparable to the total planned water use. The alternative breaks from this practice, and treats dispersed local storages as the main storage element. Consequently, the dam is seen as diverting water for filling and/or refilling local storages while the storage behind the dam mainly becomes a regulatory storage. This greatly reduces behind-the-dam storage and consequently submergence behind the dam, while at the same time allowing utilisation of larger quantities of water.

Integration of Local and Exogenous Water

The alternative has a similar approach to local and exogenous water. In fact, the alternative includes the development of local water resources and their integration allowing a doubling of the planned service area. This is a lesson from the so-called 'system tanks' in Tamil Nadu which are rainfed tanks that are refilled from large sources, greatly increasing their reliability.

Dual Role of Small Systems

Thus, small systems have a dual role: they harness local water resources and act as dispersed buffer storage elements for water from the large source. Taking into consideration that local systems need to be built in their own right, this synergy is the greatest cost-saving measure proposed by the alternative. It starts with an integrated view of the large and the small, and makes the large system a supporting and strengthening for the small system.

Feeder Canals

The canal network mainly functions as 'feeder canals' planned for speedy conveyance to the local storage systems. The system becomes modular, simplifying and rationalising the arrangement and cutting down on top-heavy, centralised aspects.

Local Storage Potential is Very Large

Local storages are mainly limited by expected yield rather than actual storage potential. If exogenous water is available it is possible to build larger storages or increase the capacity of existing ones at little additional cost.

The alternative manages to improve on set patterns because it gets away from the conceptual as well as the administrative Chinese wall that separates small and large systems. It starts with local systems as the central concept, strengthens them and integrates large sources with them as supplementing and reinforcing elements.

Run-of-the-river Operation of Hydro-power Plants

The second important change proposed in the alternative is in respect of the power from the SSP. The reduction of storage behind the dam as well as the reduction in the height of the dam would mean losing the entire power benefit if the hydroplant is operated conventionally as a peak-load plant based on behind-the-dam storage. The alternative suggests a run-of-the-river (RoR) operation of the power plant during monsoons and a pumped hydro operation for peak load in post-monsoon. This fully preserves the peak-load benefit from the plant, and comes close to the new energy generation benefit of the present SSP. In fact, adding a monsoon RoR element and post-monsoon pumped storage element to all irrigation projects would lead to a truly large, dispersed generation and peak-load capacity for the power sector.

Implication for post-monsoon flows—bound and unbound

One of the important adverse effects of present planning is the post-monsoon drying up of riverbeds downstream of big dams. In the alternative, since feeder canals mainly divert monsoon flows or accumulation and power generation is RoR, the post-monsoon flows are unbound, making for greater riparian health. A similar provision for retaining minimum monsoon flow can also be built in.

Impact on rehabilitation

The alternative, first of all, brings down submergence behind the dam by almost 70 per cent and makes the problem tractable. Second, the concept proposes rehabilitation of the oustees in the upstream area itself, in a socio-cultural milieu and sphere that is familiar to them. This is made possible, first, by reduction in submergence and second, by providing allocation from the Narmada itself for irrigation and livelihood assurance in a contiguous zone of 100,000 ha upstream of the dam within which the rehabilitation is provided.

Equity and Sustainability: Conditions for Providing Water

There is a need to insist that if exogenous water is to become available, both the users and the State must fulfil certain conditions in respect of equity and sustainability. These are listed here.

The water will be available to local systems in proportion to the local resources they harvest and harness (for most regions we have proposed 1 m^3 of Narmada water for every m^3 of local resource created). This provision ensures that when exogenous water enters an area the local systems will not simply die (as happens these days), but prosper instead.

(*i*) equitable water access or minimum water assurance for all families in the service area irrespective of landholding, and protecting livelihood needs before more water is provided as an additional economic service;
(*ii*) one-third of the service area to be brought under permanent cover. Unlike 'compensatory forestry', this ensures that minimum upgrading of the environment happens in the entire service area;
(*iii*) self-management by those who benefit will ensure equitable and sustainable use. This provision is as important as the others because no top-heavy bureaucracy can fulfil these conditions. Only if the people themselves come together and exercise control can these conditions be met.

The State too has corresponding obligations (*a*) to provide requisite funds for harvesting and harnessing local resources. In the alternative, minimal watershed treatment that is twice the service area is included in the project cost itself. Provisions (*b*), (*c*) and (*d*) require that the State enables local communities by ensuring appropriate legislation, policies and incentives, so that they are sufficiently empowered to satisfy the conditions.

As mentioned earlier, this article reiterates some of the themes detailed in Paranjape and Joy (1995). The book discusses these issues in some detail and there is an ongoing discussion on many of these issues, available at various places. There is a need to continue this discussion and bring it to a point where water users, present and prospective, come to a consensus and jointly struggle for enabling policies. We would emphasise that these aspects of the alternative are relevant, not only to the SSP, but to all irrigation projects in the country, and to the water system as a whole.

A Plea for a Pause and a Dialogue

The alternative was planned with a baseline height of 90 m (and a dam height of 107 m) in mind, and we are the first to acknowledge that every extra metre means that the benefits of the SJ plan decrease, though we feel that a remodelling of the SSP along these lines is still possible. It is important that all parties concerned, especially the people of Gujarat, pause and seriously reconsider and review the project, see to it that they devise an optimal plan that satisfies the legitimate interests of people in drought-prone regions of the State as well as the *adivasis* in the Narmada valley. If this does not happen there will be no winners and the SSP will become nothing more than a monument to our callousness.

Note

*Map 1 from Narmada Valley Development Authority, map 2 from Paranjape and Joy (1995).

References

Paranjape, Suhas and K.J. Joy. 1995. *Sustainable Technology: Making the Sardar Sarovar Project Viable —A Comprehensive Proposal to Modify the Project for Greater Equity and Ecological Sustainability*. Ahmedabad: Centre for Environment Education.

Haribad Minor Irrigation Project in Madhya Pradesh: How Multiple Conflicts Overlap

Rehmat Mansuri and Shripad Dharmadhikary

The Story of Two Villages

HARIBAD AND SAKAD ARE ADJOINING VILLAGES IN THIKRI *TALUK* OF BADWANI DISTRICT IN western Madhya Pradesh. The proposed Haribad Minor Irrigation Project is to be built on the boundary of the two villages on the Kundi river that runs through them. The project was cleared in 1977 and the cost then was 43 lakh rupees, which has now gone up to 1.66 crore rupees. Sakad is upstream of the dam and will bear the full burden of submergence but will not receive direct irrigation, while downstream Haribad will be served by canals and will also benefit from the recharging of groundwater.

Map 1. *Location of Barwani in Madhya Pradesh.*

About 27.5 ha of private land and 2.5 ha of government land in Sakad will be submerged; local paths will go under water and cut off access to grazing land. There is fear that the lower *phalia* (hamlet) will also be affected. Of the thirty-five families that are affected, many are small farmers who will lose all their lands. There are conflicting reports on the area and villages that will benefit. One version is that a 2-km-long canal will irrigate 107 ha in Haribad and Surana, another view is that only Haribad will benefit, and the third version, that several villages will gain from the dam.

The land is undulating, with large, flat open areas in between. Average annual rainfall is about 650 mm and there is limited irrigation, mainly by groundwater and direct lift from the *nallahs*. Sakad is a tribal village, and 98.4 per cent of its 1,273 population are tribals. Haribad has 799 non-tribals, 633 tribals and 428 scheduled castes in a population of 1,860. And, it is the non-tribals who are socially and politically dominant. Water has been an increasingly serious problem in the area, especially in recent years. Irrigation is minimal, and while the situation is marginally better in Sakad, Haribad suffers from drinking water scarcity as well.

Haribad Wants the Dam; Sakad does not

The people of Haribad want the dam. They see it as the solution to their serious water problem and insist that several downstream villages will also benefit. The people of Sakad oppose the dam since they will not only lose much of their irrigated land, but also access to resources. They will not gain anything directly, and even indirect gains through recharge are minimal. Moreover, only the tribal families stand to lose land and this is a major element in the conflict. They propose that the dam be shifted further downstream in order to help store more water and benefit a wider area. This way non-tribal lands will share submergence; and although their own lands will not be saved as they are at the tail end of a large reservoir, they will manage one crop in the drawdown land. While Haribad has no issues with this alternative, government surveys show that it is far too expensive and not feasible. So the clash continues: Sakad claims that Haribad is using its considerable resources and political clout to force the project through. Haribad claims

Fig. 1. *Submergence zone.*

that the tribals of Sakad are using their special representation as a tribal reserved constituency to thwart the project.

Chronology of Events

1977–78: Project announced during the visit of the then Chief Minister to Haribad.
1981: First tubewell in Haribad.
1997: Administrative clearance to project accorded on December 5, 1997.
1999: Demand for the project escalates as water resources are strained.
2003: Haribad villagers resort to *chakka-jam* (road blocks) for the project.
2003: Haribad villagers pledge part of their land to those losing land.
2003: *Gram sabha* (village assembly) votes for project to go ahead.

Old Animosities

Some old animosity also underlies the conflict and may be linked to the issue of unregulated use of water resources. The first tubewell was dug in Haribad in 1981 when water was available at a depth of about 40 m. Since then about 200 irrigation tubewells and 50 borewells for drinking water have been dug in Haribad. Only one of the borewells for drinking water is now working, and the water level has dropped to 200 m and more. Clearly, this intensive exploitation is responsible for the serious water situation. In the summer, even the lone borewell stops yielding water and the village has to be supplied by tankers, spending Rs. 1 lakh for the effort when most farmers are heavily in debt. The situation has worsened since 1999 and has resulted in Haribad villagers organising *chakka-jam* to push for the project.

An Innovative Measure?

The people of Haribad have also proposed an innovative measure—they claim that many beneficiary families have pledged portions of their land to the project command to create an 80 ha land pool from which land can be given to those in Sakad who stand to lose their land. They claim that a list with details of names and land pledged was drawn up at a village meeting in the presence of the *tehsildar* and handed over to him in April 2003. This has caught the attention of the local media. Surprisingly, none of the displaced people were called for this meeting. Also, Sakad villagers are surprised by this development and deny knowledge of any such scheme. If it was a serious proposal they would have been informed.

As Sakad falls in a scheduled area, permission from the *gram sabha* is essential for land acquisition. In 2003, the *gram sabha* was called, and it approved the project, significantly, through a secret ballot. The displaced people, primarily

Fig 2. *The new site.*

from one *phalia* of Sakad, say that wealthy individuals in Haribad manipulated the *gram sabha*. The displaced *phalia* was informed only at the last moment; and other *phalias* were bribed to vote for the project. The displaced *phalia* boycotted the meeting and later protested to the collector who has reportedly declared the *gram sabha* void.

Many of the displaced families that had opposed the project earlier and wanted it shifted downstream now appear to support the project provided they are given proper compensation. They too had not heard of the offer of land by the beneficiaries and were highly sceptical of the whole idea. They say that the drought in the last few years has forced them to accept the dam. However, the change of heart may also be partly due to the severe divisions that have come up in the village after the recent *panchayat* elections.

Scope for Dialogue

Clearly, the people have paid a heavy price due to this conflict. Water resources have not been developed, agricultural production has gone down and consequently people are suffering. The conflict is responsible for preventing other water harvesting projects. The cost in terms of the social discord cannot be quantified, but is certainly significant.

There are good chances that the dispute can be resolved through dialogue. However, it will have to be initiated by someone who commands the respect and trust of both the parties. The project should be open to modification, not just in size and location, but also compensation. The pledge of land by the people of Haribad and Surana seems to be a good starting point for a dialogue.

If the conflict is still not resolved, the project can instead be replaced by a comprehensive watershed management programme and water harvesting/recharging scheme for which the terrain and rainfall offers vast scope. However, all that is in the future; today the deadlock remains.

Note

*Map 1 from SOPPECOM, Pune; figures by authors.

References

Letter according administrative clearance to Haribad scheme dated December 5, 1997.

Letter according technical sanction to Haribad Project by the executive engineer, Water Resources, Badwani, dated October 28, 1998.

Letter by the Under Secretary, Water Resource Department, Government of Madhya Pradesh, dated December 6, 1999, to chief engineer, Water Resource Department, Narmada Tapti Division, Bhopal, referring to the letter of an MLA to the chief minister regarding Haribad scheme and informing the CE of the CM's direction to expedite the same.

Letter from the executive engineer, Water Resource Department, Badwani Division, Government of Madhya Pradesh, to Maluram Chauhan, Haribad village, dated January 3, 2000, in response to petition by villagers.

Letter dated April 17, 2003, by Haribad villagers to the collector, Badwani, giving notice of *chakka jam* on April 24, 2003 for getting the work on the Haribad scheme started.

List with signatures of the people of Haribad and Surana pledging land to the displaced people of Sakad for their resettlement. The list was given to the *nayab tehsildar* and was received by him on April 26, 2003.

Public notice issued by the executive engineer, Badwani, dated March 13, 1999, annulling the tender notice called for Haribad scheme, as land acquisition could not be carried out.

Struggle over Reservoir Rights in Madhya Pradesh:
The Tawa Fishing Cooperative and the State

Vikash Singh

An Oft-heard Tale

IN 1957 A NEWS ITEM IN *THE TIMES OF INDIA* PROCLAIMED, 'THE WORLD'S LARGEST PLANNED ecological disaster is about to commence'. It was referring to the Narmada valley development project. Conceived during the 1940s, it envisaged the construction of thirty big dams, 135 medium dams and 3,000 small dams on the Narmada and its tributaries. But there was another aspect to the script that the *Times* had understandably failed to anticipate: that this gigantic undertaking would engender one of the most protracted environmental battles in the world. The Tawa, constructed in 1975, was the first big dam to be built as part

Map 1. *Hoshangabad district in Madhya Pradesh.*

of the project. The evacuees were among the first tribal people to be displaced by 'development' in central India. Their tale is part of a narrative that winds along the length of the Narmada, alongside the project; the stories are a plexus of suffering and pain.

The dam was constructed on the Tawa, one of the major left bank tributaries of the Narmada in Kesla block, Hoshangabad district, Madhya Pradesh. The reservoir, spread over 21,000 ha, submerged forty-four Gond and Korku villages. The displaced were given a compensation of Rs. 188 to 375 per ha of land.

It is worth noting here that many of these families had been displaced twice before, once by an ordnance factory and subsequently by a firing range. The people withdrew, made their huts elsewhere and took up casual labour. Then came the dam. Today it irrigates a large part of the plains and lowlands of a non-tribal district. The first phase was accomplished with consummate ease; it was in its secondary stage that the previously muffled tension exploded, almost like a postscript. The conflict that was averted at the time of displacement flared up because of a seemingly less significant issue: fishing in the reservoir.

The State Corporation Takes Over

Once the reservoir was completed the State fishing corporation took over fishing in the reservoir. It brought skilled fishermen from across and beyond the State and had them settled in the valley. With the construction of the dam the legal definition of 'river waters' changed. While fishing in the flowing waters was free and open to all—actually a license was required until 1980 but the rule was not enforced and it was finally done away with that particular year —the law guarded the reservoir zealously. Thus the displaced tribals who had lived around the reservoir and caught fish for domestic consumption suddenly became 'poachers'. What is more, the 'poaching' continued despite the state fishing corporation, despite the people being regularly caught, beaten and handed over to the police.

Beginning with 1979–80, the State fishing corporation managed the fishing operations in the reservoir for fifteen long years. The average annual fish production in the Tawa reservoir during this period was 87.5 metric tons, with the highest being 165.31 tons in 1989–90. In 1994–95, because of mounting losses the corporation leased it out to a contractor for a year. That year the catch mounted to 176.18 tons. The contractor was, however, blamed for having been indiscriminate, for attempting to maximise the catch by disregarding fishing norms (for instance, catching fish irrespective of their size) and using hired henchmen to scare and hound the 'poachers'. The armed men spread terror in the villages, abusing and beating people if they so much as came near the reservoir. This proved to be a turning point of sorts. This was also the time when seventeen villages in the area, earmarked as Bori Wildlife Sanctuary, came under threat. They were asked to vacate their land under Project Tiger. People from these settlements joined hands with the displaced, and thus began a series of protests and demonstrations. There were road blockades, rallies and hunger-strikes.

The Tawa Fishing Cooperative

One of the major demands of the people was that the Tawa reservoir be leased out to them and that they be allowed to manage the operations autonomously. They also asked the State to facilitate the effort by contributing financially and through other means. The result of this conflict was a happy one. The State responded positively and beginning December 13, 1996, the reservoir was leased out for a period of five years to a cooperative comprising displaced people. The Tawa Visthapit Adivasi Matsya Utpadan Evam

Vipanan Sahkari Sangh (TMS)[1] was given the responsibility to produce (including seeding) and market its products. The State also yielded to a number of other demands and revoked the Bori displacement plans.[2]

Thirty-three primary cooperative societies of displaced *adivasis* were formed to commence fishing operations. In addition, three primary cooperatives made up of other fishing communities were also affiliated to the Sangh. The primary cooperatives were to deposit the catch with the Sangh, which was responsible for marketing and dispersal of seeds in the reservoir. The Sangh was also to arrange nets and boats for the fisherfolk (on credit), and look after all other activities associated with the reservoir, including administration. Each society sent one representative to the Sangh and they in turn elected the board of directors. In addition, local district officials were appointed as ex-officio members of the board. The board also had advisors. The Sangh was to be answerable to the people and report to the general body, which included all the members of the primary cooperatives.

Exceptional Performance

The TMS's performance in terms of productivity was exceptional. The average annual fish production rose to about 327.63 tons, an increase of 274 per cent over the corporation's performance; the per hectare fish productivity in the reservoir rose from an average of 7.3 kg/ha/yr to 26.95 kg/ha/yr; the average annual income per worker was almost three times what it had been earlier; the number of people employed in organised fishing in the Tawa reservoir grew from an average of 137 to 422. Among other factors, one reason for the involvement of a larger number of people was the improved access to nets and boats.

In addition, the TMS took a number of other initiatives. These included greater focus on local markets, more fishing days to ensure livelihood security for the workers, use of smaller nets as against the dragnets promoted by the corporation, worker-welfare policies like subsidised loans, etc.

The highest productivity of 32.37 kg/ha/yr achieved in the year 1999–2000 was thrice the national average of 11.43 kg/ha/yr, and more than any other reservoir in the state (for reservoirs greater than 5,000 ha in size); this despite constraints like an upstream thermal power plant that casts off its residue in the river, dead trunks of trees from the submerged forest crowding the reservoir, etc. Empirical evidence based on relative fish size and species also shows that the cooperative, unlike its predecessors, managed the reservoir in a sustainable way.

The Pitch is Queered

But despite these awe-inspiring achievements the pitch was queered as the time for renewal of the lease drew close. While the TMS had been performing with exceptionally high efficiency it had alienated itself from the ruling party. They allegedly expected the Sangh to return the 'favour' by mobilising people for the party. The TMS's refusal to oblige opened a Pandora's box. Suddenly, politicians, bureaucrats, the corporation (now a 'federation'), all turned against the cooperative. The ruling party's local units began to harass the Sangh and incite factionalism. The State fishing federation—which also oversees the State reservoirs—accused the TMS of inefficient management. At about the same time a notice was slapped on the Sangh by the cooperative registrar. It was almost as if all the State departments had suddenly woken up to hound the body.

The State stalled for many months after the expiry of the lease. During this period the Chief Minister and other officials expressed their reservations about renewing the contract, often making oblique references to the TMS's refusal to support the ruling party. The TMS on the other hand, foreseeing a dispute, began to canvass public opinion to put pressure on the state. During the few months when the deadlock was at its most tense, hundreds of academics and public personalities sent letters to the Chief Minister. A national-level conference was organised. There were a series of protests and demonstrations. Months later, clearly wilting under pressure, the government offered to renew the lease. The proposal, however, came with unacceptable conditions that would have most certainly deprived the body of its autonomy. There were clauses like: all officials will be appointed only with the permission of the State government; operations will be bound by policy-related orders of the federation, etc. The TMS was in effect expected to function as a subordinate body while the federation pulled the strings. There was a stipulation dealing with higher royalty that made the rent on the Tawa the highest in the State.

Taking a Stand

The Tawa Matsya Sangh refused to agree to these conditions and continued its fishing operations in the reservoir despite the stand-off. Though the federation sent a few notices, the case has been left hanging. After almost six months the TMS was offered a new lease with more or less the same terms as the previous one. It continues to manage fishing operations in the reservoir but the renewed lease expires next year and one can expect the struggle too to be renewed.

Meanwhile the Forest Department has begun to evict people from the area, declaring it an 'eco-development' zone that is part of a World Bank funded forestry project. In order to meet the bank's conditions the department is trying to appropriately rehabilitate those who have been evicted on already occupied land, claiming that the present settlements are illegal. The Tawa valley is presently witnessing a tense situation with the TMS engaged in a struggle with yet another department of the hydra-headed organisation called the State, not to mention the omniscient entity called the World Bank.

Highs and Lows

The year 1996—just before the reservoir was leased to the cooperative—was a tense period, as was the time between November 2001 and March 2002 when the Sangh had applied for a renewal and was turned down. Things are looking bleak now—there have been instances of *lathi* charges and forcible eviction by the forest department. Demonstrations and protests have become common features.

The most peaceful period were the five years when the Sangh went about its job and the State minded its own business. This was also a very productive time for the cooperative.

What the Different Actors Say

The logic for the cooperative wanting fishing rights to the reservoir is simple: as the reservoir submerged their homes, villages and fields, they should at least have the right to fish in its waters. After all, they have given the State more in terms of royalty than it ever got before. The TMS has proved itself the most efficiently managed cooperative in the State and therefore deserves the right to run the reservoir in the way it sees fit.

The State fisheries federation claims that it is more than capable of managing the reservoir with its body of specialists; the Sangh's autonomy has been a thorn in its side.

The State government is at loggerheads with the Sangh because it is more interested in playing politics. The ruling party is influenced not only by its own workers but also by the bureaucrats, contractors and traders. The stand the government takes is determined by the pressure that others are able to exert.

An Example to be Followed

This conflict, as has been seen in the past, can indeed be resolved. The TMS is an exceptional case: where often such experiments fail, it has proved itself a successful workers' cooperative. Such endeavours need to be encouraged, analysed and made an example of. The Tawa story also offers solutions pertaining to the rehabilitation of those displaced by big dams. The government and the fisheries' federation must rise above petty politics and look beyond vested interests. The federation should instead focus on promoting such cooperatives in other reservoirs in the State.

Notes

*Map 1 from SOPPECOM, Pune.
1. The Tawa Displaced Adivasis Fisheries Production and Marketing Cooperative Federation.
2. Rupees Six lakh were sanctioned as seed money for the cooperative, half as an interest-free loan and the other half as grant. It was also decided that no one would be evicted from the Bori Wildlife Sanctuary or the Satpura National Park against his or her will. Moreover, drawdown cultivation, exempted from any charges or levies for a period of five years, was to be permitted along the sides of the Tawa reservoir. Two lift irrigation schemes were also to be set up for those displaced by the Tawa project. Moreover, the encroachment (into revenue and forest land) of people displaced by the Tawa dam, the army firing range and the ordnance factory, was to be regularised by giving them *pattas* (land deeds). Many of these concessions are yet to be fulfilled.

References

Sunil and Smitha. 1996. 'Fishing in the Tawa Reservoir: Adivasis Struggle for Livelihood', *Economic and Political Weekly*, 31(14): 870–72.

Singh, Vikas. 2001. 'Madhya Pradesh: State and People's Initiatives; Experience of Tawa Matsya Sangh'. Available at *www.narmada.org*

The Tehri Dam Project:
A Saga of Shattered Dreams

Vimal Bhai

U‌TTARANCHAL IS ONE OF THE NEWER STATES IN INDIA. CONSTITUTED ON NOVEMBER 9, 2000, it has a total area of 53,484 km² out of which 34,434 km² is forest land. The total population according to the 2001 Census is 84,79,562, of which 41,63,161 are women.

The place where Tehri dam is being constructed finds mention as 'Dhanushtirth' in the *skandha puran*. The confluence of the rivers Bhagirathi and Bhilangna—Ganesh Prayag—is just 500 m from the main gate of the dam. Tehri town was the capital of Garhwal in 1815. It was then known as Trihari, or the confluence of three rivers, and was a hub of education, literature, culture and politics. In modern times it has been the centre of peoples' movements like Chipko and prohibition. It has always been on the route map of pilgrims and is the nearest market for two hundred adjoining villages.

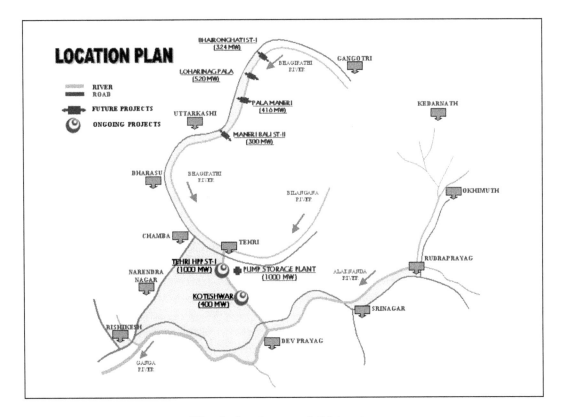

Map 1. *Location map of Tehri project.*

The main occupation is agriculture. Other sources of income are forestry, manufacture of herbal products, tourism, etc. Most of the villages in Bhagirathi valley depend on an old irrigation system. People are peaceful and the Hindu majority population consists mostly of Brahmins and Rajputs.

A Lofty Project

The Tehri Dam Project (TDP) aims to construct a 260.5 m high rock-filled dam on the Bhagirathi river. The reservoir will submerge 4,072.82 ha of forest land and 1,600 ha of agricultural land—a distance of 45 km in the Bhagirathi valley and 25 km in the Bhilangna valley.

The first phase will see the installation of a 1,000 MW power plant and a second power plant of 1,000 MW. Another smaller dam called Koteshwar dam, 103.5 m high, is under construction 22 km downstream from Tehri. This dam will also support a 400 MW power plant. The cost of construction has risen from Rs. 172 crore to more than 4,000 crore. So far Rs. 3,535.96 crore has been spent on the TDP. The government has a master plan for tourism but none for rehabilitation.

People be Damned

According to the government, the Tehri dam project will affect 5,291 urban and 9,238 rural families. About 3,810 rural families have been identified as 'partially affected' families; but the actual figures are likely to be much higher. Tehri town and thirty-nine villages (again, a number that might be revised upwards) fall in the complete submersion zone category, while the remaining eighty-six or more villages are in the partial

Fig. 1. *Tehri dam near old Tehri town.*

submergence zone. Approximately one lakh people will be affected directly. An estimated population of 80,000 on the other side of Bhagirathi–Bhilangna will lose access to basic amenities as well as to important towns.

Himalayan Blunder

There is as yet no plan in place for the people displaced by this project, which was approved by the Planning Commission in 1972. The cut-off date for project-affected families was 1976 for those living in rural areas, and 1985 for those in urban areas. Displacement started in 1978 without a rehabilitation policy. The responsibility for rehabilitation lay with the Uttar Pradesh Irrigation Department until 1989, after which it was handed over to the THDC. A half-baked programme of sorts was framed in 1995. A new rehabilitation policy was subsequently announced in 1998 that included some of the recommendations of the Dr. Hanumantha Rao Committee.

Para 3.2 of the Letter of Conditional Sanction defined rehabilitation related conditions as follows:

> The THDC will, through a reputed institution, undertake a socio-economic study of the measures needed to ensure that the standard of living of the oustees is not affected due to the project. The study will be completed by 30.6.1991. The THDC will implement such recommendations as may be made by the Ministry of Environment and Forests for rehabilitation after consideration of the study report by the MoEF. The rehabilitation package covering population affecting Koteshwar dam as well as those living on the rim of the reservoir and likely to be affected will be prepared before 31.3.1991.

Rebels with a Cause

Right from its inception, the project has faced opposition and controversy. People have protested on grounds of negative socio-economic impact, displacement, environmental hazards, seismic threat, etc. The list of people and organisations who joined the anti-dam movement reads like a roll-call. The first voice of dissent came from Queen Kamlendumati Shah of the erstwhile Garhwal State. Virendra Dutt Saklani, a prominent lawyer and freedom fighter, also questioned the logic of the project in 1965. In November 1985 he filed a suit in the Supreme Court on behalf of the Tehri Bandh Virodhi Sangharsh Samiti (Tehri Dam Protest and Struggle Committee), but the petition was rejected on technical grounds. This was a major setback that shocked many, though the 'Dam Show' carried on regardless. Organised protests were launched under the aegis of the Committee of Tehri Affected People, the Tehri Bandh Virodhi Sangharsh Samiti and the Save Himalayas Movement. The region has become the focus of national and international attention.

In 1988, environmentalist and activist Sundar Lal Bahuguna joined the battle and started working actively to mobilise public opinion and engender policy change. He linked the dam issue with the larger one of Himalayan ecology, and went on several hunger strikes, which made headline news. The movement for rehabilitation and saving the environment was further strengthened in 2001 by the participation of the Tehri Bhoomidhar Visthapit Sangthan and Matu People's Organisation.

The protest took many forms: The Save Himalayas Movement organised a cycle rally from Gangasagar to Gangotri in 1992. This 'Sea to Sky' rally made people aware of the dam's hazards and its threat to the fragile ecosystem. The administration targeted anti-dam Uttarakhand Movement activists arresting some and implicating others in false cases.

Fig. 2. *Agitation by Tehri oustees.*

Seismologists point out that the Tehri dam, which lies in a major fault zone, is likely to experience an earthquake measuring more than 8.0 on the Richter scale in the next hundred years. Incidentally, the lifetime of the dam is also estimated to be a hundred years.

Professor Vinod Gaur has questioned the design of the Tehri dam. The earthquake issue has compelled even those who are not part of the organised movement to raise their voice against this gigantic project.

After the Uttarkashi earthquake of October 18, 1992, the then Prime Minister Narasimha Rao declared that there was indeed a need to review the entire undertaking. But not much was done. In response to a public interest litigation filed by N.D. Jayal and Shekhar Singh, the Supreme Court declared on September 1, 2003:

It is made clear that the condition of *pari-passu* implementation of conditions prior to the commissioning of the project shall be closely monitored under the existing mechanism set up by MoEF and the project authorities will ensure that prior to closing of diversion tunnels T1/T2 for impoundment of the reservoir, evacuation, resettlement and rehabilitation are completed in all respects... no impoundment would be allowed until all the conditions in the Environmental Clearance Certificate of the Tehri Dam dated July 19, 1990 are complied with and stand fulfilled.

The Apex Court also transferred the case to the Uttaranchal High Court to monitor rehabilitation and ensure environmental compliance.

Nowhere to Go

The petitioners in the main case again filed an application at the Nainital High Court on December 12, 2004 seeking proper monitoring. Another petition pertaining to the cut-off area is also pending before the court. Yet another case filed on July 4, 2005 deals with the tunnel not being shut down before rehabilitation. In short, there are innumerable cases relating to the issue still awaiting conclusion in the lower courts and the High Court.

On October 29, 2005, the High Court allowed the last diversion tunnel to be closed without considering the environmental and rehabilitation issues. The old Tehri town and village up to 760 m elevation will soon be drowned. This is how actual submergence starts. It is in clear violation of the Supreme Court order dated September 1, 2005. Petitioners have therefore moved the Supreme Court.

The rehabilitation measures undertaken so far have been very poor. The new Tehri city which is part of the package is ill-suited from both the health and livelihood point of view; to top it all it is located in an active seismic zone (according to the Geological Survey of India). Assistance for house construction is grossly inadequate. Since the displaced have not been given land right deeds, people have not been able to avail of loans and other institutional facilities. This story is typical of almost all rehabilitation sites.

There are a number of villages like Khand, Bidkot, etc. that have been entirely submerged and are still classified as partially affected because the settlement itself is not part of the submergence area. How are people in such villages going to earn their living? More than one lakh people from villages like Baldogi, Kumrada and Bhaldgaon face a very uncertain future.

The District Collector confessed in March 2004 that Rs. 77 crore worth of funds meant for rehabilitation had been diverted to expenses on other heads and that the government was not in a position to pay this amount back to the THDC.

The rate of compensation has varied from village to village and sometimes even within a village. Rural markets are not part of the plan. Innumerable mistakes and inadequate land settlement have meant that many of those affected are not even being considered as 'project-affected persons'.

There has been a failure to compile proper eligibility lists in the displaced villages. In several cases either people have not been awarded land at all or if allotted, have been unable to take possession for some reason or the other; housing plots have not been awarded; agricultural land has not been considered; improper compensation for housing and failure to take shops into account—the list is endless. While the administration is in possession of all sorts of records, it is the displaced who are asked to produce records, and to run from pillar to post.

The approach roads to an alternative bridge that is to be built at Gadoliya in Bhilangna valley to serve the 'cut-off area' are nowhere in sight. The distance to the district headquarters will increase by 150 to 200 km for the people of this area. Rural rehabilitation sites like Riywala, Pashulok, Dehradun and Pathri lack basic amenities such as drinking water, electricity, health facilities and schools.

According to The Project Level Monitoring Committee (PLMC), afforestation of 6,289 ha near the dam site has failed. This was part of the Catchment Area Plan. And, 13,454 ha was only satisfactory. Till now the Government of Uttar Pradesh has not placed any plan for the command area before the MoEF.

Flooded with Recommendations

From the very beginning people have been trying to resolve the issue through dialogues with the government. Local bodies, *panchayats* and political parties have passed resolutions; rallies have been organised and public opinion mobilised all over the country; but every time flaws were pointed out some high-powered committee or the other was constituted to reject protestor's fears.

When Indira Gandhi visited Tehri in the late 1970s, she promised a task force to examine the impact on the environment. But both the S.K. Roy Committee as well as the D.R. Bhumbla Committee met with the same fate—the Comptroller and Auditor General of India termed the effort a financial burden.

Following a seventy-four day fast by Sundar Lal Bahuguna in August 1996, a five-member expert committee was constituted to study the project's safety. In September of the same year another twelve-member Expert Committee (Hanumantha Rao Committee [HRC]) came into being under the chairmanship of Dr. Hanumantha Rao to examine the environmental and rehabilitation aspects. Only 10 per cent of the committee's recommendations were finally accepted in 1998. In February 1999 it was revealed that the special group was deeply divided—the ratio being 4:1. The government accepted the suggestions pertaining to the safety of the dam design, but rejected two other recommendations.

Following the Supreme Court's decision a Project Level Monitoring Committee (PLMC) was appointed and it has submitted three reports—in November 2003, April 2004 and June 2005. All efforts made by the affected people and NGOs to meet the monitoring committee were unsuccessful.

The coordination committee on rehabilitation that came into being as a result of the HRC's recommendation has never involved the people. This has denied the displaced people an opportunity to voice their concerns through representatives. The Special Grievance Cell (also an HRC suggestion) is recognised by the government but is unknown to the project-affected people, raising suspicions over the likelihood of the displaced people getting justice. Had the inter-ministerial committee of the Central government, PLMC and the coordination committee on rehabilitation been sympathetic to some of the issues, the displaced would have found some solace.

While the government talks of paucity of resources for rehabilitation it is clear that there is a lack of political will and failure to have accountability measures in place. Officials of the rehabilitation department as well as the local administration have thus far adopted a policy of creating fear among the people or luring them to prevent them from getting organised. Needless to say corruption is rampant.

Learning from History

The broader questions raised by this project pertain to our not having learnt enough from similar experiences in the past. Public participation and transparency have to be ensured.

The Land Acquisition Act and rehabilitation policy should be scrutinised from the point of view of the environment and displaced people.

Rather than pretending to look for solutions after the problems have been magnified, efforts should be made to prevent such problems in the first place.

Important Dates and Events

1965:	K.L. Rao, the then Irrigation Minister, announces the construction of the dam at Tehri.
June 30, 1977:	All affected *gram sabhas* pass a resolution against the Tehri dam.

January 28, 1978:	Mass agitation against the dam; representatives from all political parties and affected *gram sabhas* participated in this protest.
October 26, 1986:	The task force submits its final report to the government pointing out that the construction of the dam is dangerous.
1986:	India signs an agreement with erstwhile USSR for assistance in constructing the Tehri dam.
1988:	The Tehri Hydro Development Corporation, a joint venture of the Union and State government, is set up.
1989:	Environmentalist Sunder Lal Bahuguna goes on fast on December 25, and a mass agitation takes place on December 31.
July 19, 1990:	Despite widespread apprehension, the project is cleared by the MoEF.
March 30, 2001:	Mass agitation against closure of tunnels 3 and 4; dam work halted; the *dharna* on April 24 culminates in protestors being beaten and arrested; on May 3 there is an agreement with the government.
May 2, 2001:	Chham villagers stop government officials from doing a survey.
January 6, 2002:	Villagers from Sirai are beaten and jailed for stopping trucks from ferrying mud. Their demand: land for rehabilitation.
July 7, 2002:	The Tehri Bhoomidhar Visthapit Sangthan starts a *dharna* in Old Tehri Town that lasts 900 days.
August 29, 2004:	Reservoir level rises to 649 m and Old Tehri Town is submerged. More than fifty families lose their belongings.

Note

*Map 1 from Tehri Hydro-Development Corporation; figures by author.

References

1 to 6th Document of Matu Peoples' Organisation.

Affidavits filed by the Uttaranchal government in the Supreme Court dated February 2002 and September 2002.

Affidavit filed by Tehri Hydro Development Corporation Limited (THDC) in Supreme Court of India dated March 2002.

Dr. Hanumantha Rao Committee Report, 1997.

Expert Committee Report on TDP, February 1998.

Information Department, Government of Uttaranchal.

Leaflet of Tehri Bandh Virodhi Sangharsh Samiti.

Leaflet of THDC.

Letter of the Geological Survey of India to THDC on Tehri Dam rim area, 1990.

Reports and field surveys of Matu Peoples' Organisation.

Shiv Prasad Dabral. *History of Tehri Garhwal State* (in Hindi). Parts 1 and 2.

Tehri Dam-Rehabilitation Policy 1995, THDC.

Tehri Dam-Rehabilitation Policy 1998, THDC.

The Stalled Bhilangna Micro Hydel Project: Community Protests over Infringement of Right

Pushplata Rawat and Meera Kaintura

Ideal Region for Micro-hydel Power

Tehri Garhwal district is located on the outer ranges of the mid-Himalayas in Uttaranchal (formerly part of Uttar Pradesh). The unstable hill slopes and seismological sensitivity (Zone IV) of the region has been a major concern. The district is also the melting point of snow-fed rivers like Bhagirathi, Bhilangna, Mandakini and Balganga that flow through the district.

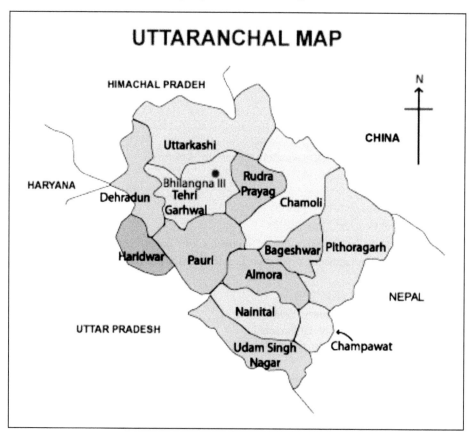

Map 1. *The location of Bhilangna.*

Considering the advantage of the sheer drops in height at many points along the river courses, there are numerous projects including the controversial Tehri dam for harvesting hydelpower in the region. One such project is being planned on the river Bhilangna that flows from the Khatling glacier at an elevation of 3,658 m. A run of the river microhydel project (2 x 11.5 MW capacity) is under construction on the river in Ghansali *tehsil* at an elevation of 1,600 m about 149 km upstream of Rishikesh, the nearest railhead.

The rural agro-pastoral economy is still predominantly subsistence-based, with about 50 per cent of rural households dependent on the village commons and forest lands. For the upland community, water shortage is a serious problem. Much of the arable land is marginal and rainfed. The perennial flow of the river Bhilangna is the major source of irrigation for the local community and existing traditional farming and water harvesting systems have evolved over a long time in response to difficult mountain terrain.

State and Company Bypass the Community

The government of Uttaranchal awarded the contract for the microhydel plant for generation of electricity to Swasti Power Engineering Pvt Ltd, New Delhi, in 1994. The project affects a total of 350 families in six villages in Bhilangna block of Tehri district.[1] Apart from the ecological threat, the villagers are affected on mainly three accounts. First, the diversion of river through tunnels will leave a segment of the river dry, so the traditional irrigation channels will not get water. This aspect has not been considered at all in the official project report (OPR). Second, the project will lead to a loss of commons and vegetation that play a critical role in meeting the fodder needs of the community, particularly women and disadvantaged sections. The extent of tree loss has also been under-reported in the OPR. And finally, the community is piqued that the *panchayat* has been given no role in the control of and decision-making on the project. They resent the entry of the outside private company and demand that the *panchayat's* right over the use of water be recognised, and that the *panchayat* be granted a share in the profits.

The conflict intensified when the company commenced the physical work in March 2004 and was faced with resistance from the affected villagers. Since then, amidst a series of protests and negotiations with the administration and company officials, the project has come to a halt and currently the State has initiated legal proceedings against protesting villagers.

Stages in the Conflict

1. During the survey carried out by the company in 1994–95, the real purpose was witheld from the community. It was only in 2001 that it became clear to the villagers that it was a hydroelectricity plant that was planned.
2. During 2001–4, the community witnessed many study teams conducting surveys. That is also when the Environmental Impact Assessment (EIA) report was made available to the community by a local NGO. This led the community to take a closer look at the project, resulting in exposure of false claims made by the company. The apprehensive villagers raised their concerns during subsequent meetings with district officials but they were ignored. The administration told the media that the community was ready for the project and that the physical work would commence in March 2004.
3. The year 2004 saw an escalation of the conflict as the State provided police protection to company employees to carry out the work, but intense opposition from the villagers put a stop to it.

4. The affected villagers organised themselves under the leadership of Dev Singh Ramola, one of the affected villagers, and formed Bhilangna Ghati Bandh Pariyojana Nijikaran Virodhi Sangathan to wage their fight against privatisation of their local resources.
5. Towards the end of 2004 and the beginning of 2005, the State intensified its authoritative measures to curb agitations by the community, and in October charged eleven people under Section 111.
6. In November 2004, a big rally was organised by Bhilangna Ghati Bandh Pariyojana Nijikaran Virodhi Sangthan. It attracted many activists and local organisations and caught the attention of the media since the people's resistance started getting support from various sources.
7. Since 2005, the conflict is at its peak, wherein the State is increasingly resorting to use of police force (including the Provincial Armed Constabulary) to help the company initiate work, along with initiating legal proceedings against community leadership.

On January 27, 2004, the company's workers, aided by the police, tried to restart work on the tunnels. They first encountered resistance from the women of Bahera village who were soon joined by the villagers of Falinda and Saruna. They staged a *dharna*,[3] forcing the company to retreat. However, after two days, additional police were sent to the project site and one of the leaders, T. S. Chauhan, was charged under Sections 107/110/151 and sent to jail in New Tehri.

A large number of people from different villages assembled at the *tehsil* office demanding the immediate release of Chauhan. The Sub-Divisional Magistrate was *gheraoed*[2] for five hours and released by the agitated crowd only after he ordered the release of Chauhan and withdrawal of the PAC force from the site. However, the confrontation between the community and the officials is intensifying without any signs of the conflict being resolved.

The Project Stands Stalled

The work at the project site has been totally stalled by the community under the leadership of the Bhilangna Ghati Bandh Pariyojana Nijikaran Virodhi Sangthan. There is a continuous *dharna* by a group of twenty-five to thirty villagers including women at the site. They have initiated a campaign against the privatisation of their river and are demanding the cancellation of the assignment to the company.

The company has backed off from its field operations in the region. Some of its equipment including bulldozers are lying abandoned at the site. Meanwhile, the state government has registered a case against the agitating community and the court recently issued summons to eleven persons to appear before it.

The community at present is also contemplating taking the issue to court, and is in discussion with some local organisation which has assured them of the requisite support in this regard.

The Three Actors

There are mainly three groups with different opinions, which has led to the conflict.

The Company

Swasti Power Engineering Pvt Ltd: the company has procured this contract from the government after following due process. They have had a meeting with the community to provide details of the project. They

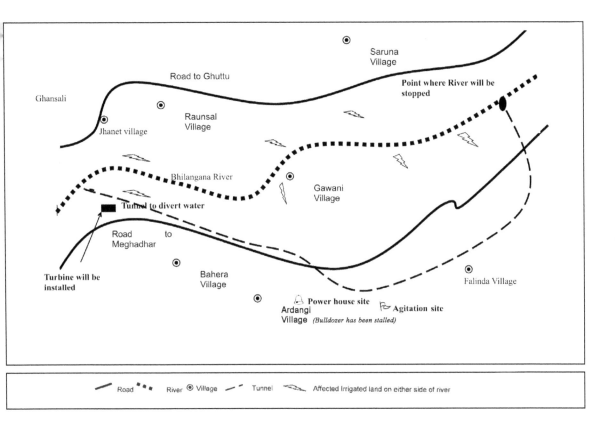

Fig. 1. *Map of the project site.*
(not to scale)

are pledging to pay compensation to the families as per the accepted guidelines. The delay in the project means a loss to them, and they are taking recourse to the State authority to provide them with all support including a police force.

The Government

The Uttaranchal government is committed to harnessing the vast hydropower potential of the region through the establishment of hydel plants and promoting private sector participation in it. The government has invited tenders for establishing of microhydro plants. Once they approve the project it is their responsibility to see that it is implemented without any obstacles.

The Community

From the start the project did not take the community into confidence. The initial survey and subsequent discussions of the company failed to demonstrate concrete benefits to the community. The project did not offer them electricity as the villages were already electrified, it offered no job opportunity to the local population, and the community's access to irrigation was threatened along with the loss of their common land. Barring trading communities and some households that would have got compensation, the entire village community felt itself at a loss. Their anxiety was further intensified when the EIA report was made available to them. They saw clearly that there were differences between the report's claims and the reality. For example, the document did not even mention the existence of their traditional *kuhls*.[4] They felt the company had tried to fool the government.

Impact of the Project

On Ecology

The area lies in seismic Zone IV and there are apprehensions about whether the company has taken due precautions in this regard. At present, an affected community organisation is compiling the information to demonstrate the actual amount of damage that has been either ignored or understated in the EIA report. For example, when the villagers—with the help of the forester—estimated the loss of vegetative cover due to the installation of the plant, they established that 557 trees would be lost as against 113 trees as reported by the company. Other vegetation like shrubs and grasses that make a significant contribution to the ecology and rural livelihood have not even been mentioned.

On Economy

At present, the irrigation needs of the affected villagers are being met by the diversion of Bhilangna water through traditional *kuhls*. The project will destroy this system. However, the report failed to even mention the existence of such irrigation channels in the vicinity of the project site.

Apart from this, the affected land has been shown as barren while according to villagers around 60 ha of the land is irrigated. This aspect has serious implications with respect to the compensation amount that needs to be paid to the community.

The project will also use the commons critical in fulfilling the community's need for fodder and fuelwood. Loss of the commons will affect the community, particularly the women and the disadvantaged whose dependence on the commons is greater.

Existing land use may not correspond with official figures and consequent tenure insecurities are a major problem in the region.

On Society

The successful struggle for Statehood has raised the aspirations of the community. In the present case, they assert that the *gram panchayat* (village assembly) should have the rights over its natural resources and should have a say in the sale of their water to any private company. They will not allow any outside private company to corner all the benefits from such deals by bypassing the *panchayat* and restricting the local community from using their own resources for their own consumption.

The community shares the State's aspiration of producing electricity; however, if this is done using local means for production they insist it should be with adequate community participation. They can set up a microhydro plant on their own the way they have done in Beef village.[5] To demonstrate that they can do things on their own, the project-affected villagers have completed a 2-km-long irrigation *kuhl* on their own.

Recognise the Local Community as a Stakeholder

The conflict has reached a very difficult stage. For a peaceful negotiation to be possible in the immediate future it is important to realise that the impasse is due to the following reasons:

- Initial miscommunication and gaps during survey, no efforts to take community into confidence.
- The entry of local activists and the consequent shift in the community's response from exclusion and betrayal to anger, and the decision not to let any private party take advantage of their resources. The failure of the State to act on the community's behalf, instead quelling community protests using coercive means. The community, which initially thought that the State had been misguided by the company, started to question the State's intentions as well.
- The first thing the government should do is to listen to the community and carry out a serious review of the claims made by the company with participation of the affected community. They are already compiling information and will be eager to share it with the officials.
- The Company should also come forward in a transparent manner, instead of putting pressure on the community through district officials. The company has already reviewed its earlier stand in response to intense opposition from the community. For example, it agrees to provide irrigation water as demanded by the community. However, the community members at present are not in a mood to heed such offers. The possibility of a compromise may exist if *panchayats* of the affected villages are made partners in the project and they get a share in the profit of the company by means of royalty. This will ensure that resources from regions inhabited by the poor are not simply robbed for the benefit of the already wealthy, and that the former become true 'partners in development'. There is experience in other places of the legal and institutional set up needed for such tripartite arrangements and the State must involve those with such expertise in the process.

Notes

*Map 1 and figure by authors.
1. The affected villages are Saruna (70 HH), Falinda (130 households [HH]), Rausal (45 HH), Jhainait (12 HH), Dabsour (50 HH) and Thayali (50 HH).
2. *Gheraoed* is a term for protest wherein the concerned official is surrounded by agitating members, thus restricting his mobility till he concedes the demands
3. *Dharna* is a form of protest wherein protesting members camp either in front of the concerned office or disputed site raising slogans.
4. *Kuhl* is a traditional irrigation canal managed and owned by the community through which water from a source is carried over long distances along the hill contours by means of gravitational force.
5. Beef village is one of the villages where a micro-hydel project is managed by the community.

Blank page

PART 7

Transboundary Water Conflicts:
A Review

Ramaswamy R. Iyer

One Part of a Vast Canvas

WATER-RELATED CONFLICTS CAN TAKE MANY FORMS, AND CAN ARISE IN RELATION TO river waters or other surface water bodies or groundwater; between different uses, areas or political units; in many different contexts; and over diverse issues. Here we deal with one part of that vast canvas, namely, transboundary conflicts over river waters. The term 'transboundary conflicts' here refers both to conflicts between States within the quasi-federal structure of India and to disputes between India and neighbouring countries.[1]

There is currently a fashionable thesis, often heard at seminars and conferences, that water-related conflicts will increase in frequency and intensity in the future and will lead to 'water wars'. That thesis has in turn been criticised as unsound and unhistorical by some thinkers.[2] 'Water wars' do seem unlikely, but as the pressure on the world's finite supplies of freshwater increases for a number of reasons, conflicts between uses, areas, groups, States or provinces, or countries may well become acute. There are already many instances of such conflicts, some resolved through treaties or agreements or adjudications, and some still simmering.

Transboundary Conflicts: Factors and Concerns

How do such conflicts emerge? Broadly speaking, and ignoring at present the deeper questions of proper water use, there are two basic factors here: first, structures on rivers such as dams or barrages that interfere with natural flows, and second, a wide divergence of perception between upper and lower riparians.

Projects as Loci of Conflicts

It is interesting to note that conflicts over river waters, whether inter-country or intra-country, seem often to arise in the context of large projects. The India–Bangladesh dispute over Ganga waters was precipitated by the Farakka Barrage Project in India. Within India, the Cauvery dispute arose because dams and reservoirs built by the State of Karnataka had the effect of reducing the flows into the Mettur reservoir in Tamil Nadu. Between Karnataka and Andhra Pradesh, there is a dispute over the formers Alamatti Project on the Krishna river. Even within a State, a large project often creates conflicts of interests (sometimes acute) between different groups or areas. It would appear that large projects tend to become the foci of conflicts. This is essentially because (*a*) they tend to alter geography and hydrological regimes, sometimes drastically; and (*b*) they involve issues of control, power and political relations, social justice and equity.

Riparian Divergences

As for the divergence between upper and lower riparians, it takes many forms and varies from case to case. However, the upper riparian generally tends to assume a primacy of rights, with only the residuary flows going to the lower riparian, and the latter tends to assert its rights to established uses and to object to any diminution in flows. The upper riparian has control over the waters and tends to exercise that power often unimaginatively and in an insensitive manner; whereas the lower riparian tends to develop an acute anxiety over the possibility or reality of control by the former, and to become excessively sensitive and touchy on this score. (Both the India–Bangladesh case and the Karnataka–Tamil Nadu case are good illustrations of this.)

The anxiety of the lower riparian becomes more acute if the upper riparian is a stronger and more powerful entity (for example, India vis-a-vis Bangladesh). However, there are also cases in which the lower riparian is more powerful and is able to impose its will to some extent on the relatively weaker upper riparians (for instance, the case of Egypt vis-a-vis Ethiopia and Sudan in relation to the Nile, though the Nile Basin Initiative, which began in 1998, seems to have ushered in a less adversarial relationship).[3]

Environmental Concerns

Apart from affecting the availability of water, the reduction of flows because of upstream dams or barrages or even because of heavy upstream water use can also have serious environmental/ecological impacts in the downstream areas within a country or beyond its borders. This was an important component in Bangladesh's complaints about the Farakka barrage, and is now a part of its objections to India's river-linking project. Environmental concerns lie at the heart of the dispute between Hungary and Slovakia over the half-completed Gabcikovo-Nagymaros Project on the Danube.[4]

Security Concerns

Lower riparian concerns about water needs can grow into serious worries about 'water security'. Given the importance of water for life support and livelihoods, a sense of insecurity in relation to water can loom very large in the consciousness of the lower riparian and become visceral. In the past, Bangladesh has felt this sense of insecurity vis-a-vis India in relation to Ganga waters,[5] and Tamil Nadu experiences it in relation to the Cauvery. India's river-linking project has given rise to acute 'water security' apprehensions in Bangladesh.

Security concerns about water can go beyond worries about availability, and become apprehensions of deliberate harm or damage, or the use of water as a weapon of war. The lower riparian can entertain fears about the upper riparian drying up the river and denying water to the lower riparian, or holding back the waters and then releasing them suddenly and in large quantities with a view to causing harm to the lower riparian through heavy flooding. Such fears may be unfounded in many cases, but they are often keenly felt. Fears of this kind lie behind Pakistan's objections to some of the proposed Indian projects on the Jhelum and the Chenab rivers in the Indus system.

Security concerns in the military sense are not the subject of this paper. It is of course possible to use a natural resource or an environmental factor as weapon of war or an instrument of political pressure. To put forward some fanciful hypotheses, a country may try to poison the water source of another; or bomb a dam in an enemy country and cause devastation and so on. These are straightforward security concerns in the conventional sense; the fact that natural resources or environmental factors, and not the usual weapons, are used as the means of inflicting damage is merely incidental. Again, where a river runs along or marks the boundary between two countries, there may be security concerns relating to that border similar to those relating to a land border. These are not the kinds of things that we are concerned with here; what we have in mind is the possibility of conflicts arising over the use—domestic, municipal, agricultural, industrial—of scarce natural resources, and over the consequences of such use.

Principles for Transboundary Conflicts

What are the principles available for obviating or resolving such disputes or conflicts? Neither the Harmon Doctrine (that of territorial sovereignty, interpreted to include sovereignty over the waters that flow through the territory), nor that of prescriptive rights (rights to historic flows) or prior appropriation, has found general approval. What has commanded a fair degree of international acceptance is the principle of equitable sharing for beneficial uses. That was the language of the old Helsinki Rules (the report of the Committee on the Uses of the Waters of International Rivers adopted by the International Law Association at the Fifty-Second Conference held at Helsinki in August 1966). The present UN Convention on the Non-Navigational Uses of International Water Courses (passed by the General Assembly in 1997, but still awaiting ratification by the stipulated number of countries) requires the watercourse states to 'utilise an international watercourse in an equitable and reasonable manner' (Art. 5, cl.1). Again, the next clause requires the watercourse States to 'participate in the use, development and protection of an international watercourse in an equitable and reasonable manner' (ibid.). What is 'equitable' has of course to be determined with reference to many criteria, and there is enormous scope for differences here. But there is some merit in a general subscription to the principles of equity and reasonableness. Similarly, a general admonition to the upper riparian on the question of causing harm to the lower riparian is unexceptionable, though the wording has changed from 'substantial harm' in the Helsinki Rules to 'significant' adverse effects in the UN Convention. The point is that if countries straddled by river systems that cross or run along boundaries wish to avoid conflicts over the use of the waters, there are enough principles and guidelines to go by. To put it in a nutshell (even at the risk of sounding platitudinous), the upper riparian, in exercising its powers of control over waters, must avoid causing harm to, or infringing the rights of, the lower riparian; and the lower riparian, in asserting its rights over the waters, must not be oblivious to the needs and interests of the upper riparian or seek to impose unreasonable restrictions on legitimate activities. Given that understanding, conflicts will either not arise at all or can be resolved without much difficulty when they do. Unfortunately, understanding of that kind often proves elusive, and is particularly true in this part of the world. Rivers here are parts of culture, history and even religion, and tend to evoke strong emotions in the minds of the people on either side of a political boundary (whether between provinces or states or countries), making rationality difficult. Moreover, disputes over river waters quickly become 'politicised', i.e. enmeshed in party and electoral politics, and this makes them extraordinarily difficult to resolve.

Water and Politics

There is a complex interaction between water issues and political relations between countries, or even between political units within the same country. It is not always a case of conflicts over water resources leading to a worsening of political relations, though that does happen on occasion; it is more often a case of a difficult political relationship rendering the water issue more intractable. This is particularly so when other issues become prominent from time to time. Water issues, in turn, can become the most dominant factor at certain times, and can have a decisive impact on the general political relationship.

The word 'politicisation' was used earlier, but a word of explanation is necessary. Water disputes are inevitably political and cannot be removed from the domain of politics. However, they become 'politicised' in a negative sense when considerations of party politics and of impacts on elections bring in adversarial attitudes and come in the way of a constructive and cooperative approach to the management of shared natural resources. The Ganga waters issue became heavily politicised in this sense (more in Bangladesh than in India); politicisation not only made the Cauvery dispute more difficult, but it also seriously impaired the working of the constitutional conflict resolution mechanism, and even led temporarily to a defiance of the Supreme Court. Politicisation also plays a part in the intractability of the differences that emerge under the seemingly successful operation of the Indus Treaty between India and Pakistan.[6] Again, if the Mahakali Treaty between India and Nepal has become virtually inoperative, the cause lies less in water-related issues than in the complexities of the political relations between the two countries.[7] (This article does not go into the 'big-country/small country' complexities in the India–Bangladesh and India–Nepal relations that tend to cloud the water issue.)

Track II Initiatives

There is, however, one positive factor that must be mentioned here, namely the constructive role played by non-official or 'Track II' initiatives in the processes that led to the Ganges Treaty of 1996 between India and Bangladesh. A similar non-official initiative has been undertaken by the Madras Institute of Development Studies to bring the farmers of Karnataka and Tamil Nadu together to promote understanding and find a way out of the impasse on the Cauvery dispute. It has generated much goodwill and helped to remove or moderate misperceptions and misunderstandings on either side, though no specific water sharing proposition or formula for difficult years has emerged from this process as yet. Such initiatives help to counter the harmful effects of politicisation to some extent.

Regional Cooperation

In the context of inter-country river water disputes, regional cooperation is often strongly urged. There is much force in that advocacy, but it tends to become doctrinaire. There are some problems and issues that are best dealt with on a national or local basis; some that call for cooperation between two countries or units; and others that demand a regional approach. The circumstances vary from case to case, and in each case the most appropriate route needs to be followed. What is called for is pragmatism rather than doctrine.[8]

Inter-State and Inter-Country Conflicts

A comment often made is that it has been easier for India to resolve issues and enter into treaties with other countries (Bangladesh, Pakistan) despite uneasy political relationships, than to resolve inter-State river water disputes within the country. This apparent paradox is easily explained. The Indus Treaty was signed at a fairly early stage in the history of India–Pakistan relations; it might have been much more difficult to enter into such a treaty after the 1971 war in the east or the worsening of the Kashmir issue and the escalation of violence in the late 1980s. Besides, mediation by the World Bank also played a role. In the Ganga waters case, there were new governments in both Delhi and Dhaka in 1995, and they were determined to improve political relations between the two countries. The perceived importance of better India–Bangladesh relations tended to override the seemingly intractable differences over watersharing. No such positive extraneous (i.e. non-water) factors operate in the domestic context to mitigate the acuteness of the water conflict; on the other hand, negative forces (namely those of party and electoral politics) do operate and make the conflict more difficult to resolve.

Integration and Segmentation

When hydrological and ecological unities are cut across by political divisions, whether within a country or between countries, there are three possibilities: (*a*) the joint, cooperative, integrated planning and management of the river as a system; (*b*) separate actions by the different political entities involved but with some degree of institutionalised coordination; (*c*) separate, segmented actions by the different entities without any integration or coordination. None of the cases referred to in this paper falls into the first or even the second category. The division of the river system into two segments in the Indus case is an extreme example of segmentation or surgery, but the 'sharing' in the Ganges case and the allocations to different States by tribunals under Indian law are not very different: they are all instances of segmentation. Each political entity (country or State or province) is given a certain allocation of waters and left free to do what it deems fit with it. That may not seem the ideal thing to do, but two points must be noted. First, a treaty or an agreement or even an adjudication is better than dispute and discord, and if the ideal is not feasible, then the second-best solution has to be accepted. Second, there are dangers even in the 'integrated' approach. The general recommendation is 'integrated river basin management', as the basin as a whole forms a natural hydrological unit.[9] That seems unexceptionable, but the hidden danger here is that such an approach carries with it a bias in favour of a centralised, technology-driven planning of big projects. That bias is of course not inevitable and can be guarded against, but we need to be aware of that possibility.

River Basin Organisations

Subject to that caution, some degree of integration or at least coordination at a basin or sub-basin level seems desirable to obviate or resolve conflicts, and this calls for an institutional arrangement or an organisation. It is curious that while such an institutional arrangement exists in international cases—the (India–Pakistan) Indus Commission and the India–Bangladesh Joint Rivers Commission—no such arrangements exist within India.[10] There is strong resistance to the idea of river basin Organisations (RBOs) on the part of the State governments.

That resistance has rendered the River Boards Act 1956 (RBA) a dead letter. The RBA, enacted under Entry 56 of the Union List, provided for the establishment of river boards with a wide range of functions. It also included provisions for conflict resolution through arbitration. Arbitration under the RBA and adjudication under the Inter-State Water Disputes Act 1956 (ISWDA) were mutually exclusive courses.[11] However, for arbitration under the RBA the prerequisite was the existence of a river board established under the Act for the river in question. No such board has been established, largely because no state government was in favour of such a course. One must hope that at some future date a more enlightened attitude will come to prevail.

Adjudication: Some Limitations

The ISWDA has of course been active, and several tribunals have been set up under it. The adjudication process seemed to be working well, and the Krishna, Godavari and Narmada tribunals can be regarded as successful instances of conflict resolution; but the process ran into difficulties with the Ravi–Beas and Cauvery cases. Apart from the procedural and operational difficulties for which solutions have been attempted through amendments to the Act in 2002, there are certain substantive points, out of which three may be noted here.

1. Adjudication may not be the best way of resolving a dispute. However, the provisions for adjudication do not rule out other means such as negotiations, conciliation or mediation; they merely offer a last-resort mechanism for the settlement of disputes when other means have failed, and such a last-resort mechanism is necessary.
2. It is often argued that adjudication promotes maximal claims on either side, and becomes an adversarial and divisive process. That could indeed happen but is not an unavoidable characteristic of adjudication. It should be possible to go through an adjudication process in a non-divisive, non-adversarial spirit. If such a spirit is absent, other routes such as negotiation, conciliation, etc. would also run into difficulties.
3. A more serious, and perverse, consequence of the adjudication process is that the States concerned might tend to build dams and barrages not necessarily because they are needed but with a view to showing that they are utilising the shares allocated to them, or creating vested rights with an eye on future reviews of the award. (There are some actual examples, but they are not being identified here.)

However, it seems clear that despite difficulties and deficiencies the adjudication of inter-State river water disputes is a necessary last-resort option, and that is what Article 262 of the Constitution and the ISWDA provide. These provisions are important features of Indian federalism in relation to water.

Root Cause of Conflicts

Conflicts over water often arise because of claims and counter-claims. In the Indus waters case, each side (India, Pakistan) wants more water than the Indus Treaty gives it. In the Ganga case, before the Ganges Treaty was signed, each side (India, Bangladesh) used to lay claim to the totality of the flows in the river in the crucial period (the leanest part of the lean season). In the Ravi–Beas case, Punjab feels that its water is being taken away by others, but Haryana and Rajasthan feel that their allocations are under threat. In the Cauvery case, both Karnataka and Tamil Nadu want a larger share of the water. In the attempts at conflict resolution, the stated claims are taken and some kind of a compromise is worked out. But how much water do the parties in question really need?

That is not an easy question to answer, but it can be said without fear of contradiction that there is substantial mismanagement of water by all parties in all these cases. Consider the incidence of waterlogging and salinity in the Indus basin. It is massive in Pakistan and is a major national problem.[12] Remedial measures are being implemented. In the Indian part of the Indus basin—in Punjab and Haryana—the same problem is being experienced, and valuable agricultural land is going out of use. We have even managed to create waterlogging and salinity conditions in the desert state of Rajasthan. This is surely evidence of mismanagement of water and land. With better water management it seems probable that each of the contending parties can make do with much less water than it thinks it needs.

Another point is that supply creates demand and necessitates more supply. The availability of irrigation water leads to the adoption of water-intensive cropping patterns (for example, paddy in Punjab where it was unknown earlier, multiple crops of paddy in the Tanjavur delta in Tamil Nadu, sugarcane in Mandya in Karnataka). More water is needed to continue with this kind of agriculture; and of course, things cannot stand still: there is a desire to expand that agriculture, creating a demand for still more water, until the demand becomes unsustainable. There is always a demand for more water and still more water. So, Karnataka and Tamil Nadu fight over the Cauvery, and Punjab terminates all water accords.

But where will this 'more water' come from? It has to be brought from somewhere. So big dams, canals and long-distant water transfers are planned. These will in turn generate new conflicts. It is clear, then, that what lies at the heart of water conflicts is 'greed' (to borrow a word from Mahatma Gandhi). Agreements, accords and treaties may temporarily bring peace, but the conflict will erupt again unless we learn to redefine 'development'. However, that is a much larger theme and will call for a separate paper.

Notes

1. This article has drawn to some extent on other writings of the author including:

    Water: Perspectives, Issues, Concerns. Sage Publications, New Delhi, Thousand Oaks, London, 2003.

    'Water and Security: A Re-examination' in Environment, Development and Human Security — Perspectives from South Asia', ed. Adil Najam, University Press of America, Inc., Lanham, New York, Oxford, 2003.

    'Water: India's Relations with its Neighbours', Chapter 16 in India and Central Asia: Advancing the Common Interest, ed. K. Santanam and Ramakant Dwivedi, Institute of Defence Studies and Analyses, New Delhi, and Anamaya Publishers, New Delhi, 2004.

    'Punjab Water Imbroglio', *The Hindu*, 26 July, 2004.

    'Punjab Water Imbroglio: Background, Implications and the Way Out', *Economic and Political Weekly*, July 31, 2004.

    'Cauvery: Disappointments, Appeals', *The Hindu*, November 9, 2004

    'Indus Treaty: A Different View', *Economic and Political Weekly*, July 16, 2005.

2. See for instance Postel, Sandra L. and Aaron T. Wolf. 2001. 'Dehydrating Conflict', Foreign Policy September–October. pp. 2–9.

3. See Global Policy Forum, 'Nile River Politics: Who Receives Water?', August 10, 2000; 'Smoothing the Waters: the Nile Conflict', by Robert O. Collins, Institute on Global Conflict and Cooperation, IGCC Policy Briefs, University of California, Multi-Campus Research Unit, 1999; 'Averting Conflict in the Nile Basin', The New Courier No.3, October 2003, UNESCO; 'The Nile Water Conflicts', Science in Africa, (Online Science Magazine), May 2003.

4. See The Danube: Environmental Monitoring of an International River, Libor Jansky, Masahiro Murakami and Nevelina I. Pachoca, United Nations University Press, Tokyo, New York, Paris, 2004.

5. See the author's paper 'Water and Security: A Reexamination' in Environment, Development and Human Security—Perspectives from South Asia, ed. Adil Najam, University Press of America, Inc., Lanham, New York, Oxford, 2003.

6. The interesting question of why divergences tend to arise under the Indus Treaty and why they become intractable is explored in the author's article 'Indus Treaty: A Different View' in *Economic and Political Weekly*, July 16, 2005.

7. See the author's 'Delay and Drift on the Mahakali', Himal, Kathmandu, June 2001, in response to an article by Ajaya Dixit and Dipak Gyawali in the preceding issue.

8. To mention one instance, the Ganges Water-Sharing Treaty between India and Bangladesh, signed on December 12, 1996, took only a few months to negotiate. Politically, it was a highly sensitive matter, and quick negotiation was essential; and it was important that it should be in place by January 1, 1997 when the 1997 lean season would begin. If a 'regional' approach had been adopted and four countries had sat at the negotiating table, one wonders how long it would have taken to reach finality, and whether it would have been reached at all.

9. A 'basin' may be a hydrological concept but it is not necessarily an ecological one. The basin approach is better than the discrete planning of isolated projects, but it may tend to focus exclusively on the river and ignore everything else: groundwater aquifers, land, the ecological system of which the river is a part.

10. The Narmada Control Authority is concerned with specific projects and is not really an RBO. The Cauvery River Authority carries a misleading name; it is also not an RBO; it is merely a high-level political committee for defusing crises.

11. Incidentally, there is no reason to believe that arbitration under the RBA would have been significantly different procedurally or substantively from adjudication under the ISWDA

12. *The Politics of Managing Water*, ed. Kaiser Bengali, Sustainable Development Policy Institute, Islamabad, and Oxford University Press, 2003, p. 17.

The Jhanjhavati Medium Irrigation Project: Inter-state Conflict Due to Submergence

R.V. Rama Mohan

Origin of a Conflict

THE JHANJHAVATI MEDIUM IRRIGATION PROJECT IS BEING CONSTRUCTED ON THE Jhanjhavati river in Vizianagaram district, Andhra Pradesh. Jhanjhavati—a perennial tributary of Nagavali river—originates in the Eastern Ghats in Koraput district in Orissa. The river enters Andhra Pradesh 15 km north of Parvathipuram and meanders for 5 km till it meets the Nagavali.

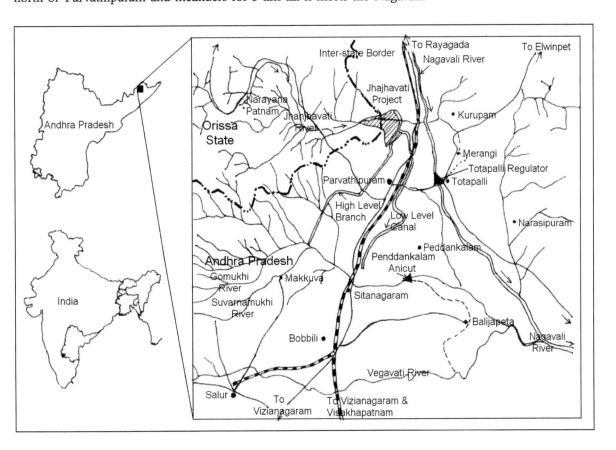

Map 1. *Location of Jhanjhavati medium irrigation project.*

The project is located at Banjukuppa village in Komaradamandal, abutting the Andhra Pradesh–Orissa border. The reservoir is located about 13 km north of Parvathipuram town and 5 km from Gangareguvalasa village on the Parvathipuram–Rayagada Road. Work on the project commenced in 1976; the idea was to utilise Andhra Pradesh's 113 Mm^3 share of the 226 Mm^3 of dependable water estimated to be available in the river basin. The total catchment area at the reservoir is 842 km^2, out of which 31 km^2 is in Andhra Pradesh and 811 km^2 in Orissa. Most of the catchment area lies within a hilly and forested terrain.

The project primarily consists of an earth dam, 3.515 km in length. A spillway regulator with six vents is located on the left and the head sluice on the right flank of the river. Two canals originate from the head sluice—a low level canal 32 km long and a 30 km branch canal traversing a higher contour. Each of these canals is intended to irrigate 4,869 ha, thereby covering 9,738 ha in seventy-five villages in five *mandals* (Komarada, Parvathipuram, Seethanagaram, Makkuva and Garugubilli) of Vizianagaram district.

Table 1 highlights the 1976 plan of the Jhanjhavati Medium Irrigation Project.

Table 1: Details of the Jhanjhavati Medium Irrigation Project

Capacity of reservoir (Mm^3)	Catchment area (km^2)	Maximum flood discharge (cusecs)	Level in metres (above mean sea level)		
			Top of bund level (TBL)	Full reservoir level (FRL)	Spillway crest level
Gross/Net 96.38/78.58	842.00	1,20,000	+149.20	+146.20	+137.20

A controversy on the area of submergence in Orissa cropped up in 1979, and has continued to remain an unresolved issue between the two states. The project cost, estimated at 15.51 crore rupees in 1976, has escalated to Rs. 120 crore in 2004. A total expenditure of 44 crore has been incurred in constructing various components of the project—earth dam, canals, surplus regulator and sluice; 8.28 crore was allocated for the project in 2004–5.

The project is set to displace approximately 1,355 predominantly tribal families in six villages of Andhra Pradesh and twelve villages of Orissa. In 1979, the area that was likely to be submerged was assessed at 971.25 ha, of which 279 ha is in Orissa. But in 2002, a joint survey by both the states revised this figure; the total area likely to go under water in Orissa stood at 464 ha. In total, 791 ha of agricultural land is likely to be affected in Andhra Pradesh and Orissa. Table 2 lists the villages that are likely to disappear as a result of this project.

Table 2: Villages Slated to Disappear with the Project

Andhra Pradesh	**Kambavalasa Panchayat**
	Rajyalakshmipuram, Kambavalasa, Lakshmipeta and Pelliguddi
	Chinakerjala Panchayat
	Banjakuppa and Boorguvalasa
Orissa	**Kappalavada Panchayat**
	Jaikota, Cheekatilova, Bothadapalli, Vepavalasa, Nagullavalasa, Tikarpadu, ChinnabankidiChekkavalasa and Eguvabothadapalli
	Alamanda Panchayat
	Bankidi, Athigada and Alamanda

Tug of War

Construction at the Jhanjhavati project started in 1978, but has remained incomplete till date for two reasons. One, as mentioned earlier, is the threat of large-scale submergence in Orissa, and the other, an inter-State border dispute. Between 1978 and 2004, a number of meetings were held at various levels—both administrative and political—between Andhra Pradesh and Orissa. Joint surveys and field investigations have also been carried out to establish the exact extent of submergence and to try and resolve the border conflict.

The project was first mooted by Andhra Pradesh in 1976 in order to utilise the 113 Mm^3 of water that was its share. However, objections pertaining to submergence cropped up almost immediately, and in 1978 Andhra Pradesh modified the design by providing a gated spillway. This was supposed to reduce the extent of affected area in Orissa from 852 ha to 280 ha. Since 1980, however, Orissa has consistently opposed the method of calculation and this figure was revised three times between 1980 and 2002. Finally, a joint survey overseen by the Central Water Commission (CWC) established that potentially 464 ha stood to be submerged in Orissa.

The dispute over the inter-State border has also been a long-pending issue between both the states. Andhra Pradesh favours the contours depicted in the 1944 Parvathipuram *taluk* map, but Orissa is not happy with this boundary and has tended to link the problems with the project to this larger dispute. It is important to mention that 2,530 ha of forest area is part of the contested land. In 2002, Andhra Pradesh set aside its claim to five border blocks equalling 2,530 ha to Orissa in order to resolve the border issue.

Chronology of Events

1976:   Andhra Pradesh moots the Jhanjhavati Medium Irrigation Project to utilise the 113 Mm^3 water of the Jhanjhavati river.

1979–80:   Andhra Pradesh modifies the project design in consultation with Orissa so as to reduce the submergence area in that state from 842 ha to 280 ha, and also compensates those affected by the project.

July 1980: Andhra Pradesh and Orissa agree that the project should proceed as planned, provided gaps are closed so that water does not enter Orissa until the extent of submergence is properly determined.

1990:   Survey of India (SoI) estimates submergence area in Orissa at 397 ha.

1994:	The Chief Ministers of both States resolve to sort out the issue at the earliest.
1997:	Studies in Andhra Pradesh estimate area of submergence in Orissa at 398 ha.
1997–98:	Joint survey by revenue and forest officials reveals that 2,530 ha of forest land are part of disputed territory.
July 1999:	The Chief Secretaries of both States meet at Hyderabad and recommend a joint survey of backwater effect in collaboration with CWC. Andhra Pradesh expresses its willingness to bear the entire cost of rehabilitation in Orissa.
2002:	The joint survey arrives at a submergence estimate of 464 ha (in Orissa).
2004:	An Andhra Pradesh delegation comprising two ministers visits Bhubaneswar to pursue the matter with Orissa.

Slow Progress

While both the problems that have slowed down construction at Jhanjhavati have ostensibly been resolved, Orissa has still not given its final nod of approval. The State is, however, expected to conduct *gram sabhas* in villages that are likely to be submerged in order to work out a rehabilitation package. The current status of the project is:

From 2003, Andhra Pradesh has been working towards a temporary plan to irrigate 2,767 ha at a total cost of 3.4 crore until the dispute with Orissa is sorted out. This plan involves raising a temporary rubber dam across the river gap to a height of +121.00 m so as to avert submergence in either state. By deflating the rubber dam, water can be diverted to a low-level canal that flows into agricultural lands because of gravity.

The two rehabilitation sites set aside for project-affected people lie vacant today, except for four or five families. These sites—one in Kummari Kunta and another in Kotipam—do not have even basic amenities like potable water or roads. Most of the people in the five tribal villages have lost the *patta* papers that prove ownership of land allotted to them at the new site.

Villagers' lives were not affected despite the gap in the river not being closed. The river continues to flow its course and allows people to carry on with their lives. Therefore villagers on both sides of the border have continued with farming activities even as they have been deriving the economic benefit of a second occupation—forestry.

The livelihood of the tribal people is entirely dependent on centres like Komarada and Parvathipuram, both of which are in Andhra Pradesh. They trek several kilometres to reach the markets in these villages to sell forest products and firewood, and buy or exchange foodgrain and groceries.

Highs and Lows

It was between 1976 and 1980 that Orissa first raised objection to the project. At the height of the controversy in 1980, though matters remained unresolved, both States agreed that Andhra Pradesh should proceed with construction provided it was careful about the gap in the river that could potentially lead to the submergence of villages in Orissa. This vital understanding between the two parties has allowed the project to remain on track.

The period from 1981 to 1999 witnessed several rounds of negotiations between the two States. Orissa continued to remain unhappy with the calculations and estimates provided by Andhra Pradesh. Consequent

Fig. 1. *Rehabilitation site in Kummari Kunta.*

Fig. 2. *Elderly villagers of Jaikota: a small village close to the Andhra Pradesh–Orissa border.*

to several meetings and mutual agreements, the two States conducted joint surveys that involved the Survey of India and the CWC. Every such attempt only added more area to the previously established estimates.

The Chief Secretaries' meeting in Hyderabad in July 1999 was a significant move towards resolving the inter-State conflict. The collaboration with CWC was a result of this meeting and the dialogue between the two high-ranking government officials underlined their effort to seek an amicable solution to the border problem so that hurdles in the Jhanjhavati project could be cleared as soon as possible.

Hope for the Future

The Jhanjhavati project is primarily meant to benefit the people of Andhra Pradesh. Orissa has nothing to gain from the programme, and because it involves submergence of land and displacement of people in that State it is essential to get Orissa's go ahead. Therefore, the onus is on the government of Andhra Pradesh to convince its counterpart across the border. Andhra Pradesh must not only provide all the information Orissa needs but also ensure that the State's concerns are addressed appropriately, and to the satisfaction of the government and the people. In this context, there seems to have been a lack of 'continuous and concerted' effort by the government of Andhra Pradesh—failure to resolve the issue in three decades points to a lack of political will.

Also, it is apparent from Orissa's requests for more information and data from time to time that Andhra Pradesh was not comprehensive or entirely accurate in its response. It thus took several decades to establish the area of submergence in Orissa. While Andhra Pradesh initially claimed that only 296 ha would go under water, the joint survey overseen by the CWC showed that the figure was considerably higher at 464 ha.

While it is hard to ascribe political motives for the delay, one has to admit that both sides remained complacent to a considerable degree. A key milestone was achieved in 2002 when they finally agreed on the extent of submergence and also managed to amicably settle the inter-State border issue. It is now up to Andhra Pradesh to persevere and pursue the matter with Orissa, at the Chief Ministers' level if necessary.

Notes

*Map 1 and figures by author; Table 1 from *Annual Report 2003–04*: Department of Irrigation and Command Area Development, Government of Andhra Pradesh; Table 2 prepared by author.

Telugu Ganga Project:
Water Rights and Conflicts

R. Ramadevi and V. Balaraju Nikku

In 1976, AFTER PROLONGED POLITICAL LOBBYING, TAMIL NADU SUCCEEDED IN CONVINCING the three riparian States of river Krishna, i.e. Maharashtra, Karnataka, and Andhra Pradesh, to contribute 142 Mm^3 each to meet the drinking water needs of Chennai. The Telugu Ganga Project (TGP) envisages diverting water from the Srisailam reservoir through an open unlined canal to the Somasila reservoir built across the Pennar; another canal connects the Somasila reservoir to the Kandaleru reservoir across the valley; from Kandaleru another canal finally carries the water into the Poondi reservoir in Tamil Nadu which is the storage point for supply to Chennai. From Srisailam to Poondi, the water travels more than 400 km traversing two river basins and states.

The involvement of different stakeholders and States in the TGP probably makes it the most conflict-ridden, redesigned river diversion project in contemporary India.[1] The project proposes utilising about 790 Mm^3 from the Krishna and 850 Mm^3 from the Pennar river to irrigate 2.3 lakh ha in Kurnool, Kadapah and Chittoor districts in the Rayalaseema region, and Nellore district in the coastal region of Andhra Pradesh,[2] in addition to supplying 425 Mm^3 to Chennai. The multi-layered conflict in the case of TGP demonstrates how water conflicts at the basin level influence local priorities and needs. We present the consultation process at local, regional and State levels and draw lessons for the management of water conflicts.

Conflicts over Design

Chennai has been facing severe drinking water problems due to the absence of perennial water sources. In its search for viable options, Tamil Nadu considered the transfer of water from the Krishna basin. Despite the fact that no part of Tamil Nadu lies in the Krishna basin, Maharashtra, Karnataka, and Andhra Pradesh agreed to give 142 Mm^3 each out of their respective shares of Krishna waters; it was hailed as a major example of inter-State cooperation.[3]

Cooperation soon turned into conflict when the Telugu Desam, then a new regional party, came to power in January 1983 in Andhra Pradesh. The new Chief Minister expanded the scope of the project to cover irrigation in the State, gave it the name Telugu Ganga, and started building a much larger canal than was needed to carry 425 Mm^3 to Chennai. The upper riparian States Karnataka and Maharashtra objected strongly to this as a deviation from the inter-State agreement on drinking water for Chennai, and an attempt was made by Andhra Pradesh to establish a pre-emptive right on Krishna waters in preparation for the ensuing review of the Krishna Tribunal's Award. As a result, the TGP was not included in the plan by the Planning Commission and was financed by Andhra Pradesh as a non-plan project.[4]

The water was supposed to reach the borders of Andhra Pradesh and Tamil Nadu on September 29, 1996, but economic and political factors have led to a serious delay in the project. Farmers all along the

Fig. 1. *Telugu Ganga canal.*

canal have been resisting the supply of canal water to Chennai. Instances of illegal tapping and diversion of water for other purposes have also been recorded. The project work has slowed down especially over the last ten years due to lack of funds. The present government of Andhra Pradesh took interest in the project and released Rs. 80 crore to speed up the completion; the project was given top priority in the Andhra Pradesh State Budget 2005–6, and was allocated Rs. 567 crore. The bottomline is that to date, the promise to supply water to Chennai remains unfulfilled.[5]

Conflicts over Storage Levels

Supplying drinking water to Chennai is feasible only when the water level touches 256 m at the Srisailam reservoir. If the water level is 267 m, it is possible to supply irrigation water as well. The total storage of the dam is 270 m. However, due to variation in rainfall, construction of new projects on the Krishna in the upper reaches, the actual storage level in Srisailam has been 256 m during the last few years. At these levels, it is possible to release some water for Chennai, but not for irrigation. The farmers, however, demand their share of irrigation water before any water is released for Chennai.

Conflict over Hydropower GO

The Srisailam Hydroelectric Project across the river Krishna is an engineering marvel. However, power can be generated only when the storage level is 260 m or more; the priority is drinking water and irrigation. But the Telugu Desam government issued a government order (GO) for power to be produced at a level of 250 m or more in order to increase power generation in the State. Upstream farmers claim that the GO is

Fig. 2. *Velugodu Balancing Reservoir.*

an excuse to release water for the Nagarjuna Sagar project and that the power generation is only a ruse. They argue that water released at 250 m only benefits coastal Andhra and Telangana at the cost of Rayalaseema farmers. Supported by political leaders from Rayalaseema, they took out processions, sat on *dharna*, and blocked traffic in August and September 2003.

The Case of the Panduru Tank Users

The Panduru tank case shows how rural food production is in conflict with urban drinking water needs and how it has become part of the larger Telugu Ganga issue. The Panduru tank supplies irrigation water for 400 ha in four revenue villages of Chittoor district. One of these, village Yanadivettu, irrigates 60 ha of wet paddy from the tank water. Yanadivettu is a small hamlet in Varadapalem *mandal;* there are 150 families of Chowdarys, a forward-caste community. The Panduru tank has been the main source of irrigation for many years for the three villages served by the stream Ralla *vagu* (riverlet). The small anicut across the *vagu* was broken and needed repairs. In 2001, the farmer leaders from the four villages made representations to government departments as well as to the local MLA. The villagers say that no action has yet been taken. Since the anicut was damaged, water has not been flowing into the tank. Last year the tank served only half the *ayacut*. In 2004, not even one ha was sown since there was no water in the tank. The villagers held a *dharna* and blocked traffic to draw attention to the issue, put pressure on the government to respond to their demand and construct a small anicut on Ralla *vagu*.

They also demanded the construction of a sluice so that water from the Telugu Ganga canal could flow into the tank. In view of the impending elections in 2004, the local politician took special interest and the sluice for the tank was sanctioned and constructed. A small amount of water was also channeled into the tank in March and farmers used that to provide one watering to the standing groundnut crop near the tank. The anicut they had demanded is yet to be built. The farmers say that they will keep agitating until the anicut is built. If they succeed, they are very sure that they will have two sources of water for the tank; that way they can grow the paddy wet crop with the anicut water and *rabi* crop with TGP canal water.

The excavation of 2L major (second distributory on the left side) on 9th branch canal is good news for the village since it means that they will also have access to TGP water. The proposed length of the major is 11 km with an estimated command of 1,203 ha of irrigated dry crop. If it is completed soon, another 20 ha in the village can also be utilised for irrigated dry agriculture. But farmers are not very sure when the canal excavation will be complete.

The example of Yanadivettu village suggests that local farmers perceive that they have a right to access the Telugu Ganga canal water. They do not object to the supply of water to Chennai, but insist that their fields be supplied water first. With the completion of branch canals, majors and minors, the demands for irrigation water will increase manyfold. Even now, farmers next to the canals irrigate their lands by siphons or diesel pumps. The practice of pumping water has been legitimised by local political leaders. These practices too are bound to continue in future. Given all these factors, the total available water in TGP too will have to increase proportionately if water supply to Chennai is to be ensured.

The Way Ahead

The 425 Mm^3 water committed to Chennai several years ago is yet to reach the city. Government records show that the maximum water released in TGP since 1996 was barely about 200 Mm^3 in the year 2000–1. The first year release to Chennai was only 5 Mm^3. The water sharing issue is intermittently dormant but comes to the fore during state elections.

The Tamil Nadu government is demanding its right to the 425 Mm^3 committed as drinking water to Chennai. The farmers from the command area do not dispute the drinking water supply but are demanding their share of irrigation water. The resolution of conflict by administrative, judicial or political means might not be permanent, as new conflicts will arise. Basin water resources do not follow administrative boundaries. Therefore, establishment of water rights is a more durable, politically tough and legally challenging option to solve the major problems of common water resources. There is scope for conflict resolution, but it has to be done the hard way by improving irrigation practices, thereby saving water, and using the saved water to protect the drinking water rights of Chennai. This will need facilitating institutional arrangements as well. These include:

1. Formation of an inter-state committee comprising all the stakeholders, namely politicians, government representatives, irrigation bureaucracy, water management specialists, researchers, NGOs and industries.
2. Formation of water user associations at the district and village levels for proper maintenance of water supply and restricting misuse of irrigation water.
3. Formation of a national committee chaired by the President of India for solving higher-level disputes, to safeguard the Bachavat Tribunal Award, and pressurise the State governments to resolve issues and take proper action.

Notes

*All figures by authors.

1. See for detailed process, Nikku, Balaraju. 2004. 'Water Rights, Conflicts and Collective Action, Case of Telugu Ganga Project, India' *Tenth Biennial Conference of IASCP*, August 9–13, Oaxaca, Mexico. Can be accessed at http://dlc.dlib.indiana.edu/archive/00001454/00/Nikku_Water_040513_Paper088.pdf
2. Amounts to 1.5 lakh acres in Kunrool, 1.67 lakh acres in Kadapah, 3.05 lakh for Chittoor and Nellore.
3. It was decided in an inter-state ministerial meeting held on October 27, 1977 that Tamil Nadu shall be permitted to draw 15 TMC of Krishna waters annually from Srisailam reservoir during the period of July to October through an open lined canal. The chief ministers of Andhra Pradesh and Tamil Nadu again met on June 15, 1978, and finalised the details for taking up the investigation of the project. The Government of Andhra Pradesh had sent a scheme report in September 1982 to the Government of Tamil Nadu.
4. The TGP case has become notorious for the bickering it involved. However, we should see the case from the point of view of legal pluralism accepting the reality that inter-state water projects are not easy to implement and are bound to face conflicts. Elsewhere it is argued that there is a need for effective partnership between state administration, polity, user groups and market mechanisms for governing basin resources.
5. Tamil Nadu's ex-chief minister Jayalalitha claims that the tragedy of Tamil Nadu is 'an artificial one, forced upon them by the upper riparian neighbour (hinting at Karnataka) which has been persistently denying them their rightful share of water.'

References

Brief Report on TGP Circle in Nandyal.

Government of Tamil Nadu. 1988. *The Krishna Water Supply Project for Madras*, Souvenir, Tamil Nadu.

Iyer, Ramaswamy R. 2003. *Water Perspectives, Issues, Concerns*. New Delhi: Sage Publications.

Mohan, Krishnan A. 2003. 'Interstate Coordination in Implementing a Mega Project for Water Supply: The Telugu Ganga Project Challenges Faced, Lessons Learnt' in Kamata Prasad (ed.) *Water Resources and Sustainable Development: Challenges of 21st Century*. Delhi: Shipra Publications.

Ramakrishnan T. 2003. *The Hindu*, Tamil Nadu edition, January 12.

Ruet, J. and M.H. Zerah. 2003. 'Providing Water Supply and Sanitation in Chennai: Some Recent Institutional Developments' in Patrice Cohen and S. Janakarajan (ed.) *Water Management in Rural South India and Sri Lanka: Emerging Themes and Critical Issues*. Pondicherry: French Institute of Pondicherry.

Saleth, Maria R. 1996. *Water Institutions in India: Economics, Law and Policy*. New Delhi: Commonwealth Publishers.

Seshadri, Hiramalini. 2003. *The Hindu*, Andhra Pradesh edition, June 25.

Status Report on Component Wise Works of TGP in Kadapa, January 2005.

www.hindustatimes.com/onlineCDA/PFVersion.jsp?ar.

The Sutlej Yamuna Link Canal:
Bogged Down by Politics and Litigation

Indira Khurana

A Project Gone Wrong

THE SUTLEJ YAMUNA LINK CANAL (SYL) WAS SUPPOSED TO BRING BEAS, RAVI AND SUTLEJ river waters from Punjab to Haryana and Rajasthan. Unfortunately, this canal has been a serious bone of contention between Punjab and Haryana. For decades, the SYL has generated hysterical propaganda against the compulsions that have motivated politicians to take decisions, leading to unpopular decisions.

In 1960, India and Pakistan signed the Indus Waters Treaty which reserved waters of the Ravi, Beas and Sutlej exclusively for India. Six years later, when Punjab was reorganised, the new State of Haryana claimed its share of water. In 1976, the Union government announced that both States would receive 3.5 million acre feet (MAF) of water from the available annual flow of 15 MAF through the construction of the SYL. This would benefit farmers in southern Haryana who could then use it through lift irrigation schemes.

The source of water for the SYL is the Bhakra dam. The canal starts from the tail-end of Anandpur hydel canal near Nangal and goes up to the western Yamuna canal from where it collects waters of the Ravi and Beas. Currently, Haryana gets only 1.62 MAF of the allotted 3.5 MAF, and the balance is to be made available through the SYL canal.

In 1978, the Punjab government moved the Supreme Court and thus started a series of litigations, with both sides remaining intractable. Meanwhile, construction of the canal started in 1981 in both Punjab and Haryana. In Punjab, construction came to a grinding halt in 1990 due to militancy and the killing of a senior officer and labourers. In 1996, the Haryana government went to court and when judgment was passed in June 2004, a huge drama unfolded. In July 2004, a special session of the Punjab assembly passed a bill—the Punjab Termination of Agreements Bill, 2004—terminating all agreements relating to sharing of waters of the Ravi and Beas with Haryana.

Up to March 1, 2005, approximately Rs. 700 crore had already been spent on the canal. Approximately 250 crore rupees is still required for completion. A case is being heard in the Supreme Court pertaining to the Act passed by the Punjab government.

On the Offensive

There are several reasons for the conflict over the SYL:

- Punjab considers the formation of Haryana under the Punjab Reorganisation Act 1966 illegal.
- The Punjab Reorganisation Act does not mention sharing of the Ravi waters while the 1976 decision of the union government does.

- Dispute over the amount of surplus water actually available based on which the allocations are made.
- Political compulsions of the governments at the Centre and the State.

Chronology of Events

- Conflict over sharing of the Beas, Ravi and Sutlej waters began in 1966, when Haryana was carved out of Punjab and the new State demanded a share under the Punjab Reorganisation Act, which in itself is not recognised by Punjab.
- In 1976, the Centre intervened—following an impasse between the two States on water sharing—and divided the unutilised water of the Beas and the Ravi between these two States and Rajasthan. Punjab found this unacceptable since this distribution allotted water from the Ravi that the 1966 Act had not taken into account. Moreover, the distribution was based on utilisation in 1960, not on actual use in 1976.
- When the two States could not come to an agreement, the Ministry for Water Resources issued a notification in 1976, unilaterally apportioning the waters of the three rivers between Punjab, Haryana and Rajasthan. This notification estimated the surplus river waters as 15.85 MAF, and allocated 3.5 MAF each to Punjab and Haryana, 8 MAF to Rajasthan, 0.65 MAF to Jammu and Kashmir, and 0.2 MAF to Delhi, cutting off irrigation water to about 3.6 lakh ha in Punjab. Until 1966, the area of Punjab, which is now Haryana, got only 0.9 MAF.
- Ground realities, however, were different. The surplus water available in Punjab was a mere 1.2 MAF. The then Chief Minister, Giani Zail Singh, asked for a review of the notification. In 1978, the Akali government moved a petition in the Supreme Court challenging the constitutional validity of the notification.
- Meanwhile, the first phase of the SYL canal in Haryana, a 75.5 km long stretch from Ismailpur to Karnal, which began in 1976, was completed in March 1982 at a cost of 40 crore rupees.
- In 1978, the Punjab Chief Minister Prakash Singh Badal gave the green signal to construction in the State. However, the government felt short-changed and moved the Supreme Court. Haryana also went to court demanding implementation of the central government's notification.
- In 1981, the next Chief Minister Darbara Singh, who belonged to a different political party—withdrew the case and signed an agreement increasing the share of Rajasthan by 8 MAF. An agreement with Haryana and Rajasthan was arrived at, wherein, based on new data, additional water was given to Punjab and Haryana. According to this agreement, 3.5 MAF was allocated to Haryana and 8.60 MAF to Rajasthan out of the surplus flow of the Ravi and Beas, then estimated at 17.17 MAF based on 1921–60 flow data. Punjab got 5.07 MAF from these rivers. This agreement created a furore in Punjab since it was believed to have been signed under pressure.
- In 1985, Rajiv Gandhi, the then Prime Minister, and Akali Dal leader Harcharan Singh Longowal, arrived at the historic Punjab Accord which recorded the resentment of the people of Punjab and a tribunal under a retired Supreme Court judge was set up. Justice Eradi was appointed to head the Ravi–Beas Tribunal and come to a conclusion on how much water Punjab and Haryana actually used so that the surplus could be apportioned accordingly. The Accord also stated that the SYL canal would be completed by August 15, 1986, allowing Haryana and other downstream states to utilise whatever share of water the tribunal would eventually allot to them. There was one stipulation: the farmers in Punjab would not have to compromise with lesser water. The Akalis, thus, endorsed the 1976 notification and the 1981 inter-State agreement.

- Justice Eradi discovered that the use of Ravi–Beas waters by farmers in the three States totalled 9.711 MAF: that is, 3.106 MAF by farmers in Punjab, 1.620 by farmers in Haryana and 4.985 by farmers in Rajasthan. This left some 6.6 MAF surplus water to be divided between these two States. Justice Eradi made an interim award: 5.00 MAF was awarded to Punjab and 3.83 MAF to Haryana. The arithmetic of this award did not add up since 8.83 MAF was allotted against the available 6.6 MAF. The water below the rim stations of the Ravi and Beas, which were the lowest points at which the data was recorded, helped make up the difference. Punjab pointed out that this water was useless since no dam or barrage could be built along the Pakistan border to store it.
- In 1987, Punjab thus contested the Eradi tribunal award on grounds that the tribunal had overestimated the free water available and underestimated the use of water by Punjab farmers.
- In July 1988, Justice Eradi adjourned the tribunal after violence broke out in the State. The tribunal began functioning again in November 1997 after a Supreme Court order. With no clear decision having been taken by the tribunal, the Haryana government again approached the apex court.
- In January 2002, the Supreme Court ordered that Punjab complete the construction of the SYL within twelve months, failing which the Centre would appoint a central agency to complete the work.
- In July 2002, the government of Haryana approached the Supreme Court to ensure that the Punjab government kept to the deadline.
- On January 15, 2003, the deadline expired. This was the seventh time that the State had missed it.
- In January 2004, the Supreme Court rejected the plea of the Punjab government to refer the controversy to a larger bench.
- In June 2004, the Supreme Court directed the Centre to construct the unfinished part of the SYL canal in Punjab to facilitate the sharing of river waters between the two States. This decision followed the Haryana High Court's petition about the Punjab government's failure to act upon the court order of January 15, 2002. The Centre was also directed to take over the project should the Punjab government fail to keep the deadline.
- In December 2004, Rajasthan Chief Minister Vasundhara Raje met Prime Minister Manmohan Singh to demand water from Punjab. The Bhakra–Beas Management Board immediately released water as per the requisition for the month. The Prime Minister was warned about unrest along the border area due to lack of water for farming.

Escalation of Events: 2004

On July 3, 2004, the conflict escalated as a result of steps taken by the Punjab government, which moved the Supreme Court seeking a review of its June 4 judgment directing construction of the remaining portion of the SYL canal in the State. The government contended that the Court did not have the jurisdiction to decide on the matter as it was a water dispute under Article 262 of the Constitution, which fell within the exclusive jurisdiction of the Inter-State River Waters Disputes Tribunal. The petition came after the centre directed the Central Public Works Department (CPWD) to construct the canal on a directive issued by the Supreme Court. Nearly 60 per cent of the 112 km long canal had been constructed before grinding to a halt in 1990.

The Punjab government threatened to stop releasing water to neighbouring States. The Rajasthan assembly passed a resolution authorising the State government to initiate legal and administrative steps to ensure that the state got its full share of water from the Ravi–Beas system, as per the 1981 agreement.

Punjab decided to bring a bill in the State Assembly to counter the obligation of handing over the SYL project to a central agency in accordance with the Supreme Court directive. The bill was drafted with the help of former Solicitor General, Soli Sorabjee, with the aim of nullifying the agreement with retrospective effect.

To retain control over the SYL, the Chief Minister brought up the Northern India Canal and Drainage Act, 1873, for amendment. The proposed amendment would make it mandatory for any work on a canal—maintenance, repair or construction—that ferried water beyond the borders of Punjab to be sanctioned by the assembly.

On July 12, 2004, a special session of the Punjab assembly was held which unanimously passed a bill terminating all agreements relating to sharing of the waters of the Ravi and Beas with Haryana and Rajasthan, two days ahead of the Supreme Court's deadline to the Centre to take up construction of the canal in the state. This Punjab Termination of Agreements Bill, 2004, also abrogated the Yamuna Agreement of May 12, 1994 between Punjab, Haryana, Rajasthan, Delhi, and Himachal Pradesh, and all other accords for sharing river water. Under the Yamuna Agreement, Haryana was allotted 4.6 MAF water, which would further be augmented by the SYL.

The bill decreed that the Indus system that existed before partition had become irrelevant after the event since only three east-flowing rivers, Ravi, Beas and Satluj, out of the six that constitute the Indus River System remained in India. All these rivers flow through Punjab. Neither Haryana nor Rajasthan are part of these river basins. The diversion of these waters was contrary to the national water policy.

Haryana termed the Act unconstitutional and lawless. Its implementation would lead to the destruction of cooperative federalism and disintegration of the country.

On July 15, 2004, the Centre filed a petition in the Supreme Court, saying that fresh directions were needed as a result of the controversial Punjab Government Act.

On July 20, 2004, the government of Himachal Pradesh also decided to move the Supreme Court against the Punjab Termination of Agreements Act, 2004, to protect the interests of the State in Thien dam and other projects affected by the Act.

On July 22, 2004, President Abdul Kalam referred the controversial law passed by the Punjab assembly to the Supreme Court.

And, on August 2, 2004, the Supreme Court agreed to examine the validity of the Punjab Act and issued notices to the Centre, and to Punjab, Haryana, Rajasthan, Himachal Pradesh, Jammu and Kashmir, and the National Capital Territory of Delhi to file written submissions on facts and the question of law formulated under the presidential reference under Article 143 (1) of the Constitution. It sought opinion on:

- Whether the Punjab Termination of Agreement Act, 2004, and its provisions are constitutionally valid.
- Whether the Act and the provisions are in accordance with the provisions of the Inter-States Water Disputes Act, 1956, Section 78 of the Punjab Reorganisation Act, 1966, and the Notification dated March 24, 1976, issued thereof.
- Whether, in view of the provisions of the Act, the State of Punjab is discharged from its obligations flowing from the judgment and decree dated January 15, 2002, and the judgment and order dated June 4, 2004, of the Supreme Court.

August 24, 2004: The Supreme Court upheld its June 4, 2004 order, directing the Centre to construct the remaining portion of the SYL canal in Punjab, dismissing a petition by Chief Minister Amrinder Singh for a review. The Punjab government was asked to provide security to the Central team. Amrinder Singh claimed that even though the Central Congress leadership was annoyed with him for getting the Act approved in the assembly without their consent, he would defend it tooth and nail in the apex court.

Evading Firm Decisions

Successive governments, irrespective of which party they belong to, have consistently evaded taking a final decision on the issue. Matters have come to such a stage that no government in power will have the courage to take up construction.

Since 2002, Punjab has been arguing that it has no surplus water to release. According to them, between 1981 and 2002, river flow data shows only 14.37 MAF against the 17.17 MAF believed to be available. The transfer of water would affect 9 lakh acres of irrigated land in Ferozepur, Faridkot, Mogha and Muktsar. The recharge of groundwater in Punjab would be affected.

Since 1990, when militancy escalated in the State, the 1,700 strong staff that Punjab had employed to build the canal have not had serious work. Engineers and labourers are unwilling to work on the project ever since militants killed a chief engineer in July 1990 in his office in Chandigarh.

The Punjab part of the canal built during the 1980s at considerable expense has fallen into disuse. The sand clogging in the canal will have to be manually removed—a time-consuming job. Land needs to be acquired along the path of the canal. Alignment will be a problem since it passes through Ropar district, a place of worship.

Since the commencement of the project several deadlines have been missed—December 1983, August 1986, December 1987, March 1988, June 1988, November 1989, January 1991, and January 2003.

As on March 1, 2005, approximately Rs. 700 crore had been spent on the project. Approximately Rs. 250 crore is required for completion. Meanwhile, the case continues to be heard by the Supreme Court.

Unbearable Burden

Far from being a welfare project, the SYL canal has become a millstone, an unbearable burden no one wants any part of. It has become a tool for politicians who, depending on which State they belong to, vow to bring/not allow the river waters to leave the State.

The root of the problem lies in the various awards by committees, commissions and other agencies appointed by the Centre from time to time. The delay in finding an effective solution has heightened emotions. The Constitution gives full and exclusive powers to the States over water and hydelpower. However, when Punjab was bifurcated into Punjab and Haryana, the Punjab Reorganisation Act, 1966, gave all powers to the Centre *ultra vires* to the Constitution. While this Act makes the waters of these three rivers distributable by the Centre, it gives Haryana full control over Yamuna river water.

Punjab is predominantly an agricultural State and one that has been living off the harvests of a single wave of reform in the 1950s and 60s which produced the Green Revolution. This process is approaching exhaustion as the application of increasing doses of water and fertilisers into land under hybrid cultivation is producing declining returns. Land is a limited commodity and continued fragmentation has reduced holdings and increased the number of marginal farmers. On the streets, the issue has taken an ugly turn and become communal.

In the midst of this, the main issue faced by farmers in Punjab and Haryana remains unanswered: that of inefficient irrigation policies and practices and increasing cultivation of water-intensive crops like paddy and sugarcane. Both States regularly record up to 30 per cent transmission losses in irrigation canals. In 1955, Punjab had 34.8 MAF of water in all forms. In 2004, the reserve is a mere 12 MAF. Groundwater levels are rapidly falling in areas like Malerkotla and Sangrur in Punjab owing to excessive groundwater use. Border areas in the State are experiencing waterlogging due to seepage from the ill-maintained canal and overuse of water. Neither State is taking water conservation seriously. Nor are they looking ahead and shifting to less water-intensive crops that are more suitable to the ecology.

References

'CM Sets Terms for Sutlej Canal, Punjab Row on Water Sharing Surfaces Again', *The Telegraph*, October 12, 1993.

'Canal that fails to link', *The Hindu*, February 13, 1994.

'Captain Digs up 131 Year-old Act to Retain SYL', *The Indian Express*, July 7, 2004.

'Centre Files Plea in Top Court over Water Issue', *The Asian Age*, July 16, 2004.

'Kalam Refers Water Crisis to SC', *The Statesman*, Delhi edition, July 23, 2004.

Personal communication. 2004. Irrigation department, Government of Haryana.

'Punjab Annuls all River-sharing Accords', *Business Line*, July 13, 2004.

'Punjab Plans Legislation to Escape SYL Handover', *The Indian Express*, July 8, 2004.

'Rajasthan to Urge Punjab to Honour Water Agreement', *The Pioneer*, July 7, 2004.

'SC to hear Haryana plea on SYL canal', *The Hindu*, January 21, 2004.

'Sutlej-Yamuna Canal: Murky Waters', *Economic and Political Weekly*, June 12-18, 2004, p. 2417.

'SYL Issue: SC Notice to Centre, 6 States', *Business Line*, August 8, 2004.

Two Neighbours and a Treaty:
Baglihar Project in Hot Waters

Rajesh Sinha

Stalled Dam

This PROJECT WAS CONCEIVED IN 1992, APPROVED IN 1996, AND CONSTRUCTION BEGAN IN 1999. The Baglihar Hydropower Project (BHP) is located on the River Chenab in Ramban *tehsil* of Doda

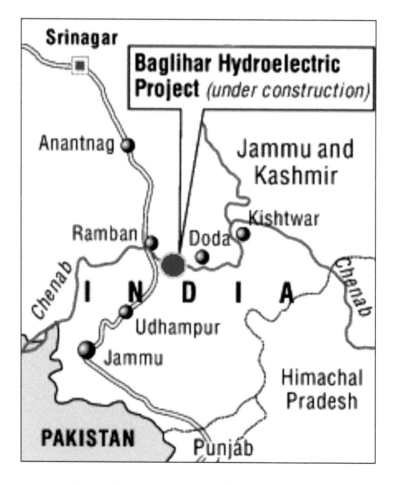

Map 1. *The location of the Baglihar hydroelectric project.*

district, Jammu and Kashmir (J&K). It is about 150 km from the nearest railhead, Jammu, near Batote on the Jammu–Srinagar Highway-1A on the Nasri bypass road.

The BHP will have an installed capacity of 450 MW during Phase I and 900 MW during Phase II. It is a Rs. 4,000 crore venture of the government of Jammu and Kashmir and about Rs. 2,500 crore have already been spent on it. The project is targeted to be completed by 2007.

Pakistan had objected to several features of the project, contending that they violate the Indus Water Treaty (IWT) that lays down the rights and obligations of India and Pakistan for the use of waters of the Indus system of rivers. India disagreed. Pakistan approached the World Bank, which had brokered the IWT, to appoint a neutral expert (NE) to resolve the differences.

Drawing Boundaries on Water

When India and Pakistan became independent in 1947, the boundary between the two countries was drawn right across the Indus basin, with Pakistan in the lower riparian region. Two vital irrigation headworks—Madhopur on the Ravi, and Ferozepur on the Sutlej—on which the irrigation canals of Pakistan's Punjab had been dependent—were in Indian territory. The resultant dispute was resolved when negotiations facilitated by the World Bank led to the signing of the IWT at Karachi in 1960. The signatories were Mohammad Ayub Khan for Pakistan, Jawaharlal Nehru for India, and W.A.B. Illif of the World Bank.

Fig. 1. *A view of the Chenab river, Jammu and Kashmir.*

The treaty has worked well for years. But since the 1980s the differences that cropped up over several projects have not been sorted out despite protracted talks. Pakistan has not only opposed the Baglihar Hydel Project on the Chenab in Doda, it has also opposed the Wullar barrage/Tulbul Navigation Project on the Jhelum, the Swalakote Hydroelectric Project and the Dul Hasti Hydroelectric Project on the Chenab, and the Kishanganga Hydroelectric Project on Kishanganga in Jammu and Kashmir.

Suspicious Neighbours

The dispute over the BHP centres around design specifications. Pakistan has raised six objections relating to project configuration: free board, spillway (ungated or gated), firm power, pondage, level of intake, inspection during plugging of low-level intake, and whether the structure is meant to be a low weir or a dam. The argument is based on Paragraph 8 of Annexure D to the IWT.

India maintains that the conditions at the Baglihar site make a gated spillway necessary but Pakistan insists that an ungated spillway will do just fine and that the plan to provide a gate contravenes the provisions of Paragraph 8 (e) of Annexure D to the IWT. Pakistan also contends that the pondage in the operating pool, at 37.722 Mm,[3] exceeds the level agreed upon in the treaty and that the intake for the turbine is not located at the highest level as required by the treaty. It believes that the height of the dam, at 470 feet, is excessive and that the reservoir created at the site will be more than is required for power generation needs. It might also block the flow of the river for a period of twenty-six to twenty-eight days during the low season (January–February). This, it is argued, will cause a drop of about 200 cumecs in the river flows during this period at the point of entry into Pakistan.

Fig. 2. *The site of the Baglihar project.*

Based on all these objections, Pakistan insists that India should stop all work on the project until the issue is resolved. India has refused, saying there is no such provision in the treaty and that past experiences in trying to find solutions by stopping construction have not been productive. The IWT gave India exclusive rights over three eastern rivers—Sutlej, Beas and Ravi—leaving the Chenab, Jhelum and Indus to Pakistan. But the treaty does allow India limited use of their waters for agriculture, domestic purposes and development projects provided there is no obstruction to the flow of waters into Pakistan.

The Permanent Indus Commission (PIC) set up under the IWT by the two countries has been meeting regularly to sort out any differences that arise. This is the first time that the committee has failed to resolve a crisis, forcing Pakistan to invoke the provision to approach the World Bank. The World Bank scrutinised Pakistan's record of actions taken before the request and acknowledged that it has the mandate to appoint a neutral expert, who would not be a guarantor to the treaty and therefore will not directly participate in any discussion or exchange on the subject. Professor Raymond Lafitte was appointed the neutral expert and the arbitration clause was put into operation for the first time in the forty-five-year old history of the treaty.

Pakistan's View

Underlying the dispute are suspicions and apprehensions resulting in much rhetoric. Claiming that the project will affect the flow of river waters to its territory in violation of the IWT, an upset Pakistan says that India planned the dam and began construction without its approval as mandated by the treaty. The gates are another big issue. Pakistan feels the closure may adversely affect irrigation in its territory. It argues that 450 megawatts of electricity can be generated even with the gates open.

Pakistan fears that tampering with the flow of the river may create floods or drought downstream. The allegation is that India has not been taking them into confidence about the project's technical details and has been adopting evasive tactics right from the start. Some sharing of data took place 'at a pretty late stage of construction', that is in 2003. Pakistani engineers believe that calling Baglihar a 'hydroelectric' project is a misnomer. The structure will create a reservoir at the site and hence should be properly termed 'dam'. And, as per the terms of the treaty, a dam is not allowed.

At a briefing after the decision to approach the World Bank was taken, Pakistan Foreign Office spokesman Masood Khan said, 'Pakistan was left with no choice but to go to the World Bank. Pakistan tried every channel provided by the treaty, but India did not change its stance and refused to meet Pakistan's legitimate concerns.'

Pakistan alleged that New Delhi denied Islamabad's repeated demand for an on-site inspection by its members of the Indus Waters Commission for four years. The treaty provides for an inspection tour once every five years or on request. Only when threatened with approaching the World Bank did India allow an on-site inspection in October 2003.

Among the reasons given for Pakistan's concern was a drought that had compelled it to economise on water. With insufficient storage capacity and inadequate rainfall, water shortage in Pakistan reached critical proportions and was a major source of inter-provincial disharmony. The Pakistan government's inability to make India bend on the Baglihar dam and Wullar barrage projects was severely criticised not only by the farming community but also by politicians in the ruling coalition.

India's Stand

India claims the BHP is a fully legal scheme that involves no water storage. It denies allegations that it violates the IWT or that it will affect the flow in the river since the IWT allows power generation projects to be built on any of the three western rivers of the Indus river system, as long as they benefit the local people and do not interrupt the flow of the river. It also says that reduction in the height of the dam will impact the power generation capacity of the project and render it worthless. It further argues that the statistics provided by Pakistan on decrease in the river water flow are faulty.

The rejoinder over the gate issue goes thus: An ungated spillway will require a higher crest and the silting level will also increase. The gates will enable the flushing of silt. India asserts that it made all efforts to dispel Pakistan's apprehensions. It claims to have consistently expressed its willingness to engage its counterpart in technical discussions in PIC meetings, arranged the Pakistani team's visits to Baglihar, and provided necessary explanations/clarifications to the queries raised by them during the 100th PIC tour. It claims that information on this project was sent to the Pakistan Commissioner in 1992 as required by the IWT. Since Pakistan feels that the design of this project contravenes the IWT provisions, the matter was discussed during the 84th, 85th and 86th meetings of the PIC held in 1999, 2000 and 2001. India thinks that Pakistan has been taking a rigid stand despite being informed about changes in the design of the plant and that they should have first conveyed their views/observations on these changes to India. Instead, they chose to invoke relevant articles in the treaty and sought the intervention of a neutral expert.

Meanwhile, India has refused to stop work on the project on the basis of its experience with the Tulbul Navigation Project (Pakistan calls it Wullar barrage), proposed to be built on the Jhelum at the mouth of Wullar lake near Sopore in Kashmir valley. After India stopped construction of the Tulbul project in response to Pakistan's objections a decade ago, it has not been possible to take it up again. India says that Pakistan's objections to the BHP are based on apprehensions rather than technical reality. India's Foreign Secretary Shyam Saran observed, 'The 1960 Indus Water Treaty under which the reference was made couldn't deal with suspicions of this nature.'

IWT and Jammu and Kashmir

To Kashmiris, the BHP is a project for and by Jammu and Kashmir, a state in dire need of power. They believe that Pakistan wants to deny J&K the right to use its own rivers, citing the situation in Pak-occupied Kashmir where they believe people have no rights over the Mangla dam on the Jhelum, built to meet the power and water needs of Punjab and other parts of Pakistan.

London-based moderate separatist leaders Syed Nazir Gilani and Shabir Choudhry reacted sharply to Pakistan's appeal to the World Bank to resolve the controversial Baglihar project, and questioned Islamabad's 'legal or moral right' over the natural resources of J&K. Gilani even justified Chief Minister Mufti Mohammed Sayeed's claim that the IWT had greatly harmed the interests of the people of the state. Sayeed had appealed to Pakistan before the secretary-level meeting between the two countries in New Delhi in the second week of January to facilitate economic growth in the state by not objecting to projects started on its own water resources. He described the IWT as discriminating against the people of J&K. But he was quick to add that the BHP had not violated the IWT.

Jammu and Kashmir has never been very happy with the IWT. On March 2, 2003, the J&K assembly passed a resolution asking the Central government to review the IWT. Public Health Minister Qazi

Mohammed Afzal said that but for the terms of the treaty, over and above the 33,000 ha which was under irrigation before 1960, the State could have increased the area under irrigation by another 40,000 ha. J&K Finance Minister Muzzaffar Hussain Beig pointed out that the farmers in the State have requested the government to take a strong stand against it. The State has been facing severe water scarcity as a result of a six-year dry spell.

Chronology of Events

- 2001: Contact between the PIC officials is broken off following tension at the border.
- July 12, 2003: PIC meeting between India and Pakistan makes little headway. A follow-up meeting is convened in August to break the logjam. Pakistan serves its first notice threatening to seek intervention.
- October 2003: A team of Pakistan's technical experts inspects the project site. Pakistan expresses its objections to the design and India refuses to make any changes. Pakistan serves a second notice to the Indian government to settle the dispute by December 31, 2003 or else it would approach the World Bank to appoint a neutral expert.
- January 18, 2005: A three-day meeting between the Indus Water Commissioners of Pakistan and India is concluded without finding a solution. Pakistan approaches the World Bank. Kashmir decides to go ahead with the construction of the dam. India terms Pakistan's decision to seek World Bank arbitration as 'premature' and 'hasty'.
- April 2005: World Bank sends a panel of three experts to both India and Pakistan.
- May 10, 2005: World Bank names a Swiss national, Professor Raymond Lafitte, as the neutral expert.
- 9 June 2005: Lafitte, in the course of his first meeting with the Indian and Pakistani delegations, turns down Pakistan's demand that India stop work on the project till he delivers a verdict. It is decided that a Pakistani team would visit the project site in July before submitting its report. India would respond with a 'counter report'. Lafitte would then visit the site in October.

Fingers Crossed

The issue has gained momentum at a time when India and Pakistan are trying to build mutual understanding and trust. Both countries have made statements offering full cooperation to the expert and have agreed to abide by his decision. The neutral expert is not concerned with any deeper issues and anxieties underlying Pakistan's technical objections. He has to answer specific questions posed to him, about whether certain features conform to the conditions laid down in the IWT. He is not expected to make any judgment or propose alternatives. Only if either of the two parties put in a request can he suggest measures 'to compose a difference or to implement his decision'. He is also free to conclude that the matters referred to him fall outside his purview or that the differences amount to a 'dispute', better handled by a Court of Arbitration. There is also the possibility that the two countries might actually reach a compromise, which makes the expert's job easier.

Diplomats fear that these developments might cast a shadow on the composite dialogue process. Seeking arbitration on the Baglihar dam means breaking out of the bilateral framework, even though it will not lead to the collapse of the dialogue process.

Note

*Map 1 from *www.rediff.com/news/2005/jan/20bagli.htm;* figure 1 from Press Trust of India, figure 2 from Lahmeyer International.

References

Ahmar, Moonis. 'The Need to Avoid a Stalemate', Director, Program on Peace Studies and Conflict Resolution, Department of International Relations, University of Karachi. Accessed at *www.jang.com.pk/thenews/jan2005-daily/11-01-2005/oped/o5.htm.*

'Baglihar Project Controversy - Separatists Question Pak Claims', *Kashmir Times*, February 6, 2005.

Baid, Samuel. 2005. 'Baglihar Power Project Politics', *Daily Excelsior*, February 3. Accessed at *www.southasianmedia.net/index_opinion4.cfm?id=53471.*

Devraj, Ranjit. 2005. 'Peace Flounders over Water Dispute', *Asian Times*, January 21.

Gupta, Alok Kumar. 2005. National Law University, Jodhpur, Institute of Peace and Conflict Studies. January 20.

India Defence Consultants. 2005. 'What's Hot? — Analysis of Recent Happenings, India-Pakistan Dispute over Water - An IDC Analysis', New Delhi, January 24.

'India Fail to Resolve Baglihar Dispute', *Xinhuanet*, Pakistan, January 22, 2005.

Narayanan, K. S. 2005. 'The Water Row - The Controversial Baglihar Hydel Project', *Deccan Herald*, New Delhi, April 24.

'Pak Accuses India of Putting Indus Treaty in Suspension', Press Trust of India, February 17, 2002.

Parliamentary questions accessed at *http://loksabha.nic.in* and *http://rajyasabha.nic.in.*

'Separatists Question Pakistan's Right over Kashmir Resources', United News of India, New Delhi, February 10, 2005.

Sud, Hari. 2005. 'Indus Water Treaty and Gas Pipeline from Iran to India'. South Asia Analysis Group, Guest Column, February 7.

'To Baglihar with Distrust', Editorial, *www.bharatdaily.com.*

http://wrmin.nic.in.

http://www.hinduonnet.com.

http://www.newsline.com.pk/NewsJune2005/view2june2005.htm.

http://www.stimson.org/southasia/?sn=sa20020116300.

http://web.worldbank.org/WBSITE/EXTERNAL/COUNTRIES/SOUTHASIAEXT/0,contentMDK:20320047~page PK:146736~piPK:583444~theSitePK:223547,00.html.

http://web.worldbank.org/WBSITE/EXTERNAL/NEWS/0,,contentMDK:20485918~menuPK:34463~page PK:34370~piPK:34424~theSitePK:4607,00.html.

The Indo–Bangladesh Water Conflict: Sharing the Ganga

Sumita Sen

The Problems of Being a Big Brother

THE SHARING OF GANGA WATERS CONSTITUTES AN IMPORTANT ASPECT OF INDO–Bangladesh relations. The Ganga basin, with its rich natural wealth, is a densely inhabited region with a combined population of about 500 million—440 million in India (2001), 37 million in Bangladesh, and 23 million in Nepal. An estimated 43 per cent of India's population lives in the Ganga basin. With the free-flowing water as an accessible common property resource, agriculture evolved as the dominant occupation of the local population. Technological innovations in the agro-farming sector have proven the Malthusian equation wrong; a vast majority of the people live in abject poverty, illiteracy and poor health. Rapid urbanisation, arbitrary drawing of water from the upland flow for irrigation by the *kulaks*, industrialisation along the river course and direct discharge of pollutants into its waters have led to the politicisation of the water issue across the country.

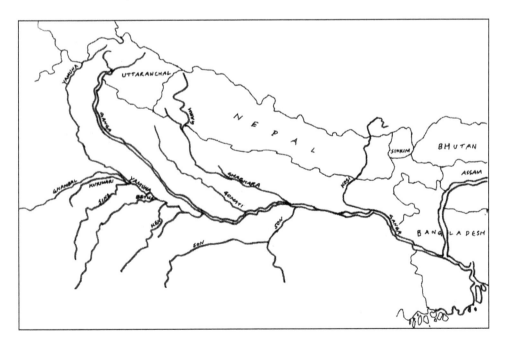

Map 1. *Location of the study area.*

The Ganga Basin: One of the Largest in the World

The Ganga rises from the Gangotri snout at an elevation of about 4,000m above mean sea level (MSL). The total length of the Ganga is about 2,500 km and the Ganga–Brahmaputra– Barak/Meghna basin is one of the largest river basins in the world. Of the total drainage area of 1,087,300 km², about 860,000 km² fall within India, 46,300 km² in Bangladesh and 1,47,480 km² in Nepal. The major trans-Himalayan rivers, that is, Karnali, Saptgandaki and Saptkosi which are the major tributaries of the Ganga, contribute about 70 per cent to the Ganga's dry season flows and about 41 per cent to its total annual flow.

The river flows through five States in India—45 per cent of its basin spans Uttaranchal, Uttar Pradesh, Bihar, Jharkhand and West Bengal—before discharging into the Bay of Bengal. Below Farakka in West Bengal, the river bifurcates into Bhagirathi–Hugli and Padma. And, about 18 km below Farakka, the Ganga enters Bangladesh near Rajshahi. Studies indicate that the Bhagirathi–Hugli received a bulk of its flow before the river's gradual diversion towards the east, along the Padma, brought down the flow in the Bhagirathi–Hugli system.

A Barrage of Woes

The river port of Calcutta was the major hub of mercantile activities and commerce during the British period. The port faced serious problems due to siltation that resulted in the deterioration in navigability. Successive reports by Sir Arthur Cotton (1853), Harcourt-Vernon(1896), the Stevenson-Moor Committee (1916–19), Sir William Wilcocks (1930), and Sir Cyril Radcliffe suggested the construction of a 'barrage' across the Ganga to keep the navigation channel free from shoaling as well as providing upland flow into the Bhagirathi–Hugli to help maintain adequate water depth of the port. Walter Hensen's report to the Government of India in 1957 too recommended the construction of a barrage across the Ganga to augment the flow of the Hugli–Bhagirathi and save the Calcutta port. Nevertheless, negotiations over sharing of river waters between India and Pakistan, and later with Bangladesh, continued before the Farakka barrage was commissioned. The Farakka barrage, commissioned in 1974, was to augment the waters of the Bhagirathi–Hugli through a 41.5 km long feeder canal. It became operational in 1975. The plan to divert water via the barrage to the Bhagirathi–Hugli became a source of political tension, mistrust and apprehension between Pakistan and India, and later between Bangladesh and India. Issues such as control of barrage gates, quantity of water to be released, sharing mechanism, augmentation measures, etc. became the focal points of tension and conflict between the nations.

The Treaty Trail

Pakistan first disputed India's decision to construct the barrage in 1951. From 1971 the issue was taken up by Bangladesh. Following the developments in Bangladesh, the short-term water sharing agreement of 1975 could not be renewed. Bangladesh raised the issue of unilateral withdrawal by India at the UN in 1976. Indo–Bangladesh relations were strained after 1988, and the water sharing issue could not be resolved between the two sides. In the absence of any agreed mechanism, Bangladesh again brought up the Ganga water issue at several international forums in 1993, highlighting the difficulties it had been facing due to India's unilateral withdrawal of water. The dissatisfaction over water sharing continues even after the 1996 agreement.

The Indian Side of the Coin

Sharing of river waters has become a contentious issue among the States within India and has taken centrestage in defining many inter-State relations. The Ganga water is no exception. When India negotiated the sharing of Ganga waters with Bangladesh, it stated:

- The annual sediment load of the Bhagirathi–Hugli is about 40 per cent of the total load of the Ganga–Brahmaputra system in Bangladesh. Hence, saving Calcutta Port from silting-related problems and maintaining navigability remains the prime objective for diversion of waters by India.
- The Ganga–Brahmaputra–Meghna basin is a single unit, and as part of the augmentation plan it proposed a link canal across the Brahmaputra that could augment the water-stressed Ganga for stable flow of the river as well as to regulate water levels in Bangladesh. Moreover, India maintained that the construction of a reservoir in Nepal to augment the flow was not ecologically feasible as the region falls within a seismic zone. The proposed reservoir site would also entail displacement.
- The Farrakka issue is a bilateral one and not a joint development plan to which Nepal can become a party.
- Increase in population and the consequent demand for water for irrigation, power generation, industrial and domestic use cannot be ignored. The basin-dependent population has grown steadily over the years and is likely to grow further (as shown in Table 1), though the catchment area of the Ganga basin remains the same.

Table 1: Ganga Basin States and the Catchment-Dependent Population

Ganga basin states	Share of catchment and population in the Ganga basin		
	Catchment (per cent of total)	Population 2001 (in million)	Projected population 2051 (in million)
Uttar Pradesh (undivided)	34.2	174.6	337.0
Madhya Pradesh	23.1	60.0	100.0
Bihar (undivided)	109.8	109.8	188.0
Rajasthan	13.0	56.0	80.0
West Bengal	8.3	80.2	121.9
Punjab and Haryana	4.0	35.0	55.0
Himachal Pradesh	0.5	6.0	8.0
Delhi	0.2	13.8	25.0

The population in the catchment area is subject to various problems related to livelihood and sustenance:

a) Delhi abstracts water at an estimated 450 Mld (million litres per day) from the upper Ganga canal in Muradnagar, Uttar Pradesh, to meet its requirements. Farmers have protested against this move since it affects a command area of 9.3 lakh ha of agricultural land in western Uttar Pradesh.

b)   The Hugli estuary is an abandoned part of the deltaic plain of the Ganga–Brahmaputra basin. The region faces shoreline shifting and beach profiling due to inter-alia diversion of the Ganga towards the east along the Padma river. In West Bengal, there are a few areas in Murshidabad district where the shifting course of the Ganga has taken its toll on erstwhile fertile agricultural lands and left the local population reeling under the dual blow of food scarcity and loss of livelihood.

c)   The water in large areas in West Bengal is affected by arsenic toxicity due to lowering of the water table in most of the Gangetic districts. At the same time, given the importance of the river in the Indian economy, there is no apex statutory body to monitor these issues in the country. The Ganga Flood Control Commission has a limited mandate and merely confines itself to studying problems related to flood control. The Ganga Action Plan was initiated to address the issue of pollution of the river. The problem of water is so great that the topic has been included in the Concurrent List of the Constitution. The debate over sharing of river waters is an ongoing one and many experts deliberate interlinking the country's rivers.

...And That of Bangladesh

Obviously the water sharing issue has been a thorn in Indo–Bangladesh relations. Bangladesh needs a minimum flow of 55 to 60 thousand cusecs during the critical months of March and April to meet irrigation and industry demands as well as to maintain the ecological balance, navigability, and control salinity ingress. Also, Bangladesh does not recognise the Ganga–Brahmaputra–Meghna as a single system. It stated that:

Fig. 1.   *Bank erosion and spur remnant, Dipchandpur, Murshidabad.*

- The construction of the barrage has deprived it of a larger share of water. The diversion of more than 11,000 cumec through Farakka combined with increasing upstream withdrawal for irrigation purposes in Uttar Pradesh and Bihar has drastically reduced the quantity of water that reaches Bangladesh.
- After the construction of the Farakka barrage, dry season water levels dropped in western parts of Bangladesh and increasing salinity levels affected the groundwater. The sudden spurt in water toxicity in the region is believed to be a consequence of declining water levels.
- The supply can be augmented through tributaries of the Ganga in Nepal by way of construction of storage dams on the Kosi, Gandak and Karnali. This would be a better solution than proposing a link canal from the Brahmaputra through Bangladesh to the Ganga. Besides, the second option will displace thousands of people and have adverse ecological consequences as well.
- There are apprehensions that India's agreement with Bhutan on the Sankosh and Manas rivers, tributaries of the Brahmaputra, will affect its water regime.

If the increased abstraction leads to lowering of the water table and salinity intrusion, then Bangladesh's ability to fulfil its demand for drinking water—which constitutes almost 90 per cent of its demand—would be severely affected. This necessitates proactive strategies for securing adequate waters for Bangladesh. Reports (*India Today*, January 1–15, 1997) suggest that the availability of Ganga waters has gone down to around 15,500 cusecs in the months of March and April during the last five to six years. Contrary to India's view, the upstream dam option would not only augment the flow, but would also help generate huge amounts of hydroelectricity to meet power demands and regulate water levels in Bangladesh as well as in the north-eastern parts of India.

Even after the 1996 treaty was signed, reports in different publications indicated that Bangladesh received a lower volume of water than it had after the 1977 agreement—the quantum of decrease varied between 117 and 144 cusecs during the critical period from February 21–28 to May 11–20. On the other hand, India gained from the 1996 treaty as it received more flow than it did through the 1977 agreement, which varied between 1,459 and 4,180 cusecs in the same period. Since the time of the signing of the treaty, Bangladesh tended to lose a large volume of water that could irrigate about 4.8 lakh ha. There is no provision in the treaty to augment the flow through regional cooperation, unlike in the 1977 agreement. The treaty also did not take into account the 'quality' of water released by India—full of toxic chemicals and heavy metals. Under such circumstances, Bangladesh will definitely not be able to redress the problems of environmental degradation, salinity in surface and groundwater, navigability, fisheries and deterioration of the river regime.

Publications and reports from Bangladesh also iterated that although the treaty has been signed for thirty years, India may not fulfil its obligations, and abstraction of Ganga waters will increase in future. Further, it will reduce Bangladesh's share, about which it can do nothing, as the 1996 treaty did not contain any arbitration clause for dispute settlement.

Adequate Mechanism for Dispute Settlement

The growing number of disputes over water sharing has resulted in the evolution of international norms and principles applicable to transboundary water disputes. They include the geography of the basin area, hydrology of the basin, utilisation practices of the basin States, economic and social needs of the basin States, percentage of population dependent on the river, comparative costs of alternative solutions, stakeholders' interests, equity, etc. Negotiations on transboundary water resources between and among co-riparian States have always been dominated by national interests, particularly between small and big countries, as in the case of Bangladesh and India.

There have been attempts to work out a mechanism to address the conflict between India and Bangladesh. After the first disagreement in 1951 over the Farakka barrage issue, India and Bangladesh have continued to negotiate since 1972 to find a durable solution to the problem. In the interim period, they have entered into a few short-term agreements in the form of treaties, agreements and memorandums of understanding (MoUs). In 1975, before the barrage was commissioned, there was a test run to release 311 to 453 cumec through the feeder canal from April 21 to May 21 that year and let out the rest into Bangladesh. The Ganges Water Agreement was signed on November 5, 1977 for sharing water during the dry months from January to May each year for a period of five years. The agreement stipulated 34,500 cusecs for Bangladesh and 20,500 cusecs for India, and guaranteed a minimum flow of 27,600 cusecs for Bangladesh during the lowest ten-day average flow period between April 21 and 30. The two sides formulated a schedule for sharing the flow, which guaranteed a minimum of 80 per cent of the flow for Bangladesh as a protection against withdrawals from upstream reaches. The schedule also provided for the formation of a joint committee to monitor the sharing arrangements. The two sides formed the Joint River Commission in 1972, but the arrangement was not free from complications because of changes in the political scenario in both countries.

The Governments of Bangladesh and India also signed an MoU on October 7, 1982, that provided for submission of feasibility studies on augmentation schemes. Another MoU, signed on November 22, 1985, included provisions for negotiations on the water of all common border rivers. This MoU expired in May 1988, after which the heads of the governments of Bangladesh and India, the Bangladesh Secretary of Irrigation, and the Indian Secretary of Water Resources tried to work out an integrated formula for a permanent settlement of the conflict. The treaty of 1996, based on the principle of reasonable and equitable sharing of water and river basin approach, followed the Non-Navigational Laws and the Helsinki Rules of 1966 which state that 'each Basin State is entitled, within its territory, to a reasonable and equitable share in the beneficial uses of the waters of an international drainage basin'. According to this treaty, the total availability of water will be measured at Farakka on the basis of the previous forty years of ten-day average flows. It calls for both parties to discuss options to augment the Ganga flow during the dry period, guided by the principles of equity and fairness without harm to either party. The Joint Committee was designated to set up suitable teams at Farakka in India and Hardinge bridge in Bangladesh to observe and record water flows. The committee was made responsible for following the arrangements contained in the treaty as well as examining difficulties arising out of their implementation. Any difference or dispute not resolved by the group was to be referred to the Indo–Bangladesh Joint Rivers Commission.

A Deeper Look

Domestic compulsions often assume primacy in diplomatic dialogues and complicate the issue of sharing an increasingly scarce resource, thereby making it difficult to arrive at a peaceful and viable solution. The Indo–Bangladesh treaty of 1996 apparently resolved the issue of sharing Ganga waters, although it appears that many loopholes need to be plugged. While India cannot claim an absolute right to divert water, Bangladesh also cannot harp on historic flows (Verghese and Iyer 1994: 196–97).

The reduced flow of the Ganga has both long-term and immediate effects (Wallerstein 1997: 1–12). The two countries must settle the issue on a permanent basis with built-in review provisions covering all aspects like ecology, hydrology and river dynamics, and above all, social, economic and political considerations. As a means of settling the vexing issue, renowned engineer Kapil Bhattacharya advocated the idea of instituting hydro-engineering work in the lower reaches of the Hugli. Construction of a reservoir between the Hardinge bridge may be an effective means of augmenting the flow, rejuvenating dry rivers, as well as facilitating irrigation activities in the catchment areas. Building a satellite reservoir in Jalangi, Murshidabad district, West Bengal, will provide an ecologically viable alternative that might help meet Bangladesh's water demand during the lean season as well as controlling erosion and rejuvenating those rivers in West Bengal that have dried up. This is an option the scientific community is seriously looking into. Another solution might be to utilise the Teesta River Project to supply water to Bangladesh.

The issue of Ganga water sharing between Bangladesh and India needs scientific intervention more than political and emotional rhetoric. Both countries would do well to learn from the experiences of peaceful water sharing like those of the Danube in Europe and the Mekong in Southeast Asia.

Note

*Map 1 and figure by P.S. Chakrabarti, Principal Scientist, Department of Science & Technology, Government of West Bengal; Table 1 from 'India's Water Wealth', Census 2001.

References

Abbas, B.M.A.T. 1984. *The Ganges Water Dispute*. Dhaka: University Press Limited.

Bhattacharyya, K. 1959. *Bangla desher Nad–Nadi O Parikalpana*. Second edition in Bengali. Calcutta: Bidyoday Library.

Biswas, A.K. and J.I. Uttio (eds). 1996. *Sustainable Development of the Ganges–Brahmaputra–Meghna Basins*. Tokyo: United Nations University Press.

Chakrabarti, P. 1998. 'Changing Courses of Ganga, Ganga Padma River System, West Bengal, India—RS Data Usages in User Orientation', *Journal of River Behaviour and Control*, Vol. 25.

Crow, B. 1995. *Sharing the Ganges; The Politics and Technology of River Development*. Dhaka: University Press Limited.

Elhance, Arun P. 1999. *The Ganges–Brahmaputra Basin, Hydropolitics in the Third World: Conflict and Cooperation in the International River Basins*. Washington: United States Institute for Peace.

Krishnamurti, C.R., K.S. Bilgrami, T.M. Das, R.R. Mathur. 1991. *The Ganga: A Scientific Study*. New Delhi: Ganga Project Directorate.

Nishat. A. 1996. 'From Ganges–Brahmaputra to Mekong' in T. Hashimoto and A.K. Biswas (eds.), *Asian International Waters*, pp. 60–80. Water Resources Management, Series: 4. Bombay: Oxford University Press.

Ramakrishnan, T. 2004. 'Sharing Water Resources', *The Hindu*, October 8, Online edition.

Salmon, M.A. 1998. 'Co-management of Resources: The Case of the Ganges River'. Accessed at *http://gurukul.ucc.American.edu/maksoud/water98/present1.htm*.

Sen, S. 1997. *The Ganga Water Treaty: From Uncertainty Towards Stability in Asian Studies*. XV (1), Calcutta.

—— 2000. Common Water Resources of South Asia: Experience of Cooperation and Development. In Upreti, B.C. (ed.), *SAARC: Dynamics of Regional Cooperation in South Asia*, Volumes 1 and 2. New Delhi: Kalinga Publication.

Singh, I.B. 1987. 'Sedimentological History of Quaternary Deposits in Gangetic Plain', *Indian Journal of Earth Sciences*. 14 (3–4).

Subramanya, K., 1994. *Engineering Hydrology*. New Delhi: Tata McGraw-Hill.

Upreti, B.C. 1993. *Politics of Himalayan River Waters: An Analysis of the River Water Issues of Nepal, India and Bangladesh*. New Delhi: Nirala Publications.

Verghese, B.G. and R.R. Iyer. 1994. *Harnessing the Eastern Himalayan Rivers: Regional Cooperation in South Asia*, New Delhi.

Wallerstein, P. and A. Swain. 1997. *Comprehensive Assessment of the Fresh Water Resources of the World*. Stockholm: Environment Institute.

PART 8

Privatisation:
A Review

Sunita Narain

PRIVATISATION AND PRIVATE–PUBLIC PARTNERSHIPS HAVE BECOME BUZZWORDS IN WATER management circles. The government says industry has the answer to the problems of water scarcity, pollution and waste. It believes industry will bring the needed investment and expertise to manage water in the interests of poor people. Industry sees it as a new and lucrative business opportunity. I would also argue that the debate on this issue has been riddled with generalised and polarised positions that are not leading anywhere.

I am not against private sector involvement in water *per se*. But given the political economy of water and sewerage in the country, I believe their role will be extremely limited. The simple assumption of private sector proponents is that if water is correctly priced—what is known as full-cost pricing—it would facilitate investment from the private sector and provide a solution to the water crisis facing vast regions of the developing world. Protagonists say private partnership is the magic bullet that will deliver safe water for all. Antagonists insist the private sector is interested only in profit and not in public good.

The truth lies somewhere in-between.

Urban Water Supply

We have to realise that urban areas—in the developed and developing world—are extremely water-intensive. They are powerful and their greed for water will make them draw it from surrounding areas. India's flatulent cities get their water supply from further and further away—Delhi gets Ganga water from the Tehri dam; Bangalore is building the Cauvery IV project, pumping water 100 km to the city; Chennai water will traverse 200 km from the Krishna river; Hyderabad from Manjira and so on. The point is that the urban–industrial sector's demand for water is growing by leaps and bounds. But this sector does little to augment its water resources, it does even less to conserve and minimise its use. Worse, because of the abysmal lack of sewage and waste treatment facilities, it degrades scarce water even further. But even after all this, its greed for water is not met. Groundwater levels are declining precipitously in urban areas as people bore deeper in search of the water that municipalities cannot supply.

This is one part of the water crisis. The larger, much more serious problem, is in rural areas where water scarcity on the one hand, and water pollution on the other, is leading to an unbearable toll on human lives.

Health Impact and Need for Standards

Worse, dirty water is the key health problem in our world. It is amazing that so many years after independence, India has not legislated water quality standards for clean water. We have guidelines for safe and clean water, which are not mandatory. Municipalities argue that they do not have the wherewithal to treat and clean water for supply. The enormous cost of poor health is borne by the poor in society. This is clearly not acceptable. Therefore, it is clear that we need major reform and restructuring of the water supply systems of the South. We cannot argue that the *status quo* of failing municipal services and increasing water scarcity is acceptable. As yet, the public sector is by and large failing people. But what needs to be understood is why this is happening. Only if we ask the right question will we get the answer we need.

The Forms of Privatisation

A French multinational, Degremont, has been awarded a contract by the Delhi Jal Board, a State-owned water authority, to treat raw water for supply to Delhi's residents. This water will come from the river Ganga. At tremendous cost, pipelines will be laid over long distances to meet the city's water guzzling needs. The cost of building the pipelines and transporting water remains with the State. But as it is a company that will build and run the water treatment facility, it becomes a privatisation model. In this build–own–operate scheme, the State still sets tariffs, collects revenues and manages overall water services.

In the second model, the responsibility shifts to the private entity. The State plays, at best, a regulatory role. The problem arises when the 'entity' sets tariffs—to pay for operating costs and to rake in profits—for a public service like water, in rich and not-so-rich cities of the developing world. For instance, in Metro-Manila in the Philippines, a successful privatisation venture has run into trouble. Water multinational Suez SA has encountered local politics, which is blocking its moves to increase the rates that its local franchise could charge consumers. Water is a business and if not profitable, the company moves out. Even so, the State technically remains custodian of the resource.

In the third model, 'ownership' shifts to private enterprise. This is the river-leasing model, so much in the news lately. The State of Chhattisgarh had entered into a contract with a private company to invest in a barrage on the Sheonath river and provide water to the local industrial estate. In this build–own–operate contract, activists alleged that a 23 km stretch of the river has been leased to the company for twenty-two years. Contract details remained unclear, but it was feared that people's access to this stretch would be curtailed. The project was scrapped because of public protests. This is not an entirely new approach. In other parts of the country, small stretches of rivers have been leased to specific industries as their water sources.

The fourth model is more unregulated. Here the water resource is free for all. An example is the use of groundwater by bottled water or beverage companies. Under the existing legal framework, companies can simply bore a hole in the ground, extract water and make profits. Nothing extraordinary, you could say. All industries, institutions, house-owners and farmers that consume groundwater are part of this 'private' army of users. Though clearly the use of this free resource by a superbrat profitable company—national or multinational—cannot be equated with, say, a sugarcane farmer, rich or poor. For instance, there are at present protests brewing in different parts of the country against Coca-Cola, Pepsi and packaged water companies for extracting groundwater at the cost of local community needs.

In Kerala, the license given to Coca-Cola is being reviewed by the local *panchayat*—village council—and the company is trying to 'work' on the government to pressurise local people.

Privatisation?

Is 'privatisation', then, a solution or a disaster? Before this, let us understand why privatisation is happening at all. Is it only because the World Bank and water companies want to 'commodify' water and push water services in our part of the world? These agencies clearly smell lucre. We have dirty and scarce water, incompetent and bankrupt municipal agencies and growing populations. Our desperate need for clean water provides a fantastic business opportunity and great grist to their greed.

But I would argue that it is the rich and middle classes of developing countries that are actually responsible for the privatisation of water. I would, in fact, go so far as to argue that water scarcity and pollution are the outcome of the fact that water has for too long been considered a free good. A free good that benefits not the poor, but the relatively rich of the developing world. For the poor, even today, there is no free lunch. They pay—through their labour or with cash—for the meagre stinking water they get. In truth, they pay for it through worsening health. The relatively rich, in stark contrast, are grossly subsidised for the water they receive. Take Delhi: it costs the city public utility between Rs. 9 and 10 per m^3 to treat and distribute water in the city. Its citizens pay 0.35 paise per m^3—less than 4 per cent of the cost. Bangalore citizens pay the most: Rs. 5 for m^3. But their water cost is Rs. 40 for the same quantity, so they pay 12 per cent of the cost. Compare this to bottled water, where we pay Rs. 10,000 for each m^3 for the clean water.

But this is only half the story. The main cost is not in providing clean water, but in taking back the flushed dirty water in the sewage systems and treating it before discharging it into rivers. We know that sewage and drainage costs can be as high as five to six times more than the cost of water supply. And with increasing chemical pollution, water treatment costs are only going to increase. We have to recognise that current water and sanitation technology based on the flush toilet and sewage system would make full-cost pricing of water and sanitation services unaffordable by most in the urban South. It is important to recognise that private sector involvement cannot be only in the water supply business. This is just one small and profitable part of the water business. The real cost is in taking back the sewage and treating it to the quality needed for disposal in water bodies. This is the real 'dirty' business.

Urban populations do not even think about this, let alone pay. Therefore, literally, they are subsidised to defecate in convenience. India treats less than 7 per cent of the sewage it generates. No wonder we have massive water pollution problems. Let us be clear. Privatisation or not, the subsidised middle class of the developing world cannot and will not pay the true costs of water and sewage. Therefore, is the issue more about the reappropriation of this natural monopoly by the poor, and not about privatisation *per se*?

The political economy of defecation is such that no democratic government will accept the hard fact that it cannot 'afford' to invest in modern sewage systems for its citizens. Instead, it will continue to subsidise the users of these systems, in the name of the poor, who would not be able to afford the systems otherwise. It is important to realise that almost all users of the flush toilet and its sewage system are the rich in our cities. Our political system today literally subsidises the rich to excrete in convenience. In fact, we get a double subsidy. The pollution of rivers and water bodies caused primarily because of domestic sewage disposal is cleaned through expensive river action programmes. And the cost to build sewage treatment plants is externalised in these environmental programmes.

The logical policy would be to accept the cost and then to impose differential pricing so that while the rich pay for the cost of the capital and resource-intensive sewage and waste disposal technology, the poor pay for the cost of their disposal system, which is invariably unconnected to the sewerage system and hence low cost. But this is easier said than done.

The democratic framework in our countries would force political leaders to keep water and waste pricing affordable for large sections of urban populations. In this situation, private investment looks for an easy way out. The answer is to invest in water services and to leave the costly business of cleaning up the waste to government agencies. In most parts of the developing world, the water industry is bidding and securing contracts primarily for the profitable water business. This will lead to a distortion in the prices, as profits will be creamed off, while costs will be left to the already strained public exchequer.

In India, industry has been lobbying for private investment in the water sector. But it would like to focus on the water supply business. Or at best it would like to build and operate the treatment plants and leave local governments to price and recover costs from consumers.

Full-cost pricing?

The mention of the term 'water pricing' gets activists up in arms. Perhaps they are justified as companies tend to think of their profits and therefore, price becomes an instrument for greed, not payment of basic operations. But on the other hand, we have to understand that it is necessary to account for the cost of the resource and to develop a public policy for the rational and minimised use of water. Full-cost pricing is important for the relatively rich of the developing world. It is evident that urban and industrial sectors in the developing world do not even begin to pay for the water they use; indeed, they misuse it in their toilets and factories. Therefore municipalities are bankrupt and this, along with all other inefficiencies, makes them terrible service providers. The private sector is, in this circumstance, given a messianic role, largely because it can bring in financial investment due to higher credit ratings and profitability. Proponents of the private sector constantly talk about the billions needed to meet the target of safe water for all in the world.

The question, then, is the kind of contract that is signed between the private entity—interested in profits—and the public entity incapable of raising profits in the first place. Clearly it will be unfair. The municipality or local government will either see the private sector as the *instrument* to recover money from subsidised consumers, or simply see it as a way to provide some efficiency even as the government continues to subsidise its consumers and also pay the private sector its pound of flesh.

This is what I would call backdoor privatisation. This is the most-favoured model. Delhi, as explained earlier, has contracted Degremont to run its water treatment plant. The city municipality recovers less than 4 per cent of the cost of water it supplies to its residents. But it will have to pay Degremont full costs, if not more. It cannot raise water tariffs because electoral politics will not allow it. The same opposition politicians—it does not matter of which party, since opposition is a stance in itself—who will make a song and dance about the unfair terms of Degremont—quite correctly—will also not allow a single penny to be increased on water tariffs. They will definitely not allow the metering of water to take place so that large users can be billed accordingly.

All this is done in the name of the poor in this strangely 'socialist' country of ours. So, a few months ago the key opposition party in Kolkata went on strike against the attempts of the Marxist State government to install water meters in each household. This, when all these politicians realise that it is the rich, or at least the relatively rich, who guzzle water and that it is they who are being subsidised in the name of the poor. The poor get a few bucketsfull; studies show clearly that they pay much more for the little they use.

It is the same with industry—another holy cow—for whom private sector services are now being procured. They, too, pay a pittance for profligate extraction. Therefore, clearly, more than the issue of privatisation, it is the issue of payment for water by the rich and relatively rich that must be resolved first.

In all this, it is important to realise that the private sector will have little to offer to large numbers of urban poor. Most poor urban dwellers are illegal occupants—living in slums and highly congested areas. The cost of reaching and maintaining services to these groups is expensive and there is uncertainty about recovery of dues. The risks are high. The profits are low. In this situation, private investment is rarely available.

Therefore, even if full-cost pricing is done—which should be done—and the costs are recovered from the relatively rich in the developing world, there is no guarantee that the same companies will invest in sectors where there are no profits. Therefore, the problem is not with the idea of pricing or recovery of costs, but the framework that would provide the incentive for water for all.

Framework for the Partnership

There is also the key issue of: safe water for whom? The private sector has no answers for the rural poor in the developing world. They are not markets. They are scattered. They cannot pay. Therefore, anyone who argues for private sector involvement in the name of the billions of rural poor is, putting it bluntly, a fraud. But worse, governments also do not care. It is the resource of the poor that gets appropriated—surfacewater extracted and returned as sewage, or groundwater piped out and away—from their villages and neighbourhoods, exclusively for cities and industries.

Given all this, I would like to step beyond the simple 'with-privatisation-or-against-it' argument. I would certainly not argue that the private sector can solve the water problems of the world. But I would also not argue to exclude it from playing a role in water services. But a contractor in the private sector can only work within the terms society sets for it. It cannot own the resource. It certainly cannot be its custodian. It can only be contracted to deliver water and take back sewage. Sewage must be included in all water contracts because this is the real and dirty business of water.

The private sector may also be asked to set the price and recover dues. But setting the tariff must be completely transparent about the full costs of treating and delivering water and waste. Therefore, the decision of governments to subsidise its middle-class electorate must not be hidden behind socialist rhetoric. I expect that once some measure of the real price is paid for water and waste, this dependence on the private sector will also magically disappear. The public utility sector could become more, or at least equally efficient as the private company as its returns on equity—profits—can be fully reinvested in the system.

In the water management framework, in which the private sector is but one player, it is equally important to resolve the issue of 'ownership' of the resource. This is important not only because the rights of the poor must be safeguarded in present and future agreements, but also because, as the State and the private sector will not and cannot provide water for poor urban and rural communities, the rights of these communities to control and manage their natural asset must be secured.

It is also equally important to recognise that the private sector with its belief in full-cost pricing does not even begin to have answers for the millions of people living in the rural South. These communities already pay an enormous amount for water. In fact, here the community sector has an enormous sum to offer. Given the State-dominated water supply systems, little effort has been made to get rural communities to develop and manage their own water supply systems. But where done, it has shown outstanding results, including the willingness of rural communities to contribute substantially (labour in a big way and materials to a lesser extent) to the construction and maintenance of the water supply systems. This reduces the cost of water supply to the public exchequer and gives ownership to the stakeholders of the water supply projects. Community-based water management has the potential to become the world's biggest cooperative enterprise.

Rural communities need financial support for creating conditions that lead to self-management of water sources. The answer is not full-cost pricing but political decentralisation and empowerment. It is this community-industry-government collaboration that we must build in urban areas as well. Public participation and political process that pushes for good governance in water management are the key prerequisites for change; not another contractor.

Therefore, to my mind the water issue is not about privatisation. It is about the governance and regulatory framework to secure the rights and access of all to clean water. It is about the right to life. It is also about the rights to water for all. Let us not lose sight of this. Not even for an instant.

Coke Versus the People at Plachimada:
The Struggle over Water Continues

C. Surendranath

The Plachimada Sit-in at Coca-Cola

FOR OVER THREE YEARS, WOMEN, MEN AND CHILDREN OF PLACHIMADA VILLAGE IN PALAKKAD district, Kerala, have been holding a sit-in in front of a Coca-Cola factory in the village, demanding that the factory be shut down because it has depleted and contaminated the drinking water sources in the village. Though the courts have allowed the plant to operate, the local people and the village *panchayat* are determined to continue their fight against the company.

Plachimada is a remote village nestled in the Palakkad Gap of the Western Ghats in Kerala. It is situated in the low rainfall Chittur block, 30 km east of Palakkad and 5 km from the Tamil Nadu border. The average rainfall (from 1994–95 to 2003–04) is 1,413 mm as against a State average of over 3,000 mm. It has an undulating topography with predominantly black cotton soil underlain by weathered rocks. Groundwater had been the mainstay of the agricultural economy in the village, and farmers in the region had competitively dug and deepened a large number of borewells even before the Coca-Cola factory started intensive extraction of groundwater in 1999–2000.

Who is to Blame?

Today, women in Plachimada and neighbouring Pattamchery walk nearly 5 km to fetch water for drinking and household needs. Many farmers have watched their wells go dry, leaving their fields unirrigated. Most villagers squarely lay the blame on Coca-Cola. However, Coca-Cola, its 300-odd workforce, the labour unions and some political parties blame poor rainfall and the farmers' own borewells.

Availability of Groundwater: A Tricky Balance

Despite several studies, how much water the Coca-Cola plant consumes remains a matter of contention. The Anti-Coca-Cola Agitation Council claims the company extracts around 1.5 Mld. A Peoples' Commission chaired by environmentalist Dr. A. Achuthan and a team from Jananeethi, a Thrissur-based human rights NGO, found eight borewells in the factory premises with a total extraction capacity of 1 Mld. They contend that the water the company siphons off each day would meet the basic needs of 20,000 people. But Coca-Cola officials resolutely maintain that the factory consumes 0.5 to 0.75 Mld depending on the season, and that globally Coca-Cola's philosophy is to 'reduce, reuse, recycle and recharge' water.

The State Ground Water Department (GWD), the Central Ground Water Board (CGWB), the Kerala Legislative Committee on Environment, and the Kerala State Pollution Control Board (PCB) have conducted

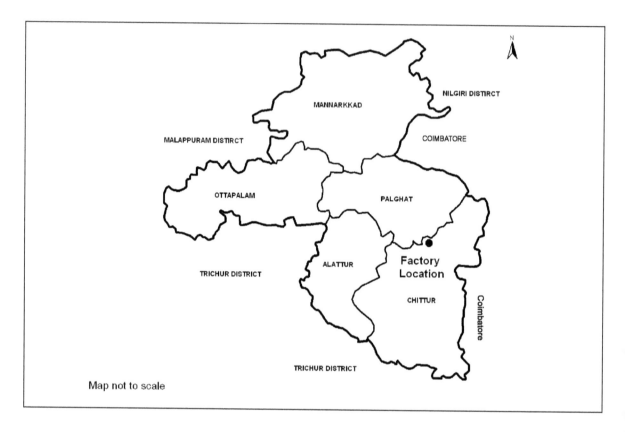

Map 1.

various studies on the depletion and contamination of water in the area. They have so far found no 'conclusive evidence' linking these problems to the extraction of groundwater and the discharge of waste water by the company. In 2002, the GWD attributed the depletion of water in some open wells neighbouring the Coca-Cola plant to poor rains, reduced groundwater recharge and high density of irrigation borewells. The report of the Expert Committee led by the Centre for Water Resources Development and Management (CWRDM), constituted on the orders of the High Court of Kerala in December 2003, also said that under normal rainfall conditions, Coca-Cola could safely withdraw 0.5 Mld of groundwater without adversely affecting local requirements.

Many other agencies challenge these conclusions. The Centre for Science and Environment (CSE) says the experts' conclusion is based on 'gross overestimation of the natural recharge potential' and availability of groundwater as well as an 'enormous depression' of real demand.

The Expert Committee calculated recharge at 20 per cent of the total annual rainfall, whereas according to CSE it would hardly be 5 to 10 per cent as found in similar geomorphologic regions. As a consequence, whereas the Expert Committee finds the groundwater availability in Chittur block to be 66.7 Mm^3 (million cubic meters mcm) and a 4.2 Mm^3 surplus, thus justifying allocation to Coca-Cola, CSE finds it to be between 16.6 to 33.3 Mm^3, and a deficit of between 47.51 to 30.91 Mm^3. According to the official water policy, drinking water and irrigation get priority. Therefore, the large deficit implies that water cannot be allocated to commercial users like Coca-Cola.

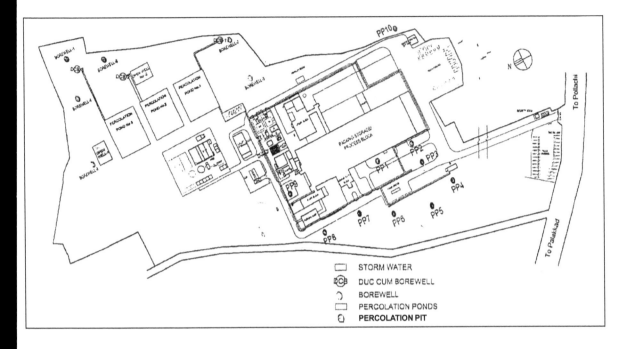

Fig. 1. *Location map of the water extraction and conservation structures within the permises of Coca-Cola Company.*

The Quality of the Groundwater

As scientists and officials keep arguing over the quantity, social activists are trying to give voice to the principles of precaution. The perceptibly worsening water quality in the village is an important reason for people's anger against Coca-Cola. The deterioration has impaired the health of many in the village. Independent researcher V.T. Padmanabhan cites low birth weight among children born after the inception of the plant. The local women have more down-to-earth complaints. Mylamma, the elderly tribal woman leading the agitation says that the water has turned bitter, and after a bath it leads to itching and swelling and causes a burning sensation in the eyes.

Official Agencies Do not Find Pollution ...

Despite these findings, most official agencies have stoutly maintained that there is no significant contamination of water in Plachimada, and that which is detected could not be attributed to the company. The GWD found only three out of twenty observation wells in the area to have quality problems—TDS (total dissolved solids) above 2,000 ppm (parts per million), hardness above 600 ppm, and magnesium content over 100 ppm, and concluded that 'pollution, if any, in the area was not due to M/s Coca-Cola factory discharge'. The CGWB found that water in Plachimada was 'suitable for all purposes' because 'water in 78.3 per cent of the wells in the area contained only low levels of TDS and was free of any heavy metals.'

...Independent Studies Do

Earlier, independent studies also found that the water in a few open wells in the neighbourhood of the Coca-Cola plant contained high amounts of chlorides and TDS. In March 2002, commenting on the lab analysis of water samples from a well in the Vijayanagar colony near the plant obtained by independent journalist Nityanand Jayaraman, Dr. Mark Chernaik of ELAW, in the US, warned that the water had become 'unfit for human consumption, domestic use (bathing and washing) and irrigation'. Subsequently, a Jananeethi team found the chloride content in the drinking water to be as high as 540 mg/l as against the recommended allowable limit of 250 mg/l. The Integrated Rural Technology Centre (IRTC) and the medical officer in Palakkad also found high values for chlorides. The medical officer of the local public health centre too cautioned the Perumatty *gram panchayat* against drinking water from these wells.

BBC's Channel Four analysed some samples at the Greenpeace Lab at the University of Exeter, UK, and found high levels of cadmium and chromium in the sludge the Coca-Cola plant sold as fertiliser, and heavy-metal contamination from sludge dumps in the water samples. Later, the New Delhi-based CSE found high concentration of harmful organochlorine and organophosphorous pesticides in the beverages Coca-Cola marketed all over the country. These revelations gave a boost to the people's struggle in Plachimada and did considerable damage to Coca-Cola's reputation and sales.

The Struggle Widens

The Plachimada struggle has by now gained a wider dimension. For many, the struggle has become a symbol of grassroots struggles against globalisation. Coca-Cola was a major target of protesters at the January 2004 World Social Forum in Mumbai. The Indian Parliament and several State governments have banned Coca-Cola products in their cafeteria.

A milestone in the course of the struggle was the April 7, 2003 decision of the Perumatty *gram panchayat* to refuse the Coca-Cola plant at Plachimada an operating license in order to 'protect public interest' and to charge the company with 'causing shortage of drinking water in the area through over-exploitation of groundwater sources'. Vandana Shiva, environmental activist, welcomed the decision saying, 'The *panchayat* taking on a mighty global corporate in the court of law is real democracy at work'.

No Agitation Against Pepsi?

Pepsico India Holdings Pvt Ltd has a bottling plant at Kanjikode in Pudussery *panchayat* in the Palakkad district, but within a declared industrial area and hence relatively removed from human habitation as well as the zone of control of the *gram panchayat*. Being located in the industrial area, the Pepsico plant had obtained its relevant operating license from the Single Window Clearance Board constituted under the Kerala Industries Single Window Clearance Board and Industrial Township Area Development Act, and not under the Kerala Panchayat Raj Act. Nevertheless, the Pepsico plant too had faced popular political protests, though sporadic and not as intense as in the case of Coca-Cola. Like Coca-Cola, Pepsico too had its license cancelled by the Pudussery *panchayat* (dominated by the Communist Party–Marxist [CPM]), but later managed to get a reprieve from the court.

The relative lack of intensity of the struggle against Pepsi has more to do with the lack of resources of the civil society organisations and their strategy of focusing the limited weapons on a single, most symbolic, enemy. Of course there was initially a game of political one-upmanship between the Janata Dal (which dominates the Perumatty *panchayat*) and the Communist Party–Marxist (which is dominant in Pudussery) that boosted the anti Coca-Cola struggle. The CPM, which had brought in both Coca-Cola and Pepsi, could have been more hesitant initially to take up an openly 'anti-development' stand. But the Janata Dal could do this because of its less rigid ideological moorings and party structure. But as the anti-Coca-Cola struggle gained local and international popularity, it created an ideological sphere of its own which no political party with its roots in mass appeal could ignore.

Legal Twists

The legality of this order has now come to be a contentious issue. Coca-Cola has approached the High Court of Kerala against the *panchayat* order. Under the Indian Constitution (Article 243 G), the legislature of a State may, by law, endow the *panchayat* with necessary powers and authority to function as an institution of self-government. In December 2003 a single bench of the court had observed that 'groundwater was a national wealth that belonged to the entire society' and needed to be protected. On April 7, 2005, however, a division bench of the same court ruled that it was permissible for 'an occupier of land to draw water out of his holding', and that the *panchayat* 'had no legal authority to cancel the license' issued to Coca-Cola. The court also permitted the company to draw 0.5 Mld of groundwater, as suggested by the CWRDM Expert Committee. The court, on June 1, 2005, categorically asked the *panchayat* to renew the license within a week.

Forced to comply, the *panchayat* went on to issue a license, but subject to strict conditions and only for three months, which Coca-Cola promptly rejected. Both parties have once again petitioned the court. Coca-Cola had hardly restarted production trials on August 8, 2005 when, on August 20, the PCB refused to renew the consent to operate and ordered Coca-Cola to close down since it could not adequately explain the use of hazardous cadmium either in production or in effluent treatment.

Justice Delayed Remains Justice Denied

The sit-in strike in front of the plant crossed a milestone when it completed three years on April 22, 2005. Many political parties and even the state government are now shifting their stance. The then local self-government minister Kutty Ahmed Kutty announced that the government would appeal against the Kerala High Court's verdict allowing the plant to reopen. Kodiyeri Balakrishnan, one of the leaders of the opposition, that is from the Communist Party of India (Marxist), said that even if Coca-Cola gets a favourable order from the Supreme Court, they would not be allowed to function, signaling a turnaround in his party's views since it was they who had brought Coca-Cola to the state in 2000.

The rains have come and gone. In spite of politicians, the State, the courts and many official committees having promised immediate supply of drinking water to the people, there is little that has been brought to parched Plachimada. The battle between Coca-Cola and the people has to be fought not only in the Supreme Court but also in the 'people's court'—in their minds and on the streets.

Chronology of Events

March 2000:   Hindustan Coca-Cola Beverages Ltd (HCBL) begins operations at the Plachimada plant.

March 2001:	First protest march demanding clean drinking water.
March 2002:	Analysis shows contamination of well water around the Coca-Cola plant.
April 2002:	Sit-in in front of the Coca-Cola plant begins.
June 2002:	Workers in the factory form a Labour Protection Forum to counter the agitation of the villagers.
April 2003:	Perumatty *gram panchayat* decides not to renew the license granted to HCBL.
May 16, 2003:	HCBL suspends operations.
July 25, 2003:	BBC Radio reports toxic heavy metals in effluent treatment plant sludge distributed to farmers by Coca-Cola.
October 2003:	State government quashes *panchayat's* decision and orders scientific investigation.
February 23, 2004:	PCB suspends consent on account of non-compliance of Hazardous Waste Rules.
March 2004:	HCBL suspends operations.
November 2004:	PCB directs HCBL to supply clean drinking water to people in Perumatty *panchayat.*
June 1, 2005:	The division bench of the High Court orders the *panchayat* to issue license to the company.
August 8, 2005:	Amidst protest marches and police retaliation on them, the Coca-Cola plant begins trial production.
August 18, 2005:	PCB orders closure of the plant.

Note

*Figure from Project Proposal, Investigations on the Extraction of Ground Water by M/S Hindustan Coca Cola Beverages Private Limited at Plachimada, filed before the High Court of Kerala, prepared by the investigation team constituted as per the directions of the High Court of Kerala (Order No. WA/2125/2003 December 19, 2003), coordinated by Dr. E.J. James, Director, Centre for Water Resource Management and Development (CWRDM), Kozhikode.

References

Bitter Drink, a documentary film by C. Saratchandran and P. Baburaj. Production: Third Eye Communications, North Fort Garden, Thrippunithura, Kochi, Kerala.

Coca-Cola Quit Plachimada, Quit India: The story of Anti-Coca-Cola struggle at Plachimada in Kerala, Published by Convenor, Coca-Cola Virudha Janakeeya Samara Samithi.

Report of the Joint Parliamentary Committee on Pesticide Residue in and Safety Standards for Soft Drinks, Fruit Juices and other Beverages. New Delhi: Lok Sabha Secretariat.

Resisting Cocacolanisation. New Delhi: Research Foundation for Science, Technology & Ecology.

Stolen Water, a documentary film by Daya and Prasanth. Keraleeyam, Thrissur.

Various Issues of *Keraleeyam*, Thrissur, Kerala.

Various reports accessed at *http://www.indiaresource.org.*

http://www.fian.org/

anticocacolastrugglesupporters@yahoogroups.com

People's Struggles for Water Rights over Kelo River Waters

Prakash Kashwan and Ramesh Chandra Sharma

CHHATTISGARH, THE TWENTY-SIXTH STATE OF THE INDIAN UNION, WAS CARVED OUT OF sixteen districts of Madhya Pradesh (comprising ninety-six of its *tehsils*, 146 of its community development blocks, and 30.47 per cent of its area) on November 1, 2000. It has a geographical area of 135,194 sq km and three distinct zones: in the north are the Satpura ranges; in the centre, the plains of the river Mahanadi

Map 1. *Location of Chhattisgarh in India.*

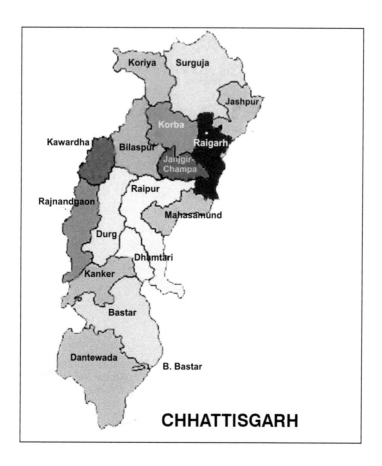

Map 2. *The district map of Chhattisgarh.*

and its tributaries; and in the south, the plateau of Bastar. The average annual rainfall in the State is about 1,500 mm and much of its yellow sandy soil is suitable for the cultivation of rice. It is one of the poorest States of the country, where an estimated 38 per cent live below the poverty line and weaker sections comprise more than 63 per cent of the population: scheduled tribes (ST) 32 per cent, scheduled castes (SC) 12 per cent, and other backward castes (OBC) 20 per cent.

The Kelo river is a minor perennial tributary of the Mahanadi, and is 102 km long. It originates in the Pahad–Lude hills near Lelunga and flows south-east through a wide middle catchment north-west of Raigarh town (Chhattisgarh) and meets the Mahanadi about 19 km south-east of Raigarh at village Mahadevpali Ghat, Jharsuguda district, Orissa. The banks of a narrow stretch in the middle portion of its catchment are marked by a range of thickly forested hills that trap and later release an appreciable amount of rainwater into the river by way of numerous small springs located on both sides of the river.

The Issue: Jindals Get Permission to Draw Water from the Kelo

At issue is the permission granted by the Chhattisgarh government (GoC) to Jindal Strips Limited, now called Jindal Steel and Power Limited (JSPL), to extract water through a combination of stop dams and intake wells from the Kelo, downstream of Raigarh town. According to the local population and activists, fourteen villages of Raigarh district in Chhattisgarh and nine villages in Orissa are affected, causing loss of livelihood, pollution of drinking water and increased drudgery for women. A writ petition against JSPL has been filed in the High Court of Chhattisgarh.

Located just outside Raigarh, JSPL is the world's largest coal-based sponge iron plant, and the largest open cast coal mine in Asia provides it with the necessary fuel. JSPL came to Raigarh in 1989 and over a decade acquired hundreds of acres of land for its industrial operations. It constructed and maintains a road that links the plants to Raigarh. The road is open to public against a toll fee of Rs. 12. JSPL offers a broad range of steel products, including the longest rail sections in the world, and also produces electricity. It is a member of the US\$ 1.5 billion steel conglomerate JSPL Organisation, a company with plants in Haryana and other Indian States. It uses coal and water to produce electricity in its captive thermal power plant to run its steel plant and the surplus is sold to the Chhattisgarh Electricity Board.

JSPL pumps out about 0.88 Mm^3 from the Kelo every month, which is about a tenth of the flow in the river. Considering that half of a river's flow is required to maintain the ecological balance, JSPL consumes 20 per cent of the available water. They also have an unknown number of borewells and some surface ponds.

'Save the Kelo' Movement

The Kelo Bachao Sangharsh Samiti (KBSS) was constituted in 1996–97 by some of the leading activists of Raigarh. In 1998, KBSS launched a public campaign to save the Kelo by asserting people's rights to natural resources, especially water. In 1998, a public interest litigation (PIL) was filed by KBSS activists and Goldy M. George. That year, tribals from the affected villages sat on hunger strike to protest against industrial exploitation of Kelo waters. A tribal woman Satyabhama died in somewhat mysterious circumstances after she was forcibly taken to hospital by a police team without any of the activists accompanying her. After an agitation, the GoC ordered a magisterial enquiry but the report has not been made public. This in some ways was the most intense period of the conflict.

Eight months later, Satyabhama's husband Gopinath, who had also participated in the hunger strike, filed a case against four activists accusing them of having caused her death by forcing her to sit on strike against her will and despite her alleged ill health. Today Gopinath regrets having filed the case, but his reasons for filing a suit are not clear. A partial explanation is that it is one of about twenty-five different kinds of cases that were slapped against leading activist Ram Kumar Agarwal. Similarly, most of the activists, journalists and some of the lawyers who were with KBSS to begin with have either been lured or forced into the JSPL camp and those who have not, including tribal villagers, are now facing several court cases.

- 1989: The Jindals enter Raigarh with a lot of public expectation and support.
- 1995: The Jindal steel strip factory is established. The Jindals subsequently establish two more plants and townships, a coal-based thermal power plant and recently another one manufacturing rail sections.
- 1996–98: The Jindals establish one open cast and one underground coal mine in the upstream of Kelo river catchment in Lelunga *tehsil*.
- 1996–97: The Kelo Bachao Sangharsh Samiti (KBSS) is constituted.
- 1998: A tribal woman, Satyabhama, dies on January 26 under mysterious circumstances after she is picked up by police from a fast-unto-death agitation.

Fig. 1. *Chhattisgarh Padyatra organised by Ekta Parishad passes through Kelo river.*

- 1998 onwards: The Jindals and their associates establish four additional sponge iron plants upstream of Raigarh town with independent water supply arrangements. One of them, Taraimal, is situated on the banks of the Kelo.
- 2000–5: Water supply in Raigarh town has to be augmented by digging more borewells. Today there are more than 175 borewells with energised pumps in Raigarh town. Both in downstream and upstream areas affluent farmers install energised borewells for irrigation, including summer irrigation.
- By 2005: Jindal and its associates establish four additional sponge iron plants upstream of Raigarh town with independent water supply arrangements. Taraimal is situated on the banks of the Kelo.

Not Quite Sufficient Evidence?

Tribals and the KBSS argue that there is not enough water in the Kelo river to fulfil the needs of Raigarh town, the downstream villages and JSPL. In support, they cite minutes of meetings of water authorities and their letters denying JSPL the right to take water from the Kelo river, and newspaper clippings reporting water scarcity; they have attached these to the PIL as evidence. In response, JSPL and the GoC argue that there is enough water and that the needs of the inhabitants have been properly taken into consideration before granting permission. They explain that permission has been granted to JSPL to draw water downstream of Raigarh town instead of upstream as it first requested, thus ensuring sufficient water for Raigarh. As far as downstream villages are concerned, the GoC claims that there is enough water left and cites expert reports and recorded water flows during the critical months from February to June. JSPL also produced letters from twenty-seven villages certifying that they had enough water and had no objections to JSPL drawing water from the Kelo, and minutes of other meetings and also letters from the state water authorities allowing the intake of water.

An expert team constituted by Ekta Parishad, a national mass-based movement fighting for rights of tribals, Dalits and women, concluded:

> The induction of the intake well and stop dam by JSPL Steel and Power Ltd. has differential impact in the three ecological regions, namely in the plains villages, the villages adjacent to hills and forest, and the upstream village. The most adverse impact on various sources of livelihood is felt in the villages in the plains as they are adjacent to the intake well and stop dam. . . . [A]ccess to water from the Kelo River for agriculture, drinking water, cattle requirements and fishing has been seriously hampered in these villages after the induction of the intake well and stop dam. Wells, ponds and lift irrigation, which were the main source of irrigation before induction of the intake well, get dried up during summer months completely. The riverbed of the Kelo was utilised by marginal and small farmers for growing vegetables, which has now become a thing of the past. It has severely affected their source of livelihood. River Kelo remains bereft of water in the areas adjacent to villages in the plains during the months of April–June. The landless labour, marginal and small farmers completely depended on Kelo river for their everyday requirements of bathing, drinking and cattle cleaning.

Apart from the water drawn from the Kelo for Raigarh and JSPL factory, a substantial amount of groundwater is also extracted from the river basin. The total groundwater extracted from the JSPL factory premises is unknown. The municipal administration of Raigarh town extracts 0.67 Mm^3 every month for its needs. The number of borewells drilled for irrigation and installed with electric pumps is increasing; for example, in Rengalpali village there are four borewells irrigating 8 ha of land and in Netnagar twenty-two borewells irrigating 18 ha of land. Obviously, the cumulative effect is reflected on the ecology of the area affecting livelihood processes. The exact impact can be revealed only through systematic detailed studies but some indicators are already visible: for example, drying up of river flow at the confluence of the Kelo and the Mahanadi, and fluctuating flow in the river as one traverses the basin from Libra village in Lelunga *tehsil* in the north-west to Rangalpali village in Raigarh *tehsil* in the south-east.

Heeding Public Perceptions

Though it is not clear how much JSPL has contributed to in this cumulative impact, one thing is certain: that people in fourteen Raigarh villages downstream of the JSPL pumping site perceive that loss of flow in the river is causing changes in their lifestyle and that it is due to the water that JSPL draws every day, 365 days a year. Unless the process is halted and this perception is dealt with, it is difficult to see the conflict being resolved. Instead, in complete disregard of these perceptions, JPSL has now mooted fresh expansion plans with investments of over Rs. 1,200 crore. A public hearing was scheduled on January 4, 2005 to consider public opinion and grievances with respect to the new plans. More than 2,500 people turned up at the hearing but it could not be held because of the chaos that ensued at the venue. As the District Collector admitted, '[The authorities] were not prepared for such a turnout for the hearing because it was just a formality for getting the proposal cleared from the Ministry of Environment and Forests'. The turnout reportedly included people from Haryana, called in by the JSPL. The hearing was postponed to January 29. On January 7, the GoC signed yet another MoU with JSPL for investments totaling Rs. 2,595 crore. According to the company's press release, 'Under the MoU, the state government of Chhattisgarh will, among other things, provide all necessary help in seeking clearances, approvals, permissions from state/central government departments/agencies for implementation of these projects'.

Eyewitnesses say that over 3,000 people turned up at the January 29 hearing and submitted more than 1,900 applications providing accounts of water and air pollution and degradation of natural resources caused by JSPL operations. A 4 km long banner was signed by 45,000 people of the area demanding complete rejection of JSPL's expansion plans.

Encouraging Signs

The Chhattisgarh Chief Minister Raman Singh is said to have instructed officials that none of the expansion plans of JSPLs could be approved until it undertook proper pollution control measures at its existing facilities. Similarly, the environment officer of Raigarh has been issued a notice to investigate public grievances of over-exploitation of water resources and various kinds of pollution. Therefore, there seems to be some sort of an attempt at accountability at least as far as the government is concerned. This could lead to some sort of dialogue if the pressure is kept up on the JSPL. Given past experiences, however, most of the activists find it hard to believe that the company will make a serious effort to address public grievances. However, the State-wide movement against the pollution and environmental degradation caused by industrialisation is an encouraging sign. The government and JSPL may be forced to heed people's voices.

Note

*Maps from SOPPECOM, Pune; figure from Ekta Parishad.

References

Gimelfarb, Léonor. 2004. 'Industrialisation, Environmental Degradation and Social Dislocation in Raigarh—A Legal Perspective'. Study on the JSPL Case for Ekta Parishad. Mimeo.

Fouzdar, Dilip, K. Gopal Iyer and Raghuvir Pradhan. 2003. 'Independent Inquiry on Water Sharing: Observations in Kelo River Basin on Changing Ecological Environment Influenced by a Rapidly Changing Water Use Pattern and Study on Villages Affected by the Kelo Shutter-cum-Stop Dam and Intake Well Installed by JSPL Steel & Power Ltd'. Mimeo.

The Source of Life for Sale, a documentary film by K.P. Sasi. Distributed by Visual Search.

Acknowledgements

This case quotes extensively from Léonor Gimelfarb's work (2004). Useful comments from Jill Carr-Harris, Ramesh Sharma, and P.V. Rajgopal from Ekta Parishad, K.J. Joy and Shruti from SOPPECOM, and an anonymous reviewer are acknowledged with gratitude.

A River becomes Private Property: The Role of the Chhattisgarh Government

Binayak Das and Ganesh Pangare

Whose River is it Anyway?

SHEONATH RIVER FLOWS THROUGH BORAI IN DURG DISTRICT, CHHATTISGARH. THIS CASE IS about the handing over of a stretch of the river near Borai to a private firm for supplying water to the region lying between two district headquarters, Durg and Rajnandgaon. Borai is a newly developed industrial hub, promoted by the Chhattisgarh State Industrial Development Corporation (CSIDC). It is 45 km from Raipur airport; on National Highway 6 and the main Mumbai–Howrah railway line. The region is rich in natural minerals and at a reasonable distance from Bhilai (10 km); surplus power is available at a reasonable cost. Surrounding the Borai region is a cluster of villages that have traditionally used the river water for irrigation and fishing.

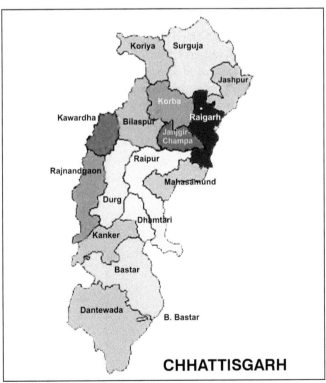

Map 1. *The location of Raigarh in Chhattisgarh.*

Sheonath river, a semi-perennial tributary of the Mahanadi, has been contracted to Radius Water, a division of Kailash Engineering Limited, for a period of twenty-two years. Radius Water is based in Rajnandgaon near Borai and has been managing the water distribution from the river. The build-own-operate-transfer (BOOT) project was commissioned in 2001 by the Chhattisgarh government. Actually, the Madhya Pradesh Industrial Development Corporation initiated the project in 1998 before the birth of the new State. Under the scheme, water from the river was supplied to the industries in bulk as part of an agreement with the CSIDC. Most of the industries located here are water-intensive by nature—distilleries, sponge iron units and thermal power plants—and CSIDC is trying to make it the hub for all water-based industries. Radius is not only responsible for supplying water, but also for operating a common effluent treatment plant (CETP).

Radius Water built a four-metre-high dam through a technique called the Flood Regulating Barrier System along a 3.5 km stretch of the Sheonath at a cost of Rs. 4 crore. The total cost for the project is Rs. 9 crore for 30 million litres per day (Mld). At present, it has a supplying capacity of 6 Mld, but the CSIDC is buying only 4 Mld now due to lack of demand. With new industries coming up in the area, however, the scenario is likely to change. The cost of water is Rs. 6.60 for 1,000 litres of water. For CETP, the price has been fixed at Rs. 3.60 per m³. The project is feasible for 180 Mld capacity through an investment of Rs. 156.70 crore, excluding the cost of CETP; CSIDC is the regulating authority for the project. One of the clauses in the agreement was that the villages downstream would get water free of cost; the clause also mentions that 'under any circumstances, the industry will be provided 30 Mld of water'. The water is stored in the reservoir and has the capacity for a 180 Mld supply.

The conflict did not start immediately. Initially, the locals were not aware that a private firm managed the new barrage that had sprung up across the river. No prior information was provided about this contract. After a few months, however, Radius Water informed the local fishermen that they were no longer permitted to fish in the 200 m zone from the barrage (on both sides) for safety reasons. There were a few skirmishes and employees of Radius Water allegedly destroyed some of the fishermen's nets. The latter complained that their catch had dwindled after the construction of the barrage. Farmers who owned land near the river were also barred from lifting water from the river with motor pumps. This ban had the endorsement of the district administration, which also banned the installation of tubewells. The CSIDC enforced this rule within Borai Industrial Area too. People from downstream villages started complaining that the groundwater table had plummeted. Many villagers from Pipalcheda, one of the surrounding hamlets, insisted that the water level in their wells had plunged since the construction of the barrage. Other complaints had to do with the company not adhering to the terms and conditions of the contract. According to the written agreement, the project

Fig. 1. *The barrage built by Radius Water on the Sheonath river.*

cost and maintenance is the responsibility of Radius Water. However, the construction was not completed on time because the company did not have sufficient funds. A decision was taken to the effect that of the total cost of Rs. 9 crore, 6.5 crore would be loaned to the firm and the remaining Rs. 2.5 crore would be generated in the form of equity.

Ironically, this hush-hush business about a private firm managing a stretch of the river came into the limelight when the media covered a publicity event organised by Radius Water to showcase their project. With the murmur of complaints rising, many activists and members of the public launched a campaign against the project, highlighting the fact that by handing

Fig. 2. *Factories in Borai that stand to benefit from this barrage, as they will be supplied with bulk water from the Sheonath.*

over the river to a private firm the State government had privatised the river. When the news spread across the country the resultant media glare and activism forced the State government to look at the deal afresh. A huge rally took place on November 1, 2003 under the banner of the Sheonath Nadi Mukti Andolan. Members of the youth federation went to fifty villages to conduct awareness campaigns. The pressured government ultimately decided to scrap the deal with Radius Water but, according to reports, despite the supposed termination of the contract the private firm continues to manage the barrage and supply water to the industries. Since the water was meant for bigger factories than the current customers, the pressure on the river has not been felt as yet—HEG remains the main consumer with a 1 Mld demand, and the river has been flowing at its normal pace. The government is also moving slowly because they will have to deal with a loss of revenue if the contract order is altered.

Greedy Government

The protesters have been questioning the very concept of privatisation of the river. They wonder how the Industries Department signed a contract for a river that legally falls in the purview of the Irrigation Department. The conflict—one of profit to industry versus livelihood for the community—is as old as time. Activists and lawyers argue that the deal violates the Madhya Pradesh Irrigation Act of 1931 and the National Water Policy, which prioritise agriculture over industries. Natural resources cannot be signed over to individuals without taking all the stakeholders into confidence.

Radius Water, on the other hand, insists that the upcoming industries at Borai will boost the State's economy and that they were merely ensuring that water was supplied to them at a low price. According to them the industrial water tariff in Borai is the lowest in the country. The comparative rates are: in Mumbai Rs. 50/m^3, and in Nagpur, Boregaon and Dewas it is Rs. 16/m^3 as against Rs. 6.60/m^3 in Borai. The company also argues that the construction of the barrage has helped the water table rise by 8 m in upstream villages, which is sure to help the farmers. With the threat of cancellation of contract looming large, Radius Water has said that they will withdraw if they are paid a compensation of Rs. 400 crore since the agreement is valid for another twenty years; they have also been threatening legal action.

Where Do We Go from Here?

This is an interesting case that has actually thrown open the debate over the rights of communities versus the rising demand from industry. Governments have to ensure a balancing act that leaves both sides satisfied. It will not be correct to name Radius Water as perpetrators—like any private firm they pounced on an opportunity to make profit. The main agency responsible for creating friction is the government, which went about the deal in a cloak-and-dagger fashion. By not respecting the National Water Policy and the Irrigation Act, they violated laws and are liable to be taken to court. The contract was signed without setting up independent regulatory authorities that could establish guidelines under which a private firm could manage a common resource. In this case, the conflict cannot be resolved by dialogue between stakeholders or by providing compensation to farmers and fishermen; the case has shown that a long lease without regulatory mechanisms can lead to unforeseen consequences. The contract will set a bad precedence and has to be cancelled for justice to prevail. The affected parties should be properly compensated for the losses.

The conflict goes to show that perhaps the country needs innovative guidelines if it is to keep up with growing demands for water across all sectors. It highlights the need for adequate guidelines and regulatory mechanisms. A dialogue that involves all parties has to be initiated to resolve the larger problem and the media could play a key role in bringing the players together: the local people who use the rivers, lakes and groundwater, the industry, and other experts, governments and politicians. This can lead to a better understanding of their requirements and constraints, and it will allow the government to build a consensus and draw up a proper framework for a just distribution of water.

Note

*Map 1 by SOPPECOM, Pune; figure by G. Pangare.

About the Contributors

Raju Adagale works with the Society for Promoting Participative Ecosystem Management (SOPPECOM), Pune. His areas of interest include watershed development, participatory irrigation management and renewable energy sources. His e-mail is rajuadagale@yahoo.com

Sara Ahmed is an independent researcher working on gender, water governance and vulnerability to disasters in South Asia. She was the Co-Project Director for the IDPAD project involving UTTHAN and other partners. Her e-mail is sarahmedin@yahoo.co.in

Paul Appasamy is Professor and Member-Secretary, Centre of Excellence in Environmental Economics, Madras School of Economics, Chennai. His research areas include water policy and urban environmental studies. His e-mail is ppasamy@hotmail.com

Gaspard Appavou is a Research Fellow at the Foundation for Ecological Research, Advocacy and Learning (FERAL), Pondicherry. His e-mail is g.appavou@feralindia.org

K.H. Amitha Bachan is with the River Research Centre of the Chalakudy Puzha Samrakshana Samithi, Thrissur, Kerala. His e-mail is rrckerala@rediffmail.com

Satyasiba Bedamatta is a Ph.D. Research Scholar at the Agricultural Development and Rural Transformation Centre, Institute for Social and Economic Change (ISEC), Bangalore. His e-mail is *satyasiba_b@yahoo.co.in*

Vimal Bhai works with the Matu Peoples' Organisation, Delhi, and has been associated with Tehri Dam issues since 1989. He has been writing on displacement and environment issues for the last two decades and published a compilation of articles entitled *Band ek Jid Nahi Hota (2005; in Hindi). His e-mail is* vimal_bhai@vsnl.net

M.J. Bhende is Professor of Economics and Trade at the Indian Institute of Plantation Management, Malathalli, Bangalore. He works in the areas of risk, insurance, production economics, credit and natural resource management. His e-mail is mjbhende@yahoo.com

Malavika Chauhan has been associated with various studies at Keoladeo National Park over the last fifteen years. Her e-mail is malavikachauhan@gmail.com

Ashim Chowla is a development sector professional and travel entrepreneur based in Bhopal. He has worked with a variety of organisations, including the UK government's DfID as a Representative for Madhya Pradesh, and for several years he was involved with action-based work at the grassroots level. He now runs his own firm providing travel solutions and development sector consultancy. His e-mail is ashimchowla@rediffmail.com

Binayak Das is a researcher with the World Water Institute, Pune. He is currently pursuing his Masters in Water Management from UNESCO-IHE in Delft, The Netherlands. His interests include water and environment, and his e-mail is phixod@yahoo.com

N. Deepa works as Senior Research Assistant at the Ecological Economics Unit, Institute for Social and Economic Change (ISEC), Bangalore. Her e-mail is deepan@isec.ac.in

Datta Desai is with the Academy of Political and Social Studies, Pune. His e-mail is svacademy@hotmail.com

R.S. Deshpande is Professor at the Institute for Social and Economic Change (ISEC), Bangalore. He has studied the Chilika in detail. His e-mail is deshpande@isec.ac.in

Shripad Dharmadhikary is with the Manthan Adhyayan Kendra, Badwani, and is a researcher and activist working on water-related issues in Madhya Pradesh for the last 15 years. His e-mail is shripad@narmada.org

H. Diwakara is Policy Officer (Economic & Social Assessment), Water Economics at the Department of Natural Resources and Water in Brisbane, Australia. His e-mail is diwakara.halanaik@nrw.qld.gov.au

D. Dominic is Programme Officer at Oxfam India/Svaraj, Bangalore, and has been in charge of the Arkavati sub-basin initiative of Oxfam India/Svaraj. His e-mail is dominic_d@svaraj.info

Sujeetkumar M. Dongre is Programme Associate with the Centre for Environment Education (CEE), Goa. His e-mail is ceegoa@ceeindia.org

Abey George is currently Assistant Professor at the Kerala Institute of Local Administration, Thrissur, Kerala. As a Fellow of Lead India, he is Honorary Coordinator of the Decentralised Resource Centre (LEAD-DRC), Kerala. His e-mail is abey@leadindia.org

Vinod Goud is Project Coordinator of the World Wide Fund for Nature (WWF) International, Gland, Switzerland implemented project 'Dialogue on Water, Food and Environment' in India, and is based at the International Crops Research Institute for Semi Arid Tropics (ICRISAT), Patancheru, Andhra Pradesh. His e-mail is v.goud@cgiar.org

Biksham Gujja is Policy Advisor of the Global Freshwater Programme at WWF-International, Gland, Switzerland, and Special Project Scientist based at the International Crops Research Institute for Semi Arid Tropics (ICRISAT), Hyderabad, Andhra Pradesh. He is associated with several global initiatives. His e-mail is b.gujja@cgiar.org

Ramaswamy R. Iyer was formerly Secretary, Water Resources, Government of India. He is Honorary Research Professor at the Centre for Policy Research, New Delhi, and the author of *Water: Perspectives, Issues,*

Concerns (2003; rpt 2005). He has written extensively on water policy and water governance. His e-mail is ramaswamy.iyer@gmail.com

Jasveen Jairath is General Secretary of the Society for Participatory Development, Hyderabad, and Regional Coordinator, CapNet South Asia, Hyderabad. She works on issues of water management, policy and practice in the context of political economy. Her e-mail is capnet_southasia@spdindia.org

S. Janakarajan is Professor at the Madras Institute of Development Studies, Chennai. He has done extensive research in agrarian relations, water, environment, and conflict resolution. His e-mail is janak@mids.ac.in

N. Jayakumar is Research Associate at the Madras School of Economics, Chennai. His e-mail is jayakumar71@hotmail.com

(Late) G.D. Joglekar retired as engineer from the State Irrigation Department, Maharashtra and was working as an independent irrigation consultant till his death in 2006.

K. J. Joy is Senior Fellow with the Society for Promoting Participative Ecosystem Management (SOPPECOM), Pune. He has actively participated in people's movements for equitable water distribution and has a special interest in people's institutions for land and water management. He has written extensively on water issues. His e-mail is joykjjoy@gmail.com

Meera Kaintura is the coordinator of SAMBANDH, a network of voluntary organisations in Uttaranchal, and has extensive field experience in community mobilisation. Currently she is working as United Nations Volunteer to facilitate disaster preparedness and response within Uttaranchal. Her e-mail is meera_utp@rediffmail.com

Prakash Kashwan is with the Workshop in Political Theory and Policy Analysis, Bloomington, USA. The cases of Lava ka Baas in Rajasthan and Kelo river in Chhattisgarh were documented while he was an independent consultant, before he enrolled for his doctoral program (Public Policy) in the USA. His e-mail is pkashwan@indiana.edu

Namrata Kavde-Datye is an independent researcher and Founder of 'Identity Foundation', an NGO based in Pune, which works with marginalised children. Her e-mail is namrata@identityfoundation.org

Indira Khurana is Documentation Consultant with Water and Sanitation Management Organisation and co-editor of *Making Water Everybody's Business: Practice and Policy of Water Harvesting (2001)* and *Scaling Up Sector Reform: Looking Ahead, Learning from the Past (2006).* Her e-mail is indirashok@vsnl.com

Jyothi Krishnan is a Research Scholar at the Department of Environmental Sciences, Wageningen Universiteit, The Netherlands. Her e-mail is jyothikr@sancharnet.in

Pankaj Lal is Program Officer at Winrock International India. He works on community-based natural resource management and institutional innovations for sustainable resource conservation and improved livelihoods. His e-mail is pankajwinrock@gmail.com

Benjamin Larroquette is on the Board of Directors, Foundation for Ecological Research, Advocacy and Learning (FERAL), Pondicherry, and Program and Livelihood Adviser at Agence pour la Cooperation Technique et le Developpement (ACTED), Dana, Auroville. His e-mail is benlar75@yahoo.com

S.N. Lele has retired as Chief Engineer from the Ministry of Water Resources, New Delhi, and is currently with the Society for Promoting Participative Ecosystem Management (SOPPECOM), Pune. His e-mail is shripadlele@yahoo.co.in

C.G. Madhusoodhanan is with the River Research Centre of the Chalakudy Puzha Samrakshana Samithi, Thrissur (Kerala). His e-mail is rrckerala@rediffmail.com

Anjana Mahanta is a freelance social science researcher. Her current projects are on cultural, developmental and socio-economic issues, focusing particularly on Northeast India. Her e-mail is kuki_iit@yahoo.com

Chandan Mahanta is Associate Professor at Department of Civil Engineering, Indian Institute of Technology, Guwahati. His research interests include environmental hydrology, environmental management, environmental impact assessment and sustainable development of water resources, fluvial ecosystems, and regional catchments cascade in the Brahmaputra system. His e-mail is mahanta_iit@yahoo.com

Mihir Kumar Maitra is a Watershed Management Specialist and is presently working as the Senior Project Consultant in India Canada Environment Facility (ICEF), New Delhi. His e-mail is mmaitra@icefindia.org

S. Manasi works as Research Officer at the Centre for Ecological Economics and Natural Resources, Institute for Social and Economic Change (ISEC), Bangalore. She is interested broadly in Natural Resources Management with special interest in water issues. Her e-mail is manasi@isec.ac.in

Rehmat Mansuri is a researcher-activist working on water-related issues in Madhya Pradesh for the last 15 years. His e-mail is rehmat@narmada.org

Jennifer McKay is the inaugural Director of the Centre for Comparative Water Policies and Laws, University of South Australia, and is Natural Resources Commissioner in the Environment Resources and Development Court in South Australia. She has over 60 publications on Natural Resources Management to her credit. Her e-mail is jennifer.mckay@unisa.edu.au

Ajit Menon is with the Madras Institute of Development Studies (MIDS), Chennai. His e-mail is jitumenon@yahoo.com

R.V. Rama Mohan works as Project Coordinator at the Centre for World Solidarity (CWS), Secunderabad. He is involved in initiatives addressing community rights over natural resources, displacement in irrigation projects and bio-watersheds. His e-mail is rvrm2@yahoo.com

Srinivas Mudrakartha is the Director of VIKSAT Nehru Foundation for Development, Ahmedabad. He has three decades of experience in natural resource management, livelihoods and people's institutions. His e-mail is srinivasm@viksat.org

R. Murali is Secretary of Modern Architects for Rural India (MARI), Warangal. He is a well-known social activist and development professional, and his e-mail is mariwgl@gmail.com

Sunita Narain is Director of the Centre for Science and Environment and the Society for Environmental Communication, and also the publisher of the fortnightly *Down to Earth*. Her e-mail is cse@cseindia.org

Bala Raju Nikku is Head, Department of Social Work at Kadambari Memorial College of Science and Management in Kathmandu, Nepal. He is currently a doctoral research scholar at the Irrigation and Water Engineering Department at Wageningen Universiteit, The Netherlands. His e-mail is nikku21@yahoo.com

Ganesh Pangare is with the World Water Institute in Pune. His e-mail is gpangare@hotmail.com

Suhas Paranjape is with the Society for Promoting Participative Ecosystem Management (SOPPECOM), Pune. He has been working on land and water management issues and has been closely associated with various mass organisations in this area for the past several years, and has written extensively on water issues. His e-mail is suhas.paranjape@gmail.com

Bharat Patankar is a leading activist of the Left-wing Shramik Mukti Dal, and of the peasant movement (especially the dam-oustees and drought-affected) in south Maharashtra. He is also one of the architects of equitable water distribution movement in Maharashtra. His e-mail is gailomvedt@rediffmail.com

Iswaragouda Patil is with the Centre for Interdisciplinary Studies in Environment and Development (CISED), Bangalore. His e-mail is iswar@isec.ac.in

R.K. Patil retired as Professor from the National Institute of Bank Management (NIBM), Pune, and is currently with the Society for Promoting Participative Ecosystem Management (SOPPECOM), Pune. His e-mail is rk_patil2004@rediffmail.com

Anita Paul is Director (Community Initiatives) of Pan-Himalayan Grassroots Development Foundation, Almora, and is involved with spearheading holistic mountain development programmes in the central and western Himalayas. Her e-mail is apaul@grassrootsindia.com

Kalyan Paul is Executive Director of Pan-Himalayan Grassroots Development Foundation, Almora, and has been involved with water sector reforms and establishment of a major river basin organisation in the central Himalayas. His e-mail is kpaul@grassrootsindia.com

Anant Phadke is a founder member of the Lok Vigyan Sanghathana, Pune, and its Health Committee. He has been writing in the local and national press since the 1980s about the peasant movement in south Maharashtra. His e-mail is amol_p@vsnl.com

Ravi Pomane works with the Society for Promoting Participative Ecosystem Management (SOPPECOM), Pune. His interests include rural development with special reference to water and agricultural issues. His e-mail is ravi_pomane@yahoo.com

Govind S. Poteker is a freelance journalist. He is Member of the Association of Friends of Astronomy, Goa, and Nisarg Nature Club, and regularly contributes to the CEE Goa publications and programmes. His e-mail is gspafa@rediffmail.com

Anjal Prakash works with WaterAid India, New Delhi. He holds a Ph.D. degree in Environmental Science from Wageningen Universiteit, The Netherlands. He has worked and researched in Gujarat for 8 years on issues of groundwater irrigation and government's rural drinking water supply project. His e-mail is anjal@wateraidindia.org

Eklavya Prasad is an independent development practitioner. He works on community-based integrated water management systems in the flood-affected communities of north Bihar. His e-mail is graminuatti@gmail.com

Kapildeo Prasad is a resident of Ghatkuraba village and has been instrumental in the formation of community institutions responsible for the maintenance of Barkop dam. His postal address is Pathargama Block, Godda Ghatkuraba Village, Distt Godda, Jharkhand, India. Tel.: (00 91) (0)641 2403110

T. P. Raghunath is Chairperson of Centre for Ecology and Rural Development (CERD), Pondicherry, and served as Nodal Officer of the Tank Rehabilitation Project. He led the negotiations and discussions with the conflicting sides from the beginning of 2002 to the end after which the tank association was formed and registered. His e-mail is tprmenon@gmail.com

A. Rajagopal is Senior Fellow/Executive Director at SaciWaters, Hyderabad. He has wide experience in water and related areas of research such as water management, agrarian conditions and poverty analysis. His email is rajagopal@saciwaters.org

K. V. Raju is Professor and Head, Ecological Economics Unit, Institute for Social and Economic Change (ISEC), Bangalore. His e-mail is kvraju@isec.ac.in

R. Ramadevi is a Consultant with District Rural Development Agency in Cuddapah district, Andhra Pradesh. She is involved in capacity building, community health, and drinking water-sanitation issues since last two decades. She is also a member of Asian Health Institute, Japan. Her e-mail is ramanayam_ramadevi@yahoo.com

M. Chandrasekhara Rao is a Consultant with water resource management working with Watershed Support Services and Activities Network (WASSAN), Hyderabad. His e-mail is cmulukuri@yahoo.com

M. Pandu Ranga Rao retired as Professor from the Department of Civil Engineering, National Institute of Technology, Warangal. His e-mail is profraomp@yahoo.com

J. Rama Rao is Chairperson, Forum for Better Hyderabad, Hyderabad. His e-mail is captjrrao@eth.net

Pushplata Rawat is engaged in providing support to local institutions working with issues of natural resource management and livelihood in Uttaranchal. Currently she is working as United Nations

Volunteer to facilitate disaster preparedness and response within Uttaranchal. Her e-mail is pushplata@rediffmail.com

P. B. Sahasranaman is a lawyer based in Kerala specialising in Public Interest Environmental Cases, and has written extensively on environmental law. His e-mail is sahasram@dataone.in

R.K. Sama retired from the Indian Forest Services in 1997 after 38 years of service in Gujarat, and has contributed significantly to the watershed programme. Since 2002 he is associated with Water and Sanitation Management Organisation (WASMO). His e-mail is board-wasmo@gujarat.gov.in

S.B. Sane is a retired engineer from the State Irrigation Department, Maharashtra. Currently he works with the Society for Promoting Participative Ecosystem Management (SOPPECOM), Pune, and his e-mail is sanesb@vsnl.net

Sanjay Sangvai is a freelance journalist writing mainly on development politics. He is an activist of the Narmada Bachao Andolan (NBA), and Editor of *Movement of India*, a magazine of the National Alliance of People's Movement (NAPM). His e-mail is sanjsang@gmail.com

Sumita Sen is Professor and Head of Department of International Relations, Jadavpur University, Kolkata. Her areas of interests include legal regime and policy issues relating to environment and development especially in water and biodiversity in South and South-East Asia and their impact on foreign relations. Her e-mail is senmita@yahoo.co.in

Esha Shah is a Fellow with the Institute of Development Studies (IDS), University of Sussex. Her research explores various aspects of relationship between science, technology and development. Her e-mail is o_she@hotmail.com

M. P. Shajan is Secretary, Janayogam, and a member of the Puthenvelikkara Grama Panchayath in Ernakulam district, Kerala. His e-mail is rrckerala@rediffmail.com

Batte Shankar is Convener, Musi Parirakshana Samiti, Edulabad, Ranga Reddy District, Andhra Pradesh. His e-mail is capnet_southasia@spdindia.org

Ramesh Chandra Sharma is associated with Ekta Parishad for the last 8 years as a Campaign Coordinator. He is mostly involved in grassroots research, advocacy and action. His e-mail is ektaparishad@gmail.com

Jatin Sheth is Advisor at VIKSAT, and has 27 years of experience with Industry. He is currently promoting micro-enterprise activities, and is involved in the Sabarmati Stakeholders Forum in VIKSAT. His e-mail is mail@viksat.org

Kamaldeo Singh is a resident of the Ghatkuraba village, Godda District. He has worked for the Development Alternatives group for two years. At present, he is pursuing PGP-ABM from IIM, Ahmedabad. His e-mail is kamaldeo@rediffmail.com

Nandita Singh is a freelance writer interested in the field of social transformation and development. She has been a witness to the conflict in Paschim Midnapur District, West Bengal for a year. Her e-mail is nandita_28@hotmail.com

Praveen Singh has been working on issues related with water, especially floods, in eastern India. He has worked with Eco Friends in Kanpur on 'De-polluting the Ganga in Kanpur', and his e-mail is nispraveen@yahoo.co.in

Vikash Singh is an independent researcher based in Bhopal. He works in development and cultural studies, and writes as a supporter and friend of the Tawa struggle. His e-mail is hi2vikash@gmail.com

Chandan Sinha is a practitioner and student of development administration, with a special interest in participatory community development and capacity building. As District Magistrate from March 2004 to April 2005, he was associated with the dispute over the Gravity Dam in Paschim Midnapur (West Bengal) and with the recent attempts to resolve it. His e-mail is sinhadear@gmail.com.

Rajesh Sinha is a Delhi-based journalist, and his e-mail is sinha.rajeshsinha@gmail.com

J. Srinath is Programme Coordinator at the M. S. Swaminathan Research Foundation, Chennai; previously he was a Senior Programme Officer with VIKSAT, Ahemadabad. His e-mail is srinath@mssrf.res.in

M. Suchitra is an independent journalist. Recipient of the 2005 Appan Menon Memorial Award for her study on the 'Socio-Economic Impact of Special Economic Zones in India', she was named Development Journalist of Asia at the Developing Asia Journalism Awards (DAJA) 2006. Her e-mail is msuchitra@journalist.com

C. Surendranath is a freelance journalist, and Editor of The Quest: Features and Footage, an independent media alliance based in Kochi. His e-mail is suran@vsnl.com

V. K. Ravi Varma Thampuran is Chief Sub-Editor, *Malayala Manorama*. He has been reporting regularly on Kuttanadu, and is the author *Kuttanadu Kanneerthadam* (2004; in Malayalam). His e-mail is thampuran2005@sify.com

P. Umesh works with the NGO 'Gamana' in Hyderabad which supported by Oxfam India/Svaraj Bangalore. His areas of interest are conflicts arising due to competing claims between different users/sectors of water, and urban water supply policies. His e-mail is umesh_varma@yahoo.com

UTTHAN: UTTHAN (meaning 'progress' in Gujarati) is a team working in four water-stressed districts in Gujarat for the past 20 years, has been facilitating self-reliant village-level institutions to find local solutions to basic livelihood issues. Central to UTTHAN's participatory, rights-based approach is the focus on women and vulnerable communities. Their e-mail is utthan@icenet.net

A. Vaidyanathan is a well-known scholar and has contributed significantly to the discourse on the water sector in India. A former member of the Planning Commission of India, he has also headed various governmental committees including the Irrigation Pricing Committee set up by the Government of India in the early 1990s. His e-mail is avnathanster@gmail.com

R. Vasanthan is the Nodal Officer for the Tank Rehabilitation Project - Pondicherry (TRPP). He has 5 years experience in the formation and management of tank associations and an excellent track record in conflict management. His e-mail is cerd_psf@sify.com

Praveen Vempadapu is currently pursuing his Masters in Development Studies at Institute of Social Studies (ISS), The Netherlands. He has more then 3 years of experience in development sector, and his e-mail is pravee.v@gmail.com

Shruti Vispute is with Society for Promoting Participative Ecosystem Management (SOPPECOM), Pune. Her research interests are development politics, environmental activism and feminisation of international politics, and her e-mail is shrutivispute@yahoo.co.uk

9 781138 376755